Environmental Studies

Environmental Studies

(Based on the compulsory core module syllabus prepared by UGC for Environmental Studies for undergraduate courses of all branches of higher education)

U.S. HAMPANNAVAR

Assistant Professor
Department of Civil Engineering
KLE Society's College of Engg. and Tech.
Belgaum

GOURI SURESH

Reader in Botany and former Co-ordinator
Environment and Water Management
Jamshedpur Women's College, Jamshedpur

I.K. International Publishing House Pvt. Ltd.

NEW DELHI • BANGALORE

Published by

I.K. International Pvt. Ltd.
S-25, Green Park Extension
Uphaar Cinema Market
New Delhi–110 016 (India)
E-mail:info@ikinternational.com
Website: www.ikbooks.com

ISBN 978-93-80578-93-4

Reprint 2014

Published by Krishan Makhijani for I.K. International Publishing House Pvt. Ltd., S-25, Green Park Extension, Uphaar Cinema Market, New Delhi–110016 and Printed by Rekha Printers Pvt. Ltd., Okhla Industrial Area, Phase II, New Delhi–110 020.

Preface

Environmental Studies have acquired great significance in these days of environmental degradation. It is high time that man realizes the grave consequences of his actions which are damaging the environment beyond repair. Any development during which natural resources get depleted faster than they are renewed is called unsustainable development. This leads to environment degradation, global climate changes, increase in the occurrence of various diseases etc. Though industrialization, urbanization etc. are all beneficial to mankind, they are not without pitfalls. Social and psychological impacts set aside, developmental activities upset the environment as a whole. The entire biosphere is affected by these. Disturbances in any one part of the biosphere are conveyed to the other parts and cause great harm. This shows that development should not be at the cost of natural resources. We must develop, no doubt; but we must see to it that we do not disturb the nature, of which we are essentially a part. The importance of maintaining the harmonious equilibrium of this beautiful earth must dawn on each individual sooner than later. Thus, the need for creating awareness in society regarding our environment has become crucial for all round development. In such a situation, making Environmental Studies a compulsory paper for undergraduate students of all streams is a highly laudable step. Recommendation to this effect by the University Grants Commission is a welcome measure.

This book is an attempt at explaining the topics suggested by UGC in simple language for the benefit of students. The contents have been arranged as per the core module syllabus prepared by the University Grants Commission. There are 8 units in the syllabus prescribed, which have been maintained as such. The eighth unit comprises field work. Many suggestions have been provided for field work in this unit. Though conditions differ in different localities in the country, a general guideline is always helpful. Numerous questions have been given at the end of each unit. These are of four types - multiple choice objective questions, and questions with very short, short and long answers. Answers to the multiple choice questions have been given in Appendix 3. Appendix 1 comprises a list of international conventions and protocols to protect our environment, while Appendix 2 presents a glossary.

I have great pleasure in thanking the publishers for their offer and prompt publication.

Constructive comments and suggestions for improving any aspect of this book in the next edition are most welcome from students as well as fellow teachers and other readers.

– Authors

Contents

Core Module Syllabus

Unit No.	Title of the Unit	No. of lectures
Unit 1	Multidisciplinary nature of environmental studies	2 lectures
Unit 2	Natural Resources	8 lectures
Unit 3	Ecosystems	6 lectures
Unit 4	Biodiversity and its conservation	8 lectures
Unit 5	Environmental pollution	8 lectures
Unit 6	Social issues and environment	7 lectures
Unit 7	Human population and environment	6 lectures
Unit 8	Field Work	Equal to 5 lecture hours

Unit 1

Multidisciplinary Nature of Environmental Studies

1.1 INTRODUCTION

The exact origin of life is a mystery. Yet we can reason that the first formed tiny droplets of protoplasm, the physical basis of life, were neither plant nor animal. It is assumed that this happened more than a thousand million years ago. The first living things were simpler than cells. Gradually, these structures evolved into cells. These simple cells branched off along two courses of evolution. One division became *animals*, and the other division became *plants*. The third group of cells remained simple in structure and function. These were the microbes or the microorganisms of the present times.

Both plants and animals developed from a common stock. They show many similarities with each other, yet they are so different from each other. Evolution did not stop there. Diversification proceeded in many directions. This resulted in the formation of innumerable varieties of living things. The biodiversity that we see today is the result of millions of years of development and diversification.

Plants, animals and microorganisms are thus the living beings in nature. In other words, they make the biotic part of nature. The non-living or the abiotic part includes the physical and chemical components of nature. The biotic and the abiotic components together make our environment. They are in perfect harmony with each other and form a stable system. Therefore it is our duty to protect the environment, of which we are a part.

1.2 DEFINITION OF ENVIRONMENTAL STUDIES

Environment can be defined as the external conditions influencing the growth and development of living organisms in any way. Study of any one or more aspects of the environment in relation to their effect on the living organisms and also the effects of living organisms on the environment is Environmental

Studies. Unlike other streams of studies, Environmental Studies pertains to our every-day lives. It has direct effect on our health and overall well-being. It is basically an applied field of study.

1.3 SCOPE OF ENVIRONMENTAL STUDIES

The living organisms of the earth interacting with the non-living environment form a vast, gigantic system. This is the largest, almost self-sufficient biological system we know of, i.e., the biosphere or the ecosphere. In the biosphere, biotic and abiotic components are constantly acting, reacting and interacting with one another. This results in structural and functional changes in the biosphere.

It is difficult to study the biosphere as a whole. Therefore, we study nature by dividing it into smaller parts—forests, deserts, grasslands, croplands, ponds, lakes, rivers, oceans etc. These systems—the ecosystems—are separated from one another only by space and time. In other words, the plant, animal and microbe communities in different areas and at different time periods (seasons or years or centuries) are different, yet they are linked with each other through air, soil and water. The molecules of carbon dioxide being absorbed by a plant, or the molecules of oxygen inhaled by us may have come from a far-off area. They have also been used and reused countless number of times in the past. Thus plants, animals and the non-living components of nature are linked to each other functionally.

There are no practical demarcations between the ecosystems. All of them are interdependent in various ways. For example, land plants absorb water from the soil, but this water ultimately comes from rain, and rain is the result of condensation of clouds, which are formed by evaporation of water from the water bodies. Thus land and water are closely linked not only physically, but also functionally.

The plants, animals and microorganisms of any region influence and interact with one another, and together constitute a biome or biotic community. In a biome, the living beings act on the environment and are acted upon by it. There are innumerable examples of such interdependencies. Some of them are cited here. Trees provide shelter to birds and birds help in the pollination and dispersal of fruits and seeds of many plants. Animals provide carbon dioxide to the plants, which need it for photosynthesis. Plants give oxygen to the animals which require it for respiration. Microorganisms act on the dead bodies of plants and animals and convert them into carbon dioxide, water and minerals, which are thus returned to the air, water bodies and soil in the environment so that they can be used again. The minerals from the soil enter the plants through their roots. Animals get minerals mainly from plants. In this way, the living and the non-living are closely connected to each other in nature. The energy needed to support life on earth comes from the sun. This radiant energy is absorbed by the non-living as well as the living components of the biosphere, and heat energy is released into the environment by both the components.

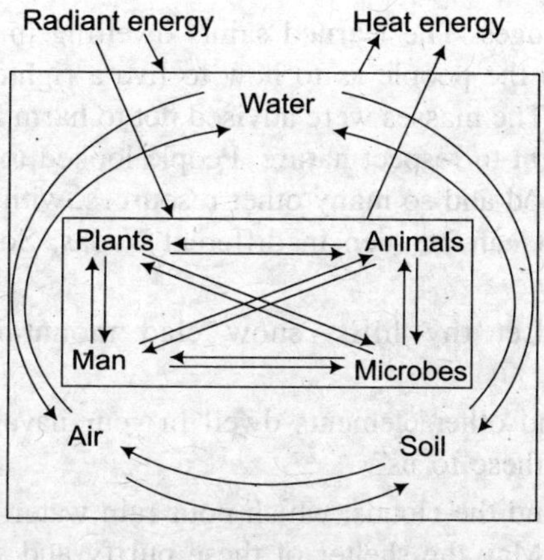

Figure 1.1 Interrelationships in the Bioshere

Man is an animal too. He depends on other animals, plants and microorganisms and all the non-living components of the environment in many ways. Therefore, if the equilibrium of the natural systems is disturbed by any means, the resulting harmful effects will have to be endured by all the components, including man.

To study the various aspects of biosphere, contributions from workers in many different fields are essential. All that we know today in this field is the result of studies in various branches of biology, chemistry, physics, geology, engineering, resource management, anthropology, sociology, disaster management, medicine, economics, philosophy, statistics, computer science, meteorology etc. Thus, the study of the environment is of a multidisciplinary nature and the scope of Environmental Studies is very widespread.

1.4 IMPORTANCE OF ENVIRONMENTAL STUDIES

We are surrounded by the environment. Whatever happens in our environment has a bearing on us. Therefore, studying all the issues concerning environment, understanding the problems related to environment, finding solutions to these problems, and implementing the solutions sincerely are of utmost importance to our well-being as well as of the living and non-living world around us.

The primary natural resources available to us are air, water, soil, forests, minerals, food, energy resources etc. Some of these are non-renewable like soil, minerals etc., while others are renewable like forests, water etc. However, even the renewable sources need time to be regenerated. If trees in a forest are cut down faster than saplings can grow into trees, the forest will be destroyed completely in no time. So we have to be very careful while taking out resources from nature.

In India, our ancestors had fully realized the importance of conserving natural resources. Indian sages of the Vedic age lived in hermitages (ashrams) in forests. The best form of spiritual, scientific and cultural researches was done in these hermitages. The learned saints dwelling in the forests provided the right guidance to the people as to how to live a righteous life and how to treat Mother Nature. The masses were advised not to harm any natural resource; rather they were taught to respect nature. People looked to nature, the provider of air, water, light, food and so many other resources, with a sense of gratitude. Prayers to this effect can be seen in different Vedas. Some examples are as follows:

1. "O Earth! Let thy hills, snow clad mountains and forests be pleasant"
2. "Minerals and other elements dwell in your naval part (central core). Please offer these to us".
3. "Your land and the clouds, which pour rain water on it, nurture us like our parents. May the shelter of these purify and strengthen us".

The Vedas also guide us as to what we must do to protect our environment. The meanings of some verses go thus:

- "We should establish harmony with water and medicinal plants".
- "It is in the interest of mankind to plant more and more trees as these safeguard the water resources".

Indian culture was thus based on harmonious coexistence with the environment. The same culture was also followed later during the times of Budha and Mahavir.

It was in the modern times, approximately two centuries ago, that man's greed overtook his needs. He became so self-centred that he stopped thinking about the environment. For example, without stopping to think of the consequences, man cleared forests on a large scale for various purposes. The immediate effects of this blatant deforestation were soil erosion, floods and droughts.

Today one of the prime concerns is continuing deforestation. Planting trees in as many places as possible will help to reduce the severity of this problem to some extent. Plantation forests cannot match natural forests in species variety of trees or other plants or wildlife. Yet, they can help to reduce air pollution, purify the atmosphere and conserve precious soil. Therefore tree plantation programmes must be organized regularly. This can be done by students in groups all over the country. While planting trees, care must be taken in the selection of species. For example, arid (dry) zones should be planted with drought resistant trees like Shisham (*Dalbergia sissoo*), Ber (*Zizyphus jujuba*), *Cassia* etc. In this way, the suitability and adaptability of the species and the needs of the different zones should be looked into during tree plantation programmes.

As population is increasing in geometrical progression, there is a resultant shortage of land for cultivation. In some parts of the country, forest land is cleared for cultivation even now. Application of chemical fertilizers in huge quantities is degrading the land. Pesticides are polluting air, water and soil. Where hybrid seeds have become popular, traditional seeds are being neglected, leading to gradual loss of many wild type genes and gene combinations. It has thus become necessary to convince the farmers of the importance of using traditional seeds to maintain the fertility of the soil. Otherwise, more and more areas in the country will have to face the problems being faced by the farmers of Punjab and Haryana today. In these two states, large tracts of land have been degraded due to unsustainable farming using hybrid seeds, large quantities of fertilizers and pesticides. Monoculture, i.e., cultivation of one particular crop year after year, has been another reason for the loss of fertility of the soil. Intense irrigation with the help of dams is bad for soil too, as it makes the soil saline and uncultivable. Green revolution has thus its limitations. It must therefore be practised carefully and only in areas suitable for such agricultural practices.

Dams of many types and capacities have been built on different rivers. They are irrigating thousands of hectares of crop land, and are providing us with electricity. But major dams have always been a heatedly debated topic. It is not that dams are not required, they are reservoirs of water, and help in irrigation and power generation and flood control. The point is that how worthy it is to construct major dams, while many minor dams can serve the purpose. Let us examine points for and against major dams. The advantages projected are that they help in.

1. storing water for irrigation;
2. flood control; and
3. power generation.

Now, these purposes can be served by other mean also. For example, rainwater conservation will be of great help in this regard. Most pasts of India get rain except some states like Rajasthan, Delhi, Gujarat etc.

1. If the rainwater is conserved properly by constructing ponds and lakes in suitable low lying areas, this stored water can be used for many purposes including irrigation of fields. The huge amounts needed for constructing one major dam is more than sufficient for constructing a few minor dams or thousands of ponds, reaching a far more number of villages that need water.
2. Rainwater harvesting will also reduce the amount of water that flows towards the rivers. The occurrence of floods can be thus automatically reduced.
3. Hydroelectric power (hydel power) can also be generated in minor dams. The quantity of power generated will of course be less, but then there will be many minor dams. Electricity from many power stations can be more easily and cheaply transmitted to surrounding areas; losses during transmission will be less too.
4. Besides all these, one major disadvantage of constructing major dams is the human factor, because the tribal and other people who are evacuated from the dam site are never given proper rehabilitation packages. Thousands of villagers are always displaced, and rehabilitation is never satisfactory.

5. Moreover, vast forest areas have to be cleared of trees for making a major dam. This loss is irreparable.
6. Whole villages are wiped out in floods when vast areas in the dam site are inundated due to blocking the path of rivers.
7. Major dams are very expensive to be built.
8. They cause flash floods
9. They change the breeding patterns of fish. In fact, all plants and animals in the river are disturbed when dams are built.

The developed countries have stopped making major dams. Therefore, World Commission on Dams (WCD) of 2000 suggested that projects on dams should be undertaken only after all other alternatives have been assessed and the consent of communities to be affected should also be obtained.

Rainwater harvesting can do wonders. What is needed is awareness. Most people do not know the potential of rainwater harvesting. It is high time different methods are popularized among the general public. It can be done with ease in rural as well as urban areas (See Water Harvesting in Unit 6 for details). In this era of rapid urbanization, rising population and growing industrialization, water scarcity is becoming more and more acute all over the world, especially the developing countries. The cities have become concrete jungles. As the green areas decrease giving way to cement, mosaic and marble, surface runoff increases greatly. Much less water is percolating underground. Whatever water is available underground is overexploited too.

In such a situation, we must guide the runoff from rooftops and footpaths and other hard surfaces into storage reservoirs or underground aquifers. Rain is the primary source of water. Forgetting this truth, we depend on secondary sources like rivers, lakes and wells for our daily needs. This trend must be reversed. Harvesting water was an Indian way of life in the distant past. We must revive our traditional methods of water harvesting and must also implement new methods.

At the Centre for Science and Environment (CSE), New Delhi, most of the rain is captured and recharged into underground aquifers or stored. It is not allowed to drain away into the streets. The Centre uses various methods for water harvesting. Annual Rain water Harvesting Potential (ARHP) at CSE is 3,66,600 litres. The Rashtrapathi Bhavan has taken many measures for harvesting rainwater. Many children in India are harvesting rainwater successfully in their respective schools. Colleges in general have large flat concrete roofs; all the rain falling on these roofs can be harvested by suitable methods.

Keeping the environment clean is the duty of every citizen. People in general keep their homes clean, but are seldom bothered about their surroundings. They throw plastic covers and pieces of paper on the streets. Men urinate in public. People spit freely in all sorts of places. They throw their kitchen garbage outside their own homes. All this must change. For managing household wastes, composting and vermicomposting can be done. (See Unit 5 for details) Other bad habits must be controlled and finally stopped.

Using plastic carry bags must be discouraged. Cloth or jute bags can be used while shopping. These are biodegradable, while plastic is non-degradable. Plastic will not decay for years and will pollute the soil. People throwing used plastic bags here and there are leading to another lesser known problem. Cows eat them and develop tumours. In the case of the small sized plastic bags, it must be realized that those black in colour are the thinnest, being the most recycled and the most dangerous. Food put directly into these bags becomes toxic; it may even be fatal. Therefore, these must not be used at all. White coloured ones are the thickest, and those of other colours slightly thinner. These differences in thickness show which ones have been recycled more number of times.

Segregation of wastes at source is a very good practice. Degradable garbage like kitchen wastes, paper, dry leaves etc. must be put in one bin, while nondegradable wastes like plastics, metal, broken glass, crockery etc. in another bin. This way, the contents of the first bin can be used at home in the making of organic manure, while the contents of the second bin can be deposited in the municipality waste collectors, or in their absence, in a common dumping area from where rag pickers can take them, sort them and send them to recycling plants. The wastes that cannot be recycled usually find a place in landfills.

Water pollution is a major concern in all countries, but more so in the developing countries. Many industries in India discharge their waste materials into rivers and other water bodies. This should be stopped. Drinking polluted water causes many diseases. In India and other less developed countries, water-borne diseases like diarrhoea, dysentery, jaundice, cholera etc. and diseases spread by mosquitoes like malaria, filarial, encephalitis (brain fever) etc. are taking a heavy toll year after year only due to the lack of proper waste management. Many cases of dengue fever and the uncommon chikungunia were reported recently from many parts of the country. In some localities, they even reached epidemic proportions. These diseases are related to sanitary conditions in the environment.

The value of natural resources must be realized fully by everyone. Just like water, other resources like food, minerals, wood, petroleum products etc. should not be wasted. Rich and developed nations like the US follow a wasteful style of living. Large quantities of food, paper, plastics etc. go waste in the US. They do have facilities for waste management; still wasting resources must be considered a crime, environmentally and also on humanitarian grounds. It is an irony that tonnes of food is wasted in some parts of the world while millions are starving in other parts. Wasting food is a practice with some people in India too. While having food, one must think of how many peoples' efforts have gone into the making of it, starting from the crop land to the dining table.

Biodiversity is precious. There are many possible gene combinations in nature; the number of chemicals produced by living organisms based on them is even more numerous. Till now, the use of only some of them have been found. There are thousands of plants, animals and microbes, whose importance we do not know yet. Therefore, any species becoming extinct is a loss for us.

As such, some species become extinct as time passes due to various natural factors. We should not add to their numbers by our activities. A species lost is lost forever.

Wild species are most essential for research work in plant and animal improvement. Wild species contain wild or dominant genes, which are used in breeding experiments. New diseases appear now and then in plants and animals. Resistance to these diseases is seen in the wild species. Such resistant genes have been used to create disease resistant plants and animals all over the world. Wild stocks cannot be maintained in parks and zoos, as the attention they receive there results in rapid loss of variability. Therefore, entire habitats must be protected. To prevent deterioration, they must be given the status of national parks, sanctuaries etc. At the same time, zoos, aquaria, botanical and zoological parks etc. are also essential, as they help urban people to know about the wild plants and animals and the variety in nature, thereby making them more environmentally oriented. Both these types of protection, known as *in situ* and *ex situ* conservation respectively, also help to maintain the biodiversity of nature.

Conserving nature for future generations is the duty of the present generation. We are duty bound to prevent extinction of species. If we do not take care, our future generations will not get a chance to even see many of the creations of nature. This will be a great injustice to them. We inherited a beautiful nature with a great variety of plants and animals from our ancestors. We are reaping all benefits from it. Now we cannot destroy the same and deny our descendents what we inherited.

Ecotourism combines conservation with development. The revenue generated by eco tourism can be reinvested in the conservation of nature. It also gives employment opportunities to traditionally remote and marginalized communities. It increases environmental awareness and creates a desire in the public mind for developing more and more tourist spots. It builds public support for conservation and encourages the participation of private sector in conservation efforts. However, care must be taken that areas opened for eco tourism do not get degraded by the activities of the tourists themselves.

Nature and natural scenes provide great inner peace. Nature has a great aesthetic value. We forget all our worries while enjoying the beauty of nature. An unpolluted river, lush green forests, a clean blue sea, a majestic mountain, multicoloured flowers, birds and butterflies—all give us tranquillity. Deformation and pollution of the beautiful nature is a crime in itself.

Depletion of natural resources, lack of water harvesting and proper waste management practices are the chief environmental problems facing the mankind today. All these can be rectified to a great extent by creating environmental awareness among the general public. Students, as an enlightened group of energetic youngsters, can play a major role in bringing about this awareness in society.

1.5 NEED FOR PUBLIC AWARENESS OF ENVIRONMENTAL STUDIES

Pollution of air, water, soil, and mindless utilization of natural resources have raised environmental degradation to dangerous heights across the globe. One of the best methods to reverse this trend, undoubtedly, is creating environmental awareness and educating the people. What is needed is a holistic approach, involving all strata of the society.

The need for creating public awareness of environmental issues has been discussed at national and international levels. The goal is to make people aware of and concerned about environmental problems, and committed to work individually and collectively towards solving current problems and preventing the occurrence of new ones.

People's attitudes towards their environment determine to a large extent how their lifestyle affects the environment. If people are aware of the environment and its significance to their lives, they will act in an environmentally responsible manner and support the environmental programmes. Education, training and supplying information to the public are the tools to boost this up. Information must be presented to different target groups so as to serve the local need for the immediate solution of the problem. Activities must be organised in every school, college, panchayat, village, town and city. It is in this context that NGO's assume a primary role as they can easily be in constant touch with the local communities.

Mass media like radio, television, newspapers, documentary films, hoarding boards etc. have a great influence on the people. Media can spread awareness with ease. More visible changes are given sufficient coverage by the media. For example, we regularly get information on population growth and disappearing forests. The media also help in giving the public information on how to construct rainwater harvesting systems, where to get eco-friendly products etc. To have a more positive impact, the media will need to understand the issues and develop a long-range outlook. Mass extinction is a very important issue, but as there is not much change in a day, it is not considered newsworthy. Such long-term changes should also receive the attention due to them.

Population explosion is perhaps the biggest problem faced by our country today. This is leading to rapid environmental degradation. According to the World Bank Report, 1984, in North India, it is the population growth that has led to severe deforestation, resulting in devastating floods. Desertification, pollution and all other environment related problems increase greatly by unchecked population growth. So, spreading awareness about the need for family planning is of utmost importance.

The need to increase literacy can never be overemphasized. Non-formal methods of education like Adult Education, Total Literacy Mission etc. have been taken up by both Central and State governments and also various NGO's in India. These have been partially successful.

Environmental awareness among the common people is seen more in places where literacy standards are high. This is but natural. Kerala can be cited as a typical example. People protest vehemently if any industry is polluting air, water or soil. Similarly, they are very much conscious of the problems of deforestation, rehabilitation etc., which can occur due to the construction of major dams. The Silent Valley project had to be dropped by the government as it would have led to the deforestation of vast tracts of forest land with great biodiversity. In November 2006, there was a statewide protest when just ten families left the premises of Mullaperiyar Dam fearing evacuation due to rising water level and a possible breach in the dam. This is undoubtedly the result of education and awareness. Therefore, literacy campaigns must be conducted on a war footing in India. It will help in creating awareness about not only environmental issues, but also the importance of sanitation, hygiene and health, family planning and many other issues, all related to environment and overall welfare of the general public.

Activities in which students can take part to create environmental awareness among the general public:

- Organize activities like seminars, workshops, competitions in essay writing, elocution, slogans, posters and quiz etc. at college and inter-college levels
- Conduct public meetings, festivals, rallies, exhibitions, demonstrations, folk songs and dances, street theatres, science fairs etc. on various themes
- Distribute pamphlets/handbills on environmental issues
- Celebrate Van Mahotsav
- Participate in local literacy campaigns
- Write relevant articles and publish them in newspapers, magazines etc.
- Help neighbours and other members of the society in setting up community waste management and rainwater harvesting projects
- Discuss with friends and relatives topics like resource management, pollution control, saving water, paper, energy etc.
- Set practical examples by using cloth and jute bags instead of plastic and walking short distances instead of using automobiles etc.
- Encourage civic sense and hygiene by not spitting on walls and roadsides, not urinating in public places, not throwing garbage on roads and inside trains and buses, not smoking in public etc.
- Join local NGO's who popularize environmental issues
- Read newspapers and magazines regularly and be up to date with global environmental issues
- Visit local zoos, parks, sanctuaries etc.
- Celebrate important days on the Environmental Calendar, like World Environment day, Wild Life Week, World Food Day, World Popualtion Day etc.

There are many institutions working for the betterment of environment all over the world. In India, the Ministry of Environment and Forests (MoEF) has many divisions and departments, with various centres and institutes working under it. Some of them are listed below:

1. Centre for Environment and Education, Ahmedabad
2. CPR Ayar Environment Education Centre, Chennai
3. BB Pant Himalayan Paryavaran Evam Vikas Sansthan, Almora
4. Tropical Botanic Garden and Research Institute, Thiruananthapuram
5. Centre for Ecology, Research and Training, Bangalore
6. National Museum of Natural History, New Delhi
7. Salim Ali Centre for Ornithology and Natural History, Coimbatore
8. Botanical Survey of India, Kolkata
9. Zoological Survey of India, Kolkata
10. Centre for Environmental Management of Degraded Ecosystems, New Delhi
11. Centre for Mining Environment, Dhanbad
12. Central Pollution Control Board
13. Wildlife Institute of India, Dehradun
14. Indian Institute of Forest Management, Bhopal etc.

These institutes and centres do valuable work in:

- Spreading environmental awareness and promoting environmental Education
- Publish magazines and journals
- Train environmental educators
- Conduct programmes on conservation and biodiversity
- Perform environment impact assessment work
- Produce research scientists
- Train personnel in eco-development, wildlife biology, habitat management, disaster management etc.
- Procure information on India's wildlife wealth
- Prepare extensive lists of the wetlands of the country
- Prepare resource material on environmental education
- Prepare books, posters, video films etc. and conduct workshops, seminars, conferences etc.
- Establish nature clubs/eco clubs etc.

Many awards are also given annually to citizens/institutions by MoEF, like:

- Indira Gandhi Paryavaran Puraskar
- Pitambar Pant National Environment Fellowship Award for Biodiversity

- Rajiv Gandhi Wildlife Conservation Award
- Indira Priyadarshini Vriksha Mitra Award
- Dr. Salim Ali and Dr. Sankala Fellowship etc.

Many projects have been set up in India to protect wildlife. Some of these are the Project Tiger, Project Elephant etc. Crocodile breeding is done in the Madras Crocodile Bank Trust. Survey of plant resources is done by BSI, while corresponding survey of animals is done by ZSI, both located in Kolkata. There is also the Ganga Action Plan, the Yamuna Action Plan etc.

Many other national organisations are also doing valuable work in the feld of environment. Some examples are:

- State Pollution Control Boards (SPCB's)
- Industrial Toxicology Research Centre (ITRC)
- National Environmental Engineering Research Institute (NEERI) etc.

In addition to this, hundreds of non-governmental organisations are also involved in various projects. Some well known NGO's in India are:

- World Wide Fund for Nature-India (WWF-I)
- Kalpavriksh, Pune
- Kerala Sastra Sahitya Parishad (KSSP)
- Bombay Natural History Society (BNHS)

The names of many individuals who have spread the importance of conservation of environment stand out clearly in our country. The contributions of Medha Patkar, famous for her Narmada Bachao Andolan, Sunderlal Bahuguna of Chipko Movement, Salim Ali, the great ornithologist (expert on birds), M.S. Swaminathan, the agricultural scientist, Rajendra Singh, the Magsaysay award winner for water harvesting, M.C. Mehta, the famous environmental lawyer etc. are unforgettable.

The good work done by them and so many others has to be carried forward with a renewed zeal. There are thousands involved in projects creating awareness and propagating environmental education in our country. Yet the task cut out for us is an immense one as we have to educate more than one hundred crore people. There is need for as many more environmentalists as possible. And it is here that young, creative, motivated students can render invaluable service to the nation.

QUESTIONS

A. Very short answer type:

(a) Multiple choice type: Tick the right answer:

1. The largest ecosystem on earth is:

(a) The oceans	(b) Rainforests
(c) Mountains	(d) The earth itself

2. Salim Ali was an expert in:
   (a) Agricultural Sciences (b) Rainwater harvesting
   (c) Ornithology (d) Environmental law
3. NEERI is:
   (a) An Indian Environmental Research Institute
   (b) An NGO
   (c) An international organization
   (d) A Pollution Control Board
4. Which of the following waste material is degradable?
   (a) Plastic (b) Metal
   (c) Glass (d) Paper
5. Which of the following diseases is not spread by water?
   (a) Diarrhoea (b) Chickenpox
   (c) Cholera (d) Dysentery
6. Botanical Survey of India is located in
   (a) New Delhi (b) Mumbai
   (c) Kolkata (d) Chennai

(b) Define the following:
   1. Environment 2. Environmental Studies
   3. Biosphere 4. Biome

(c) Name the following:
   1. Five disciplines related to Environmental Studies
   2. Five natural resources
   3. Three environmental awards given in India
   4. Three diseases caused by drinking contaminated water

B. Short Answer type:

(a) Differentiate between:
   1. Renewable and non-renewable natural resources
   2. *In situ* and *ex situ* conservation

(b) Write short notes on:
   1. Deforestation
   2. Effects of population explosion on environment
   3. Ecotourism
   4. Role of media in creating environmental awareness
   5. Functions of MoEF and the centres and institutions working under it
   6. Environmental awareness and literacy

7. Segregation of waste at source
8. Use of wild species in crop improvement

C. Long answer type:

(a) Write long notes on any two:

1. Water pollution
2. Biodiversity
3. The advantages and disadvantages of major dams
4. Rainwater harvesting
5. The advantages of minor dams

(b) Discuss in detail the scope of Environmental Studies.

(c) What can students do to promote environmental awareness in society?

(d) Give an account of the importance of Environmental Studies.

(e) Public awareness of Environmental Studies is the need of the hour. Discuss.

Unit 2

Natural Resources

2.1 INTRODUCTION

Ecology of natural resources is called Resource Ecology. A resource is a source of supply. Natural resources are substances, different forms of energy etc., which the living organisms can take from nature for various purposes. Forests, soil, grasslands, water bodies, minerals, sunlight, different gases in the atmosphere etc. are all natural resources. They may be living or non-living. Man depends on all the natural resources for various purposes. In fact, there is no existence for him in the absence of these resources.

These natural resources, on which we depend, are provided by four 'spheres':

- The atmosphere—The atmosphere is a protective covering around the earth, and is a reservoir of several elements essential for life. It is a mixture of gases in different proportions. Of all the gases, oxygen is the most important as all living organisms depend on it. Carbon dioxide is needed by green plants to make food. Without plants, there can be no life on earth. Water vapour is needed to form clouds. Ozone absorbs the harmful ultraviolet rays. Nitrogen and some other gases are also useful to us in various ways.
- The hydrosphere—All the water on the earth together is called the hydrosphere. It includes the rivers, lakes, ponds, the seas and oceans and the glaciers and the water stored underground in aquifers. Only very little of the available water is fresh water. Water is essential for drinking, cooking, bathing, washing, cultivation, industries, transportation, recreation and many other purposes man, plants and animals cannot survive without water.
- The lithosphere—The soil, the rocks and other parts of the earth's surface together is called the lithosphere. We need soil for cultivating plants. Minerals are required for various purposes. We get coal and petroleum from the lithosphere. Stones and sand for construction are obtained from the lithosphere. In this way, we depend on the lithosphere for living.

- The biosphere—The place on earth where life exists is the biosphere or the ecosphere. The biosphere includes all the living organisms on the earth and the different parts of the environment on which life depends. It is therefore the whole earth itself.

Natural resources may be inexhaustible or exhaustible. Inexhaustible resources are those which are in unlimited supply in nature. Activities of human beings do not affect these resources, because they are present in nature in very large quantities—much in excess than needed for the existence of human beings and all other living organisms. Some of these are necessary for man, but others may not be necessary directly for human life, though they can be made use of. Important inexhaustible natural resources are solar energy, rain, wind, tidal power etc.

Exhaustible resources are limited in nature and can therefore be exhausted if not used carefully. These are of two types—renewable and non-renewable.

2.1.1 Renewable and non-Renewable Resources

Renewable resources are resources that can be restored and replaced. Examples are:

- Forests
- Water
- Soil fertility etc.

These can be renewed by us or they have the capacity to renew themselves by replacement, reproduction or recycling.

Non-renewable resources are those which, once used, are exhausted forever, and cannot be restored or replaced. Examples are:

- Fossil fuels like petroleum, coal etc
- Minerals
- Metals
- Species of plants, animals and microorganisms.

There is an end to the reserves of coal, petroleum, minerals, metals etc. in nature. Similarly, the gene combination of a particular species can never be duplicated once it becomes extinct.

2.2 NATURAL RESOURCES AND ASSOCIATED PROBLEMS

The world is facing many problems associated with both these types of natural resources. Recklessness in the utilization of these resources without replacement is the main reason for this.

The important natural resources useful to man, after air in the atmosphere, are listed below:

- Forests

- Water
- Minerals
- Food
- Energy
- Land

All these natural resources are being over-exploited by human beings. Forgetting the fact that we are part of nature, we are using a lion's share of the natural resources. We are also not taking care to return to nature what we have taken out of it. This is creating many problems like:

- Imbalance in nature due to pollution
- Extinction of valuable species
- Deepening divide between the More Developed Countries (MDC's) and the Less Developed Countries (LDC's) because of:
  - (a) more population growth in the LDC's;
  - (b) more poverty, hunger, diseases, illiteracy etc. in the LDC's;
  - (c) more per capita income in the MDC's;
  - (d) more non-vegetarianism in the MDC's, leading to more land use for rearing animals;
  - (e) more energy being used by the MDC's; and
  - (f) more wastage of food, electricity etc. in the MDC's.
- Energy crisis affecting industries as well as day-to-day life
- Water scarcity, leading to social tensions
- Scarcity of available land for agriculture, resulting in food scarcity
- Depletion of minerals for industry
- Deforestation leading to desertification, soil erosion, floods, droughts, loss of biodiversity etc.
- Global climatic changes
- Increase in man-made environmental disasters
- Social conflicts and wars etc.

In short, it is high time we all pay more attention to the problems created by over-exploitation of natural resources, discuss the problems associated with it and find out and implement the remedial measures on a case-by-case basis.

2.2.1 Forest Resources

Forests are a great national wealth. A dense forest is full of trees. The forest floor is densely populated by various shrubs (bushes) and herbs (small plants), ferns and mosses. Many climbers and epiphytes (plants growing on the branches of trees, without absorbing any water or food from them) depend on the trees.

The atmosphere in such a forest is cool and humid and wind velocities are very low. Forests are a great tourist attraction in any country.

(a) Uses of forest resources

The uses of forest resources, in other words, the functions of forests, can be classified as follows:

1. *Protective uses*:
   - Trees afford excellent protection to soil erosion. The deep and spreading roots of the trees and other plants in the forest hold the soil particles strongly and prevent them from being washed away in rain.
   - Forest trees build up a thick layer of humus to prevent floods, droughts and erosion.
   - Roots of trees and other plants serve as a natural dam; they catch and hold rainwater.
   - Forests prevent early siltation of dams and lakes.
   - In the absence of trees, soil at the edges falls continuously into the water bodies, thus making them shallow and reducing their water storage capacity. This is especially bad for dams because when a dam is built, its life is usually predicted. Siltation shortens the life span of dams and also increases floods.
2. *Regulatory uses*:
   - Forests regulate various biogeochemical cycles, especially that of carbon. This is because trees of forests absorb huge quantities of carbon dioxide from the air for photosynthesis (making carbohydrates from carbon dioxide and water with the help of the green pigment chlorophyll, using solar energy for the process).
   - During photosynthesis, trees release huge quantities of life giving oxygen, which is one of the end products of the process.
   - Trees help purify the atmosphere, thereby reducing pollution drastically.
   - Acting as sponge, forests absorb and hold water which charges springs, streams and groundwater. Thus they regulate the flow of water from mountain highlands to croplands and urban areas.
   - Forests also influence local, regional and global climate.
   - Trees absorb large quantities of water from the soil and release it into the atmosphere during the process of transpiration. Transpiration is the release of water vapour from green plants into the atmosphere. Fifty to eighty per cent of the moisture in the air above the tropical forests comes from trees in this manner. This moisture soon becomes rain clouds. Therefore, if large areas of forests are cleared, rainfall will be affected.
   - Wild animals and plants are most essential for the balancing of nature. In their absence, ecosystems will be unbalanced.

- Forests buffer us against noise, absorb air pollutants and give us relaxation from tension.

3. *Productive uses*: Forests provide us with:
   - Timber for various construction purposes and furniture
   - Firewood, one of the most popularly used fuels by the rural poor
   - Wood for paper pulp
   - Alkaloids
   - Essential oils
   - Gums
   - Resins
   - Perfumes
   - Tannin
   - Charcoal
   - Dyes
   - Wood alcohol
   - Plywood
   - Turpentine
   - Drugs
   - Honey
   - Lac and many other things.
   - Many forest lands are also used for mining, grazing animals, recreation etc.
   - The wild animals and plants are used by plant breeders, genetic engineers and biotechnologists as a reservoir of wild type of genes. This is also a productive function.

An assessment of the forest cover of the country was presented as per the Survey Report, 1999 (Forest Survey of India). For preparing this report, satellite remote sensing technology was used. Thirteen states of India together comprise 63% of our forest cover. These are Andhra Pradesh, Arunachal Pradesh, Assam, Delhi, Himachal Pradesh, Madhya Pradesh, Maharashtra, Manipur, Meghalaya, Mizoram, Nagaland, Sikkim, and, Tripura. Among these states, Madhya Pradesh tops the list with 20.68% of its area under forest cover. The 7 north-eastern states together cover 25.70% of India's forest cover. Of the mangrove forests, 85% are in West Bengal, Gujarat and Andaman & Nicobar islands.

(b) Over-exploitation of forests

Most of the over-exploitation of forests is taking place in the tropical countries. Tropical forests are of different types and cover about 6% of earth's land area. All types are affected, but the worst situation is in the rainforests. Worldwide, about half the timber cut each year is used as a fuel for heating and cooking, especially in the poor countries. Some of the timber is burnt directly as fire-wood, while other is converted to charcoal, which is used in urban areas in poor countries and also by many industries. One-third of the world's annual timber is used in building materials. Forests are over-exploited in different manners. The most important methods of overxploitation are deforestation, mining and dam building.

Deforestation

Worldwide, most of the destruction happened since 1950. But as far as India is concerned, the greatest loss was about 150 years ago, when trees were cut on a huge scale by the British for making railway sleepers for England and other European countries. Forests degraded into open woodlands, savannas or deserts. However, even after the British left, the process of deforestation contin-ued. Rising population, developmental activities, industrialization and pollution have decreased the forest cover in India drastically. It is most unfortunate that this is happening in a country which gave to the world the earliest reference to forest conservation in the *Mahabharata*, the longest epic of the world. Even after the *Mahabharata*, more than 2,500 years ago, Lord Budha had pointed out strongly the necessity of forest conservation to the people of the country. Over the years, all the words of wisdom of the past were forgotten.

It has been assumed that for the health and prosperity of any country, 33% of its total land must be under forest cover. At the time of independence, India had only 25% area under forests. Now it has come down still. Forests cover 6,37,293 sq. km. of India, which is 19.39% of the country's area. The 3 types of forests that together make-up this percentage are:

Type of forest		*% area covered*
Dense forest		11.48%
Open forest		7.76%
Mangrove forest		0.15%
	Total	19.39%

Deforestation is occurring at a dangerously high rate in many parts of the world, like South East Asia, and South and Central America. Philippines has lost 97% of its original forest cover, while Madagascar has lost 84%. Of the total tropical deforestation, about 40% is taking place in the vast Amazon basin of South America alone. All these are important hotspots of the world. Thus there has been great loss to precious biodiversity due to deforestation alone. Madagascar is the world's 4th largest island, which is located in the Indian Ocean, east of Africa. Most of its species are endemic. Half of the island's birds and all its reptiles and mammals are endemic. These are the most endangered species in the world.

Tropical forests contain 50 to 90% of the earth's total species. So defores-tation of these is the most serious problem. Half of the supply of hardwood, coffee, tea, cocoa, spices, nuts and tropical fruits are obtained from these forests. Twenty-five percent of the ingredients of medicines are obtained from medicinal plants growing in tropical forests. Most varieties of rice, wheat and corn used in the world have been developed from wild tropical plants. Thousands of plants in tropical forests have not been studied yet for their food or other values.

It is often thought that tribal people destroy the forests in which they live. However, this is not true. The quantum of damage done by the forest dwellers is not much. They depend on the forests and their produce for their livelihood. Why should they do anything to diminish their own source of income? The real causes lie elsewhere.

Causes of deforestation: Deforestation is the result of many interconnected factors. It may be due to natural causes like forest fires, diseases, climatic changes etc. or due to human activities listed below.

Human activities leading to deforestation:

- Population growth and poverty: The landless poor keep on shifting to forests, where they try to do cultivation just to survive.
- Road construction: The process of degradation of a forest in fact starts with the construction of a road through it. This 1st road is usually made by the company/contractor for transporting timber. Once there is an approach to the forest, many people enter it and start clearing the forest.
- Unsustainable small scale farming: People living at the margins of the forest move into the interior, build huts and plant crops in small cleared areas.
- Shifting cultivation: When soil fertility of areas, once cultivated as farms, decreases, these areas are abandoned. New areas are cleared and made into farms. This is called shifting cultivation. Valuable forest patches get degraded by this practice.
- Cattle grazing: Villagers living around the forests leave their cattle to graze in the forests. The surface soil becomes hard under the hooves of animals and water percolation decreases. Seeds cannot germinate on such hard and dry soils. Reproduction is hampered as the tops of plants are eaten too. Forest land thus gradually turns into barren land with thorny bushes.
- Commercial logging: Wood is cut from forests for various purposes by industries.
- Forest fires: Forest fires may be natural or man made. In nature, on very hot and dry days, forests may catch fire. Man made fires are also common. Large forest areas are burnt down for cultivation of cash crops like tea, coffee, sugarcane, bananas, pine apples, different types of spices etc.
- Faulty Governmental policies: Without realizing the full ecological value of trees,

- Governments put subsidies on timber and make it comparatively cheap in the third world countries. This has speeded up deforestation.
- Mining and dam building: These also cause immense damage to forests.

Case Studies

1. *The first forest movement in India*: The first forest movement against indiscriminate tree felling was perhaps fought in India. In 1730 Amrita Devi, a village woman, along with 362 followers, all women, resisted illegal tree felling by contractors in Jodhpur. They declared that trees could be felled only after putting them to death. Thereafter, each woman stood hugging a tree. Contractors axed them all to death and then cut down the trees. It was after this gruesome incident that the ruler of Jodhpur ordered a ban on illegal tree felling. A similar movement started by tribal women in U.P. later became famous as Chipko Movement. It was soon popularised by Sunderlal Bahuguna. The Movement has now crossed all geographical boundaries, and environmentalists from various countries are meeting him to study the details of this mass movement.
2. *Sacred groves*: There was the practice of setting up sacred groves in many parts of India. For example, in Kerala, these are called 'Kavus'. These are dedicated to different gods and goddesses. Kavus dedicated to snakes are also common. A sacred grove is a patch of dense vegetation including a small shrine and a pond or stream. The vegetation is so dense, undisturbed and pollution free that many rare species of plants and animals live here. The ponds and streams never dry in the cool and humid environment and conserve precious water. The 'kavus' are owned by families who take every care to maintain them. Big sized sacred groves are maintained by a few families together or the whole community. The belief that gods lived in the groves helps the owners to look after them well. The non-believers, among Hindus themselves and also other religions, have often contested to the existence of the sacred groves. But it is a good sign that environmentalists have come to their rescue. Now the government has also decided to protect them.
3. *Forest protection by villagers*: In Amlasole, a tribal dominated village in West Bengal's Midnapore district, villagers have found a way to drive away the timber mafia from adjacent Jharkhand, increase their income, and at the same time protect the forests thereby conserving water too. Till 2008, the village was inaccessible to the rest of West Bengal due to the absence of roads and had almost turned into a dry desert. Hunger deaths became common. It was then that some villagers got together to crack down on the timber traders. No stranger is allowed to roam around in the forest. Greenery has spread. Protecting the forests has given the villagers multiple gifts. Return from trees like

mahua, coconut, jackfruit, mango etc. and various berries fetch them good money. The wells do not dry up in summer any longer. The villagers have also decided on the punishment to be meted out to those caught cutting trees. They have to pay up fines of Rs. 5000/- per tree. If they are not able to pay up the fine, they have to feed a local family for one month!

Timber extraction

Wood is cut from forests for various purposes. It is made into doors and windows, furniture, packing cases, matchboxes, paper etc. The problem is not that these are cut, but that the cutting is at a much faster rate than the natural process of regeneration. Replacements are not done by the industries either. Moreover, when one tree falls, some others trees around it are injured, and many of them die. It has been estimated that many timber exporting countries in South East Asia will not have any timber to export at all in a few years. One of the most important reasons for the dwindling forest resources in the world is timber extraction.

To cancel the effect of timber extraction and the resulting deforestation, Forest Management should be done. This includes afforestation, reforestation and forest conservation.

- Afforestation is planting trees in areas where there were no trees earlier.
- Reforestation is bringing areas, now devoid of trees, back to the original form of forests.
- Forest conservation is protecting the present forests and preventing them from degradation.

Tree planting is now done on a large scale in different parts of the world. In India, the festival of Van Mahotsav was introduced for this purpose. It is celebrated every year in July. Planting is done in wastelands, open lands, in and around cities etc. Thus new trees will come up to replace the old ones. This is a sound planning for future. However, Van Mahotsav should not become a mere ritual. Trees planted should be cared for and not forgotten after the festival. It should be made into a well planned country-wide afforestation programme. The needs of the different zones should be looked into while selecting the saplings. For example, arid zones should be planted with drought resistant trees. The new trees should blend well with the surrounding trees. Soon, one or two keystone species can be seen flourishing better than other species. These will gradually dominate the area.

A plan of action that can be followed for forest management is given below:

- Degraded land not fit for agriculture must be converted into forest land with suitable species.
- Special types of salt tolerant trees can be planted in the saline soils.

- Trees should be planted on roadsides, near railway tracks, canal banks and farm boundaries.
- Encourage planting of trees around villages, school yards, public buildings, church yards, cemeteries, market squares etc.
- Enforce the present laws against tree felling strictly.
- Force industries using wood as a raw material plant trees 20 to 50 times the number cut.
- Develop fuel efficient stoves to decrease the quantity of fire wood used in rural households.
- Make alternative energy sources like solar energy, biogas etc. popular.
- Practise selective harvesting, i.e., marking trees according to age, cutting selectively and filling gaps with saplings.
- Popularise social forestry, i.e., involve local people in afforestation and reforestation programmes.

Realising the importance of forest management to negate the adverse impact of timber extraction, the Ministry of Environment and Forest (MoEF), Government of India, formulated the Joint Forest Management (JFM) in 1988. Accordingly, the Forest Departments work with local communities. In 1990, Village Forest Communities (VFC's) were formed. New guidelines were given in 2000. At least 25% of the income from the villages, which volunteer to manage the forests, goes to the villagers. All the states in India now have JFM's and all are functioning well. This will hopefully help cancel the effects of timber extraction going on in our forests. However, it is also essential to punish the guilty to prevent further exploitation.

Mining

Mining is the extraction of minerals from the earth. There are three stages in any mining operation:

1. *Prospecting and exploration*: First of all, the areas where minerals are hidden underground have to be located. This process is called prospecting. In the early days, spots where veins of minerals were exposed on the surface were mined. But now there are sophisticated methods of prospecting. Areas where mineral ores are located are now detected with remote sensing devices. Geographical Information System (GIS) is extensively used for this purpose. (for GIS and remote sensing, see Disaster Management in Unit 5) The size, shape and extent of the spread of minerals underground can also be assessed using these methods.
2. *Development*: Once the locality is known, transportation to the target area has to be assured. For this, roads must be constructed, and all other necessary infrastructure developed. This is the development stage.

3. *Exploitation*: Minerals may be superficial or deep seated in position. Accordingly, mining methods differ. Surface or open cast mining methods are different from deep or shaft mining. The mined mineral is first processed and then transported to various industries as per the requirement. In all mining operations, safety of the miners is given due importance.

Impact of mining: Mining degrades forests to a great extent. In many countries, large tracts of forest land are cleared for mining operations. Mining disturbs the physical, chemical and biological characters of the soil. The fertile topsoil of the forests goes deep inside the mines, while the infertile subsoil, devoid of humus and organic nutrients necessary for plant growth is brought to the top. Erosion increases. The whole area becomes useless unless soil reclamation practices are not adapted promptly. Even then, it is not possible to revert the land back to the natural forest. What can be done at the most is to grow some plantation forests, which lack the biodiversity and beauty of natural forests.

Environmental impacts of mining include erosion, formation of pits and loss of biodiversity. Soil and groundwater and surface water get polluted by chemicals from mining processes. Sometimes, trees may be cut down for space to store debris and soil. In other cases, the debris and slurry produced are dumped into ponds dug in naturally-existing valleys. These waste materials may be toxic. Since mining involves removing dirt and rock from its natural location creating large empty pits, cave-ins leading to accidents occur frequently in mines. Chemicals may leak out and affect the health of the local people. There are strict laws to be followed by mining companies; however, in many cases, laws are not followed due to the slackness of the authorities.

There are thousands of abandoned mines throughout the world. Old mines are often dangerous and can contain deadly gases. Wastes, sometimes toxic, after the mined materials are taken away may be left in the surrounding areas.

Gold, copper, diamonds, and other precious metals and gemstones are important natural resources. In many cases, these are found in rainforests around the world. Extracting these natural resources damages the rainforest ecosystem and creates problems for people living nearby.

Large-scale mining operations result in significant deforestation through forest clearing and the construction of roads. After extracting building stones, slate etc. from underground rocks, wide gaping quarries are left open in many places. For stone cutting in this manner, many trees are felled. While deforestation and chemical pollution from mining can impact the rainforest environment, water bodies in the vicinity are also polluted, affecting fish and other aquatic animals and plants. Resettlement and rehabilitation of the local people, mainly tribal, is a serious issue too.

Dams and their effects on forest and tribal people

There are more than 45,000 dams in the world. More than half of the large dams are in India and China. During independence, in 1947, there were fewer than 300 large dams in India. By the year 2000, the number had grown to over 4000, more than half of them built between 1971 and 1989. India ranks third in the world in dam building after US and China. While some of these dams were built primarily for flood control, water supply, and hydroelectric power generation, the primary purpose of most Indian dams (96 percent) remains irrigation.

Construction of large dams has been a major policy of the Government of India. However, public investment in large dams in India has been criticized by many, especially environmentalists since 1980's. The economic costs of dams and their benefits are now-a-days correlated with the social and environmental issues. Dam gates are opened during the periods of very heavy rain to prevent cracking of the walls of the dam due to the pressure of water. Huge quantities of water suddenly gush out from the open gates. This is called a flash flood. If prior information is not given to the local people, flash floods can bring death and misery.

Most of the dams built in our country are for irrigation purposes. To store water upstream in rivers, walls are built across the rivers. As a result, a reservoir is made upstream. There is a system of gates in the wall. This maintains normal river flow downstream. Water from the dam is made to flow into a network of canals for irrigating regions downstream. The upstream area that stores water as well as the area submerged by the reservoir is called the catchment area. The downstream area fed by the irrigation canals is called the command area. People living in the catchment area, who lose property and livelihood, do not benefit from irrigation. Those living in the command area do not lose anything, but on the other hand, gain very much from irrigation.

Behind almost every major dam in India, there are stories of broken promises. This is because of the mishandling of the issues by the administration. Moreover most of the rural poor, mostly tribal people, are not aware of their rights. They have no idea how much compensation they are entitled to get or how much land will be submerged. Many do not know that there is a Measurement Book. Those who know get annoyed when surveyors come, just measure the lengths and breadths of houses and put very low values for them. Actually, the purpose of the Measurement Book is to help engineers to evaluate houses properly so that compensation to build similar houses can be given.

The aim of the government is to evict people in minimal time with minimal expenditure. The residents are caught in the red tape, and at each step lose more and more. Sometimes, the houses of the affected people are demolished before alternative arrangements are provided. In many cases, tribal and other people do not get the land for settlement even after the dams have been built.

At other times, substitute land given is far away and barren or rocky, and absolutely uncultivable.

In Arunachal Pradesh, 89 large dams have been cleared for construction. People of many tribes (some of them whose size is as small as 3,500) will be badly affected by these dams socially and economically. Tehri dam on River Bhagirathi is 260.5 m high; it is the 5th tallest dam in the world. Rehabilitation had started in 1979, yet, till now it has not been completed. The dam has been built on earthquake prone Central Himalayas. There is a considerably high chance of major dam causing earthquakes. Thus social and environmental issues have been ignored.

When the area under forest cover is decreased due to timber extraction, mining or dam building, it is the overall prosperity of the country that suffers, and the people most affected are the tribal people who depend on the forests for their livelihood. The tribal people live a simple life. They get most of their food in the form of leaves, fruits, roots and rhizomes etc. and the small animals from the forests. They drink water from the streams of the forests. They collect firewood for cooking and also for selling in the nearby villages. They also sell flowers, fruits, leaves, twigs for brushing teeth, honey, lac, gum, fibres and other such forest produce for a living. They make huts using bamboos and other building materials available in the forests. They use the medicinal plants available in the forests for different types of ailments. In this way, the life of the forest dwellers is dependent entirely on the forests. Therefore, the direct effect of deforestation by any method—whether it be lumbering or mining or dam building—is felt by the tribal people the most, though everybody else is affected too in various ways.

2.2.2 Water Resources

Water is one of the most valuable natural resources. All the water on the surface of the earth together is called the hydrosphere. It includes the seas and oceans and also the rivers, streams, springs, lakes, ponds, glaciers (huge masses of ice in the Polar Regions) etc. The hydrosphere covers $3/4^{th}$ of the earth's surface.

A major part of the hydrosphere is the marine ecosystem comprising the seas and oceans, estuaries, backwaters etc. Only very little is freshwater. This includes ponds, tanks, lakes, rivers, springs, streams, glaciers etc. The freshwater in the lakes, rivers, glaciers etc. is continuously being renewed by the process of evaporation and rainfall. Some of this freshwater is stored underground; these are called aquifers. This is also the part of hydrosphere. Thus hydrosphere includes all types of water resources.

Of the total amount of water on the earth, 97% lies in our oceans; only 3% is freshwater. Of this 3%, 2% is locked in the polar ice caps, while only 1% is available as freshwater. Water found below the surface of the earth is called groundwater resources, while that above the ground is called surface water resource.

Table 2.1 Distribution of world water resources

S.No.	*Location*	*% of Total water*
1.	Freshwater lakes	0.009%
2.	Saline lakes	0.008%
3.	Stream channels	0.0001%
4.	Groundwater (<0.8 km deep)	0.31%
5.	Groundwater (>0.8 km deep)	0.31%
6.	Soil moisture	0.005%
7.	Ice caps and glaciers	2.15%
8.	Atmosphere	0.001%
9.	Oceans and seas	97.2%
	Total	100%

This shows that water available for our use is very little. More than 2/3 of all countries have already started experiencing some degree of water shortage. Global population is increasing at a fast pace. Each individual needs at least 250 litres of water per day. This is in addition to the much larger hidden water demand required to produce food and other consumer items. A terrible water crisis is therefore not far. The prediction that "The 3rd world war will be fought for water" may really come true.

Ours is a country of vast water resources. We have annual monsoons, which regularly recharge the water storages maintained in:

- the great glaciers (ice blocks) of the Himalayas;
- the rivers and streams of our plateaus; and
- the enormous multilayered underground water pools etc.

Yet, the per capita availability of clean water is decreasing day by day due to population explosion, urbanization, industrialization, pollution etc.

As such, available freshwater in the world is very less in comparison to other forms of water. Therefore, every effort must be made to conserve this precious natural resource.

(a) Use and over-exploitation of water

Water is the most abundant ingredient of living beings. Sixty to 95% of the body is water. Man can live without food for a few days, but he cannot live without water even for a day. Thus after oxygen in the air, water is the most important need of man and other animals.

Uses of water:

- Most chemicals can function only in the presence of water.
- Cytoplasm, the basis of life in all living organisms, can function only in a hydrated condition. If dehydrated beyond a certain level, living cells will die.
- Water is a raw material for photosynthesis (making of food by plants) on which all organisms depend directly or indirectly for food.

- The circulation of minerals absorbed from the soil within the plants and also the food made during photosynthesis takes place through the medium of water.
- For the condensation and the synthesis of various compounds inside the living body, and outside in the soil, water is a must.
- Water is the universal solvent in nature.
- In addition to all this, many plants and animals live in water. These are called aquatic organisms.

In short, life on earth cannot exist without water.

Man needs water for various purposes. This water may be taken from underground sources like wells, or surface sources like rivers or lakes. Wells may be dug wells or tube-wells. Both depend on underground water. Surface water seen in rivers and other water bodies are used by us for different purposes like:

- Drinking
- Cooking
- Washing
- Bathing
- Irrigation for crops
- Rearing animals
- Transport
- Fishing
- Hydroelectric power
- Construction works
- Aquatic sports and recreation etc.

Maximum usage of water is in the field of agriculture. About 69% of the total water consumed in the world is for agricultural purposes. Only about 8% is used for domestic and municipal purposes. The rest, i.e., 23% is used for industries of various types like mining, textiles, cement, leather, pharmaceuticals, paper, sugar and the like, for energy production (oil and gas production and power plant cooling) and for various construction works of buildings, bridges and the like.

Over-exploitation of water: Without stopping to think of the consequences, human beings are over-exploiting water from different sources in various ways. The main culprit, one can say, is the population explosion itself. This has led to an increase in the amount of water used by human beings directly and indirectly. When the world population is more, agriculture is more, industries are more; everything is more.

At the time of independence, the population of our country was 40 crores. Now it is 120 crores! Just imagine how much we would have developed in every field and how less problems we would have had to face, including the present shortages in water, electricity, and everything else if only our population had been stable at 40 crores!

Continuous over-exploitation of water has led to many problems.

- When too much water is withdrawn from the surface resources, wetlands dry up.
- Groundwater is also depleted due to overuse and lack of recharge.
- Too much irrigation in dry areas leads to soil turning saline when the water evaporates. This results in the reduction of yield.
- If water from near sea coasts is overdrawn, sea water moves into these wells.

The worst problem is freshwater shortage. There are 4 reasons for this:

- Aridity, i.e., water shortage due to dry climate
- Drought
- Desiccation, i.e., drying up of soil due to deforestation, overgrazing etc.
- Water stress, i.e. low availability due to overpopulation

It is high time the consequences of over-exploitation are understood clearly and corrective steps taken.

(b) Floods

In many countries like India there is enough rainfall, but most of it is only during a few months. There is no or very less rainfall rest of the year. The results are very disastrous.

- Soil gets waterlogged (too much water in the soil)
- Soil nutrients move deeper down in the earth
- Topsoil is washed away
- Flowing water overflows its normal channel leading to floods

Causes of floods: Floods are due to many factors, which act in various combinations and thus become very complex. Floods are difficult to manage and control due to the complex nature of these factors.

Natural causes:

- *Heavy rainfall*: Continuous heavy rainfall for long periods or melting of enormous quantities of snow leads to a huge increase in the quantity of water in a river. When this happens upstream, floods occur downstream. In arid (dry) and semi-arid areas, rivers cannot contain large quantities of water at any given time. Therefore, in such areas, if there is a heavy rainfall for continuous periods, rivers get flooded. This was what occurred in some parts of Rajasthan in the monsoon season of 2006.
- *Winding course of rivers*: Some rivers take many twists and turns during their course of flow. In such cases, the speed of flow is reduced. This results in accumulation of water in some places, which overflows beyond the banks.

- *Occurrence of vast floodplains*: On either side of some rivers like the Ganges, there are vast floodplains. Even comparatively small increases in the quantity of water in such rivers spread over large areas, causing floods.
- *Sudden change in slopes*: Rivers flow along slopes. If the slopes are uniform throughout the course of the river, the flow remains uniform. But if there are sudden changes in the gradient of slope in the river bed, floods result during heavy rains.
- *Landslides*: If there is a huge landslide upstream (in nature, these are caused by heavy rains and the resulting loosening of soil especially on the mountain sides), flash floods occur downstream.

Anthropogenic (man-made) factors:

- *Deforestation*: The commonest and most important man-made cause of floods is deforestation. Trees in forests are felled by man for various purposes like logging, mining, dam building, cultivation, overgrazing, forest fires etc. Removal of plant cover has many grave consequences:
  - (a) Infiltration capacity of the soil decreases, i.e., less water filters into the soil, resulting in more runoff.
  - (b) When there are trees and other plants, rainwater filters through the gaps in their leaves and branches in the form of tiny streams on the soft forest litter (fallen leaves etc.). In the absence of these plants, rainwater strikes at the land with force and gets no time to seep down. So it leads to rapid runoff.
  - (c) Roots of plants hold on tightly to soil particles and keep them in place. With the removal of plants, soil becomes loose and is washed off into rivers, making them shallower. This increases floods.
  - (d) Removal of the top layers of soil leads to a decrease in infiltration of rainwater in the valleys. Therefore, rain falling in the valleys runs off into rivers, causing floods.
- *Urbanization*: When any city spreads sideways, there is an increase in the cemented and concreted areas of land, tarred roads, pavements (footpaths) laid with bricks etc. All these reduce infiltration and increase runoff. The water flowing fast through the drains of the city reach the nearby river. Constructions of bridges across the river, dumping of garbage in the river etc. also follow urbanization. All these increase floods.
- *Changing the course of rivers*: Sometimes, the course of flow of rivers is changed by man to suit his convenience. If this is not done keeping the topography (form and surface behaviour of earth) of the place in mind, rivers get flooded.
- *Construction of dams*: One purpose of dams is to control floods. However, construction of dams results in obstructing the river flow. So,

when there is heavy rain, dam gates have to be opened for the safety of the dams, to prevent the dam from cracking under the enormous pressure of water. This results in flash floods.

- *Some agricultural practices*: In many farmlands, after the winter (rabi) crops are harvested, the farms are left to dry. Rain washes this soil into rivers, leading to siltation. Rivers become shallow and floods become more severe.
- *Man-induced landslides*: Major dams store enormous quantities of water. The mere pressure of this water may lead to earthquakes. Mining activities may also lead to earthquakes. Such man-induced earthquakes result in landslides. They block the water flow in rivers and cause floods.

People in poor countries have no choice; the poor usually live in lowlands prone to floods. Therefore, it is necessary that governments, non-governmental organizations and the people themselves understand the causes of floods and take remedial steps to prevent them.

It is a myth that the proposed interlinking of rivers in India will control floods. The average flood flows in the Ganges and the Brahmaputra are about 50,000 cumecs. The maximum flows that the project will be able to divert will be 1,700 and 3,000 cumecs respectively. This will not make any big difference. On the other hand, the 1000 km of canals planned may make the flood situation worse.

Worldwide, it is being understood that instead of flood control, it is flood management that must be done. India's failure to control floods even after almost 60 years of independence and after spending crores of rupees proves this point. The flood affected areas have been increasing steadily in India, after all this expenditure.

Methods of flood control: Following are some flood control methods:

- Channelization, i.e., deepening, widening or straightening a suitable section of a stream
- Lining the stream with concrete
- Building artificial embankments (stream bank protection)
- Building check dams, even minor dams across the stream/river

Methods of flood management: Some measures practised globally are given below.

- Prepare a flood frequency curve 1st based on the history of floods in the area.
- Based on this, prohibit certain types of buildings/activities in flood-prone zones
- Increase the height of buildings
- Construct flood proof buildings
- Construct floodways that allow the flood water to flow through with minimal damage

- Make a rule that construction projects proposed on flood-prone areas will not get permission from the government.
- Arrange flood insurances for the people living in such risk areas

(c) Droughts

Drought is a period when rainfall is nil and evaporation from the soil is higher than normal. There are at least 80 dry/semidry countries in the world. Forty per cent of the people live in these countries. They experience droughts throughout the year. Drought kills thousands of people every year in the world. Millions of poor people are compelled to live on drought-prone land in the developing and underdeveloped countries. Droughts are experienced to different degrees in many countries of Africa, Gulf countries of the Middle East, northern China, much of the former Soviet Union, Mexico, parts of India etc. (see Biogeographical Regions of India in Unit 4).

Land degradation and deforestation that induce droughts are also closely related to uncontrolled population growth. When there are more people in a country, more trees are cut, more cattle are grazed, land is cultivated more, leading to loss of fertility etc. Although there are huge lives in some countries, they flow through forests and their water is not available to the people, who face drought in other parts.

Poor natural resources and frequent droughts cause high variability in production. Because of this, farmers often do not have enough money, and sometimes they have no money at all to invest for the future. This is one reason why farmers in the dry lands remain poor forever.

Causes of drought:

- Actual deficit of rainfall
- Uneven distribution of rain
- Not managing the monsoon water by proper methods
- Deforestation
- Overgrazing
- Huge wastage of water by wrong methods of agriculture etc.
- Global climatic changes (In *El Nino* years when the ocean currents reverse their direction this happens periodically. There are storms in America and drought in Australia).

One-third of India is either facing drought or is under the threat of drought. Therefore it is essential to adopt suitable measures to counter the effects of drought. These can be listed as follows:

Methods to reduce drought and its effects: These are highly varied.

- Minimize the waste of water. About 70% of the total water available is used in agriculture. Therefore, there is an urgent necessity to reduce this amount. Many farmers practise flood irrigation even today. In this method, the fields are flooded with water. While the plants absorb what they need, huge quantities are lost by evaporation from soil. This is a highly wasteful method. Use of sprinklers, drip irrigation etc. must therefore be promoted among Indian farmers who are not aware of such systems.

- Plan water resources development and management properly. Water harvesting techniques must be adopted wherever possible.
- Crops suited to dry land must be cultivated in drought-prone zones. This is called dry land farming.
- Cultivate drought resistant varieties of crops. Plant breeders have already developed such varieties. Biotechnologists are also trying their best, with some successes achieved. For example, the International Crops Research Institute for the Semi-Arid Tropics (ICRISAT), Headquarters Hyderabad, is using transgenic technology (transferring genes from one species/variety to another) to develop drought resistant chickpea and groundnuts.
- Do not overuse rivers and wells for irrigation
- Do not clear forests, because just as this leads to floods, it also leads to droughts. When trees are cut down, in dry years, the soil dries out more quickly. The result is a man-made drought. It was such a drought that caused a devastating forest fire in Indonesia in 1997-98 — the biggest forest fire recorded in the world. Tropical forests are usually moist and do not burn fast, but due to the felling of trees the forests had dried so much that the fire set to clear land spreads uncontrollably.
- Community based rainwater harvesting has the potential to drought proof the entire country. India receives most of its rainfall in just 100 hours out of the 8760 hours in a year. If this water is not captured and stored, there will be no water in rest of the year. The solution is rain-water harvesting.

Drought is a natural phenomenon in most cases, therefore we cannot prevent it fully. But we can surely reduce the severity of droughts by suitable water conservation measures and thus prevent desertification.

(d) Conflicts over water

International scene: Water is everybody's need. Water resources are becoming lesser and lesser while population is increasing fast in many parts of the world. Agriculture based on irrigation has increased greatly and so has withdrawal of water from underground aquifers. Thus there is no wonder that war-like situations have arisen in many parts of the world over water. The next war in the Middle East may be over water, not oil. Most water here comes from 3 rivers—Nile, Jordan and Euphrates-Tigris. At 3% growth rate, the population is expected to double in 20 years. Conflicts over water from Nile among Ethiopia, Sudan and Egypt are spreading fast.

Nearer home, the sharing of the waters of Ganges and Brahmaputra has always been a tricky problem between India and Bangladesh. When floods bring misery to the people of Bihar, on most occasions, it is because Nepal opens its dam gates.

The national scene: Within India, there are many conflicts over water.

Karnataka - Tamilnadu: Sharing of water from Cauvery is a big conflict even now. The case is in the Supreme Court at present. The Cauvery Water Authority has been set up under court order.

Punjab - Haryana: The Supreme Court ordered the Punjab Govt. to release part of the water from Satluj to Yamuna many times, But Punjab has not constructed a canal for the same till now.

Orissa - Chattisgarh: These two states are supposed to share the waters from River Indravati as per a water sharing agreement drawn up in 1975. But Chattisgarh is complaining that Orissa is not releasing the amount of water it is supposed to.

Tamilnadu - Kerala: The bone of contention here is the inter-state tributary of Cauvery, River Bhavani. If Kerala diverts water in its territory, Tamilnadu will not get water.

Similar water sharing conflicts exist between Goa and Karnataka, Maharashtra, Goa and Karnataka etc. The most recent conflict, widely reported in the media, has been between Gujarat and Maharashtra. This conflict is over the raising the height of a dam on River Narmada. Environmental activist Medha Patekar is still fighting her *Narmada Bachao Andolan*. Other important personalities have joined her and her thousands of followers are demanding justice in all matters, proper rehabilitation of the displaced tribals being one of them.

Dams—Benefits and Problems

Dams are planned according to their geographic suitability. Dam location is strongly influenced by the nature of the rivers concerned. Dams like Bhakra-Nangal, Damodar etc. in India have turned large tracts of unproductive lands into prosperous fields. To conserve, retain and store surplus water, many projects have come up in India, like the Gandak and Kosi in Bihar, Thungabhadra in Andhra Pradesh and Karnataka, Hirakud in Orissa, Upper Krishna in Karnataka etc. Thus, in general, dams have many uses. This is the reason why dams—major, medium and minor—are built across many small and big rivers all over the world. One example is the Aswan dam across the Nile in Egypt. It has given a boost to the economy of Egypt.

Dams built across major rivers and streams serve the following purposes.

- They store water to be used in irrigation during dry seasons.
- Aggregate productivity benefits of major dams are great. Between 1951 and 2000, India's production of food grains increased fourfold, i.e., from 51 million tonnes to about 200 million tonnes. About two-thirds of this increase was in irrigated areas.

- Major dams increase the economic activity in the catchment area through construction and economic activity around the resevoir, such as tourism and fishing
- They prevent soil erosion.
- They provide drinking water.
- Major dams help to control floods.
- Hydroelectric projects at some dams generate the much needed electricity.
- Other methods of harvesting water for irrigation, such as groundwater and check dams have not become popular or efficient in India. So these cannot be relied upon to meet the needs of India's large and growing population.
- Alternate water harvesting systems are not cost-effective.
- They do not have the added advantages of hydropower generation.
- They cannot do anything about flood management.
- In the command areas, irrigation and agricultural productivity increase.
- In this area, poverty is reduced.
- Command areas become less vulnerable to rain etc.

Though dams are most essential for the development of a country, many small dams will fully serve the purpose of one large dam. The problem of rehabilitation of tribal and other people too, if at all is there, can be tackled much more easily.

It is incorrect to attribute the entire production gains in dam irrigated areas to dams. First, the increase in irrigation coincided with increased uptake of other inputs and technologies, such as high yield varieties beginning in the 1960s, fertilizer, machinery, and multi-cropping. Second, there are other methods of harvesting water for irrigation, and so some of the dam-irrigated areas would still have been irrigated even if the dams had not been built.

Problems with major dams:

- The energy generated from large hydroelectric projects has been continuously falling over the last 11 years in India.
- As per the projections made by the Planning Commission for future electricity needs of the country, annual capacity has to be increased by 5,040 MW for the next 25 years. This will need Rs. 30,000 crore each year.
- 0.67 million people, mostly tribal people, will be displaced each year.
- 100,000 ha of land, mostly forest land, will be submerged each year.
- Surrounding land will become waterlogged, and this will also lead to salinisation of soil.

- Habitats of animal populations will be disturbed; their migratory routes will be cut off.
- Dams built on seismic (earthquake prone) zones can be very dangerous. The huge quantity of water stored exerts a tremendous pressure on the ground below. This can lead to earthquakes.
- Fishing and water transportation will be affected.
- The cost of irrigating land by big dams is much more compared to that by minor dams.
- Big dams are affected worst by siltation, which will lead to degradation of soil in the forests.
- It is easier to reduce siltation and to desilt minor dams.
- In spite of having so many major dams, water use efficiency in Indian agriculture is one of the lowest in the world as admitted by Finance Minister P.Chidambaram in Feb. 2005.
- Ground water recharge is done better by minor dams.
- Dam construction and land submergence leads to significant loss of arable farmland and forest.
- The cost of building major dams escalates unimaginably high. When the ecological cost of deforestation is added to this, it becomes all the more. Minor dams do not lead to deforestation to this extent.
- Advocates of major dams say that we need more storage capacity in the country. But storage is a means, not an end. Storing huge amounts of water and then releasing it all of a sudden whenever there is a danger of breaching of the dam leads to flash floods bringing death and misery all around. This affects the poor people, mainly tribal, as they live in the forests or lowlands.
- Resettlement and rehabilitation packages are never properly implemented. This is very evident in the 3,000 odd dams built on River Narmada. This river has the largest number of dams built on it than any other river in the world.
- The total storage capacity of many existing minor dams and small surface storages has never been calculated. This must be very large.
- The existing storage capacity of major dams themselves is not utilized fully.
- Increase in the economic activity around major dams is either temporary or depends on the ability to learn new trades, and often cannot compensate for the loss of familiar livelihood.
- Research has shown that agricultural productivity in the catchment areas is unaffected.
- Major dams built on large rivers flood vast areas of tropical forests.
- Electricity is generated using the water flowing past the dams in the developing countries. This is used mostly in mining and smelting minerals, which are exported to the developed countries.

- In the Amazon basin, many large dams have been built, displacing many indigenous people.
- The local people living near River Nile in Egypt are facing loss of soil fertility and a rise in diseases caused by snails, as predicted by environmentalists.
- Waterlogging and increased salinity reduce the agricultural productivity in the vicinity of the reservoir.
- Policies to ensure adequate flow into the reservoir sometimes prohibit water harvesting in the catchment area, reducing agricultural productivity even more.
- Large-scale storing of water increases exposure to vector-borne diseases such as malaria, filariasis etc.
- The Indian government's compensation policy towards the displaced remains insufficient in many cases.
- Since the compensation is based on the amount of land owned, landless households were typically not compensated whatsoever.
- Flash floods result from opening of dam gates when the water stored crosses the danger level. This happens during the years of very high rainfall. This leads to untold misery to people living in the adjoining areas as flood gates are opened unannounced in many cases.
- Storing too much water in dams for the purpose of hydroelectric projects and irrigation may cause much harm, because the weight of the water in the dam itself may stress the earth, causing earthquakes.

A cost-benefit analysis suggests that the dams are, on an average, only marginally cost-effective. The increase in poverty in the catchment areas suggests that even though losers are clearly identified, they are rarely compensated properly. This finding suggests that losers do not have the institutional capacity to bargain for higher compensation. NGO's advocating for the losers are relatively weak or non-existent. The scheduled tribes have been found to be the most affected.

The report of the World Commission on Dams (WCD) was released by Nelson Mandela in November 2000. Though the Commission did not say that all big dams are bad or that big dams should not be built, the bad experience with such dams came to be known to the world. Many countries around the world decided to follow the recommendations of WCD. India however rejected the report and is still planning to get many big dams built.

Large dam construction has been a major and expensive undertaking opted for by the Indian government. This shows that the departments that do the planning and implementation of such projects must be made aware of the disadvantages of the major dams and the advantages of the minor dams in the present world.

2.2.3 Mineral Resources

A mineral is an element or an inorganic compound that occurs naturally and is solid. Some minerals are made of a single element, like gold, silver, diamond (carbon) and sulphur. However, most of the 2000 minerals identified till now are inorganic compounds, e.g., salt, mica, quartz etc. All these minerals are of economic importance to man; others form rocks. A mineral resource is a concentration of a naturally occurring solid, liquid or gaseous material inside or on the earth's crust and whose extraction is profitable.

Earlier it was thought that mineral resources are unlimited, so nobody cared for their conservation, and anyway, it was impossible to preserve them (Odum, 1971). But now we know that these beliefs are wrong. As population is increasing, per capita consumption is also increasing. Total resources available remain constant. The mineral resources are non-renewable. The only hope is that new deposits may be found or new technologies discovered to extract existing minerals profitably.

Any solid natural material containing compounds of metals in the combined state along with earthly impurities and found below the earth's surface is a mineral. Those minerals from which metals are extracted commercially at a comparatively low cost and with minimum effort are called ores of metals. Hence all ores of metals are minerals, but all minerals are not ores.

Formation of minerals: Minerals are formed in many ways. In some cases, erosion of rocks leads to concentration of ores. Seawater contains such ores brought from land e.g., magnesium and bromine. Sometimes, materials from rocks dissolve in circulating groundwater, and get deposited somewhere else as veins in the cracks of rocks, e.g., gold, copper, tin, lead, zinc and molybdenum. Sometimes, dissolved materials precipitate out of solutions and form minerals, e.g, sandstone, banded iron ores etc. When water in marine conditions evaporates, salt, potassium minerals, gypsum etc. are formed. Some minerals are formed from organisms. Phosphate rock deposits are phosphates from the bones and teeth of fish and other aquatic animals. Petroleum is formed from the remains of organisms; coal is formed from dead trees under heat and pressure of the earth.

(a) Uses of minerals

Coal is a very important solid fuel needed for many industries. It is also the commonest domestic fuel. It is the ultimate product of slow decomposition of large forests of trees buried underground and subjected to high heat and pressure. The amount of carbon in coal increases as time passes. Thus, different varieties of coal are present in nature. These are, starting from the poorest quality coal based on the carbon content, peat, lignite, bituminous and anthracite. Peat has only 60 – 70% carbon. Anthracite is the most superior coal, with 90 – 95% carbon and burns with maximum heat and minimum smoke.

Petroleum is mineral oil occurring in the pore spaces of sedimentary rocks. Crude oil is used as a fuel for steam ships and other locomotives. Petrol is used in automobiles and aeroplanes. Diesel is used in automobiles and trains. Huge amounts are burnt in industries. Kerosene is used as oil for lamps, cooking, lighting and heating. Hundreds of other by-products are made from it like lubricants, vaseline, tar, wax, ink, medicines, soap, terylene etc. Another product mined with petroleum is asphalt (bitumen or pitch). It is in fact a very thick oil. It is an excellent material for road surfacing.

Iron is the basis of the modern industrial complex. About 90% of the world's iron ore is used in making steel. Many industries are based on iron and steel. Important iron ores are:

Magnetite—Richest in metallic iron content, (55 to 72%). Seen in igneous rocks. Black in colour. World's best magnetite is found in Sweden.

Haematite—Fifty to 70% iron. Seen in sedimentary rocks. Reddish in colour. Most of the iron ore in Lake Superior area is haematite.

Limonite—Commonest of the commercial iron ore. Iron content up to 50%. Impurities very much. Brown in colour.

Siderite—A low grade ore, which has only less than 45% iron. Seen in large quantities in England and France.

Manganese is used in making good quality steel. Addition of manganese makes steel tough and hard and it does not rust easily. It is also used in many industries.

Aluminum is the most abundant metal making up about 80% of the earth's crust. It is obtained from an ore called bauxite, which contains about 25% of the metal. Aluminium is not strong and does not rust. It has a wide variety of uses.

Mica is a transparent mineral found in most igneous rocks. Some varieties are coloured. Mica splits easily into thin plates. Chief use is as an insulating material in electrical goods. Indian micas are the finest in the world because of the large size and perfect plates.

Gold differs from other metals as it is not combined with any metal. It occurs as veins running through quartz or other hard rocks. This can be exploited only by using costly machines for mining and crushing rocks. Gold is also found as grains in sand on some river beds. It is mainly used to make ornaments.

Copper was probably the first metal used by man. It forms different alloys like bronze, brass, German silver etc. Copper is used mostly in electric appliances.

Tin is a very useful metal for coating cans used to preserve food materials. It is also used for soldering and making different alloys.

Diamonds of superior quality are used in jewellery. Poor quality diamonds have many industrial uses.

Uranium and thorium These are used for generation of atomic energy.

In this way, different minerals are used in different ways.

(b) Over-exploitation of minerals

Minerals show uneven distribution in the world. Some areas are very rich in mineral resources such as Canada, South Africa, Some countries formed from the erstwhile USSR etc. while large areas like the Amazon basin, Netherlands, Denmark etc. are comparatively poor. In India, mining for minerals is mainly done in Uttar Pradesh, Bihar, Jharkhand, Orissa, Madhya Pradesh and Andhra Pradesh.

Some minerals are found in great abundance in only some parts of the world – for example, more than 69% of the world's supply of gold comes from South Africa. Similarly, there are vast reserves of petroleum in the Middle East.

India is rich in about 35 minerals. The coal reserves of India amount to about 86,000 tonnes; annual production is more than 150 million tonnes. Annual production of petroleum is more than 9 million tonnes. Reserves are seen in the valleys of the Brahmaputra, deltas of Cauvery, Krishna, Godavari and Mahanadi. Coastal areas of Kerala, Andhra Pradesh, Orissa and offshore areas near Mumbai also have deposits of petroleum.

However, mineral reserves of the world are not never-ending. As human population is expanding in leaps and bounds all over the world, especially in the poor countries, the need for various articles made of minerals is increasing in geometric progression. This leads to extraction of more and more minerals from the earth. That day may not be far off when the deposits of minerals may be exhausted fully.

(c) Environmental effects of extracting and using mineral resources

Extracting and using minerals affect the environment to a great extent. Extracted minerals like iron ore, bauxite etc. are used in the native countries or exported to the developed countries. Mines of mica, manganese, coal, limestone, iron ore, copper and many other metals and minerals lead to large scale destruction of forests. In fact, extraction of large quantities of minerals is one of the major causes of deforestation.

Iron ore is mined out from many mines in India. Large scale afforestation programmes are being done continuously for regeneration of the forests cut down for mining in some cases as in Noamundi, Jharkhand. Yet plantation forests are very unlike natural forests in that they do not have the original biodiversity, either plants or animals. It is also possible that diverse species have been lost for ever.

Burning of fossil fuels like coal and petroleum adds to the level of CO_2 in the atmosphere and causes air pollution. It also results in global warming and climate change. Similarly, use of various minerals cause environmental pollution of different kinds. For example, gold mining is very damaging because mercury, which is a toxic substance, is used to separate gold from sand and gravel. Once the mining is over, the mercury is left at the site, polluting the soil and the groundwater. (See relevant parts in Units 5 and 6 for more examples)

Underground mining for minerals is more dangerous than surface mining. Accidents occur more often in underground mines. These are mainly due to the lack of ventilation, falling of mine roofs, flooding of mines, explosions, inhalation of toxic fumes etc. In the absence of adequate precautionary measures in mines of radioactive substances, like uranium, there will be the risk of radiation sickness or even cancer, in addition to mutations (changes in the genes) of reproductive cells leading to birth of deformed children in the families of workers.

Case studies

1. *Asbestos waste on Roro Hills in Jharkhand:* Asbestos is a carcinogenic substance; it induces cancer. It has been shown to cause cancer of lungs and other organs as well as breathing disorders. The World Health Organisation (WHO) estimates that at least 90,000 people die every year due to asbestos related diseases. Yet it is used extensively for different purposes, mainly in building construction, mostly to add to cement, in the LDC's. It is also used as an insulating and fire proofing agent in buildings, ships etc. Mining of asbestos is done in many places. More than twenty years ago, Hyderabad Asbestos Cement Products Ltd. (now known as Hyderabad Industries Ltd.) used to mine asbestos from the Roro mine, situated about 20 km away from Chaibasa in Jharkhand. This was one of the largest asbestos mines in India, producing about 4 to 5 lakh tonnes of asbestos annually. In 1983, when the company realized that the grade of the asbestos was no longer good, the mine was closed down and the assets taken away.

However, thousands of tonnes of asbestos waste were left on the Roro hills. This waste is poisoning land, water and vegetation around. River Roro, flowing down from the hills, joins Subarnarekha downstream, carrying tonnes of asbestos waste. A community of about 5,000 tribal people is living within a 5 km radius of the waste dump. The small paddy fields of the tribal people are covered with the waste. People cross the dump sites while grazing their cattle. When children slide down the slopes of the hill, clouds of lethal dust rise in the air. It is high time this carcinogenic waste is disposed so that the health of the poor tribal people is not affected any more.

In this case, mining has resulted in a double environmental destruction—it has resulted in deforestation of the Roro hills alongwith all the dangers associated with it; it is also causing serious pollution problems.

2. *Marble mining in Aravallis, Rajasthan:* Aravallis are one of the oldest hill systems of Rajasthan and Haryana. These mountain ranges have a major role in the control of climate and the drainage system in the entire region. These also act as a mini watershed as many seasonal streams originate from the Aravallis. A lot of mining goes on here. The mines are made of marble,

granite and other types of stones. In 2002, the Supreme Court of India put a blanket ban on all mining activities in the Aravallis. As this affected the lives of 65,00,000 lives in Rajasthan—workers in the 9,700 industrial units with their families—the ban had to be lifted after one week. However, during the 7-day period, the water level in the wells in the region went up. The people of Rajasthan realized that Magsaysay Award winner Rajendra Singh was right when he said that "Mindless mining is responsible for the depletion of water levels in Rajasthan. In fact, mining affects *jal, jangal* and *jameen* (water, forest and land)".

2.2.4 Food Resources

It has been estimated that the earth has about 30,000 plant species with parts that people can eat, but 90% of our food comes from just 15 plant and 8 animal species. Wheat, rice, corn and potato together make-up more of the world's food production than all the other crops taken together. Plants are our main food providers. Meat of animals and animal products like milk, cheese, butter etc, some types of birds and their eggs and fish form another part of our food. Thus, we get our food from agriculture, animal husbandry and fishing. However, all animals ultimately depend on plants for food. Thus we need to increase agricultural production greatly to keep pace with the ever increasing population.

In India and many other developing countries, to boost agriculture, several methods that have harmful effects on the environment are used. These are against the policy of sustainable agriculture. In sustainable agriculture, air, water and land are not polluted by any means and the natural gene pool is also not disturbed. If the aim of agriculture becomes just increase in yield without any thought of the environment, the ecosystems in nature will be harmed beyond repair.

The fact that only sustainable agriculture will ultimately pay is being realized by all. Production of all crops, including cereals, oil seeds, fruits, vegetables etc. has increased in India considerably in the recent years. After a very serious drought of 2002-03, we have made an impressive recovery in all areas of agricultural production.

Agriculture is of many types:

- Industrialised agriculture uses huge amounts of fossil fuel energy, water, fertilizers and pesticides to produce large quantities of single crops or animals for sale.
- Plantation agriculture is cultivation of cash crops like banana, coffee etc. in the developing countries mainly for sale in developed countries.
- Traditional subsistence agriculture is that done by families on a small scale. They produce only what they need in the family; some may be sold.
- Traditional intensive agriculture uses a lot of inputs including human and animal labour.

In traditional agriculture, intercropping (2 or more different crops grown in the same field), polyvarietal cultivation (many varieties of the same crop in one field), agro forestry (crops and trees planted together) etc. are done. All these are good for the soil and are being practised in India.

In the fisheries sector, coastal fish stocks are being exploited in India while resources in the deeper seas are not exploited. This is again an offshoot of poverty as most of our fishermen cannot afford to buy trawlers.

As for freshwaters, as the state governments have the ownership rights of ponds and lakes, proper development is not taking place. In general, Indian aquaculture (rearing and using fish and other water animals and plants) needs to improve a lot to be able to compete with global markets.

On the other hand, milk production in India shows great promise. A growth rate of 4.5% has been achieved by our dairy sector during the last decade. We have already overtaken the USA in milk production. We are moving further ahead. The target of doubling milk production by 2015 is definitely possible. India's ranking is good in poultry market too. The output of eggs in our country in 2004-05 was 48 billion. Our rank is among the top 6 egg producing countries in the world.

Livestock population in India is large and widely diversified. We possess the largest cattle and buffalo population in the world. Goat population is second only to China. The total production of livestock products has also increased greatly over the past 3 decades.

With all these increases in production of food resources, the reason for poverty in India is mainly due to population explosion, faulty public distribution system, lack of adequate cold storage facilities for the produce, lack of sufficient units for making preserved products, poor transportation due to lack of roads or bad quality of roads etc.

(a) World food problems

As the year 2000 began, an estimate showed that 1.1 billion people were undernourished and underweight. And the World Bank estimated that 1.3 billion people were living in poverty, at the beginning of the year 2000. Hunger and poverty go together. There are many reasons for the food problems that the world faces today.

Population explosion—This is a very big problem, especially in the Asian continent and Africa. In India, with more than a billion people, 53% of all children are undernourished, and in Bangladesh, 56%. In all such countries, most of the poor live in villages, and population growth rates are alarming.

Subdivision of farms—When a farm passes from one generation to the next, it is subdivided among the children. In India, in the last 30 years, the number of farms with less than 2 hectares of land has doubled. In Bangladesh, the average farm size has already fallen below 1 hectare. Thus in addition to the real landless, many others are 'effectively landless' as their plots are extremely small.

Scarcity of water—Another major setback of increasing world food production is scarcity of water. Of the nearly 3 billion people to be added to the world population in the next 50 years, almost all will be added in countries already facing water shortage, like India, Pakistan and many countries of Africa. In India, water tables are falling in large areas. Water is taken for granted and is wasted in many ways in agriculture and other fields.

Increasing demand for animal food—Over the last century, the world's demand for animal food has increased. There is a great economic disparity in different countries and also in different parts of the same country. Thus, the number of rich people has increased greatly in the world even when the number of poor has increased. The rich demand animal protein more. To rear animals, more and more land is converted into grazing pastures; great quantities of world's grains are also used. This adds to the shortage of grains for human consumption.

Falling of fish production—Water pollution has increased in many water bodies. Added to this is the fact that in many countries, fish are caught even in the breeding season. Due to both these, fish production is falling in many places. As fish is an important source of protein, fisheries must be given due importance. Over-exploitation using modern trawlers must be discouraged.

Over-dependence on some countries—Many countries are importing food grains from developed countries. Just 5 countries—United States, Canada, France, Australia and Argentina together export 88% of the world's exports. Thailand, China, Vietnam and the US account for 68% of all rice exports. In the case of corn, US alone exports 78% of the world's needs. In such a situation, importing countries face great risks. If the production of grains in the US falls due to any natural calamity, the whole world will suffer.

Lack of taste for new foods—There are so many plant species on this earth, but human beings use very few as food. There is a need to develop a taste for food made from plants, which have not become popular yet. Similarly, the sea has abundant food resources in the seaweeds. Yet these have not found a popular market till now, except in some countries like Japan.

Gender inequality—A social problem can also be cited here. In many countries, women are not getting their due share in anything. Food is no exception. Whenever there is a shortage of food in the family, it is the women who sacrifice it for the sake of the other members of the family. This is one reason why anaemia and other such effects of malnutrition are more prevalent among women and girls. Considering that women form 50% of the people in the world, this is not a good sign. To correct such anomalies, social changes are essential.

(b) Changes caused by agriculture and overgrazing

Wrong policies used in agriculture and overgrazing have disastrous effects on air, water and land. These are as follows:

- Monoculture (cultivation of single crop in a field) is very bad for land. It depletes soil of nutrients.

- When a field is irrigated too much, the salts in the deeper layers of the soil are brought up. This makes the soil saline and waterlogged.
- In many places, marshes are drained and made into fields. This disturbs the whole ecosystem.
- Trees in forests are burnt down to convert them into croplands. This also has many serious consequences.
- Shifting cultivation is bad for the land. Here, when a plot becomes infertile due to agriculture, the farmer abandons that plot and starts cultivation in another plot.
- By depending only on hybrid seeds only, farmers are neglecting traditional seeds completely. This leads to the loss of genetic diversity.
- When grains and other crop produce from the developed world are imported at cheap price in the developing countries, the farmers in the latter countries adopt any method, sustainable or unsustainable, to compete with the imported stuff. This degrades the quality of the land forever.
- Too much fossil fuel is burnt to run the machinery and pump water.
- Use of too much fertilizers and pesticides in agriculture pollutes the environment.
- Overgrazing is very bad for the soil. The soil becomes hard as heavy animals walk on it for long. Water cannot go down in such a hard soil. Seeds cannot germinate in hard soils. Soon, the area becomes barren as the tops of the grasses and other plants are munched off by cattle and other animals. Thus overgrazing is one way by which deserts spread.

(c) Effects of modern agriculture

The main solution to the world food problem is raising cropland productivity. All countries, including India, are trying their best to do this. Many effects of modern agricultural practices are seen all over the world these days.

- The most popular and common practice is to try to increase the yield per unit area. This process is called Green Revolution. For this, many methods are being used.
  1. Plant breeding has given rise to many better varieties than the usually cultivated ones in many aspects like increased yield, disease resistance, drought resistance etc.
  2. Irrigation is being done on a much larger scale.
  3. More artificial and biofertilizers are being used.
  4. Better weather forecasts help reduce losses.
  5. Better awareness regarding modern practices of soil testing, disease control, seed treatment, use of hormones, biopesticides etc. have helped to increase the yield.

6. Biotechnology and genetic engineering are of help too. Tissue culture plants have become very common. Transgenic plants (genetically modified crops) have also made their appearance.

- Another method is to increase the number of crops in each field by multiple cropping. The farms where monoculture was the practice are being converted into multiple cropping fields. This has led to an increase in world food production greatly.
- The third method is to utilize different crop residues for various purposes. Thus, now we have rice bran oil, pellets made of crop residues for fish ponds, biogas plants etc. Such measures increase the overall efficiency of integrated farming (farming in which agriculture, animal husbandry, poultry, fishing etc. are all integrated).

Green Revolution involves many steps:

1. Developing monocultures of high yielding varieties
2. Using fertilizers, pesticides and water in large quantities
3. Multiple cropping

This was the first Green Revolution that began in 1950.

A second Green Revolution has been taking place since 1967. It was at this time that fast growing dwarf varieties of rice and wheat, especially bred for tropical countries, were released. Between 1970 and 1992, India doubled its food production.

However, the harmful effects of Green Revolution are being felt more and more. An extract from M. S. Swaminathan's address to the Indian Science Congress held at Varanasi in January 1968 is being reproduced here:

"Intensive cultivation of land without conservation of soil fertility and soil structure would ultimately lead to springing up of deserts. Irrigation without arrangements for drainage would result in soils getting alkaline or saline. Indiscriminate use of pesticides, fungicides and herbicides could cause adverse changes in biological balance as well as lead to an increase in the incidence of cancer and other diseases, through the toxic residues present in the grains or other edible parts. Unscientific tapping of underground water would lead to the rapid exhaustion of this wonderful capital resource left to us through ages of natural farming. The rapid replacement of numerous locally adapted varieties with one or two high yielding strains in large contiguous areas would result in the spread of serious diseases capable of wiping out entire crops ..."

Thus it is most important to think of the consequences before starting new programmes in agriculture or for that matter for any scheme in increasing world food production.

(d) Fertilizer-pesticide problems

Use of inorganic fertilizers in agriculture to increase production is nothing new. These have been and in fact are being used worldwide. As far as India is concerned, fertilizers have played an important role in making the country self-sufficient in food grain production. Urea is the most widely used fertilizer.

It contributes about 80% of the nitrogen needs of the country. Other common fertilizers used are ammonium sulphate, potassium nitrate, superphosphate etc.

Some agricultural scientists argue that fertilizer use is a must in India due to two reasons.

1. As more land cannot be brought under agriculture in the already over-populated India, productivity has to be increased per unit area.
2. The main grains used here are wheat and rice, and both these give higher yield by the addition of nitrogen.

They even feel that the present prices of fertilizers should be brought down by the government as most of our farmers are small or marginal farmers. They believe there is need to increase both fertilizer consumption and productivity of agriculture.

On the other hand, some other agriculturists as well as all environmentalists are against the use of too much fertilizer. They argue that by reducing the price of nitrogen fertilizers the government is promoting inefficient fertilizer use. Even after a 10% reduction in the cost of nitrogenous fertilizers, production did not increase proportionately. This is because soils in India suffer from the deficiency of many nutrients.

Moreover, the gap between nutrient addition and nutrient removal is very high. This shows that soil fertility is being depleted. Therefore, the solution is not adding more fertilizers, but an integrated nutrient supply. Alongwith fertilizers, organic manures like cowdung, green manure, crop residue, compost, vermicompost, biofertilizers etc. must be given.

Use of too many inorganic fertilizers pollutes not only soil, but also water and air. From the soil, the inorganic festilizers reach surface water along with the runaway water. From the surface waters, they leach into aquifers also. Fertilizers, which are used in greater quantities than pesticides, are bad for many aquatic organisms. It has been found that some amphibians are very sensitive to the nitrogen compounds leaching out of the fields. For example, the Oregon spotted frogs are poisoned by even very low nitrates or nitrites—so low that it passes the drinking water quality standards set by the US Environmental Protection Agency (EPA).

One main problem is that farmers apply too much nitrogen to the soil for better yields. It has been proved that in China, crops use only 40% of the given nitrogen. The rest gets dissolved in water and moves down to join the aquifers. In many parts of Europe, US and India also, farm runoff is polluting surface and ground water.

Some of the extra nitrogen from the fertilizers is released into the air in the form of nitrogen oxides. These are converted into nitric acid (HNO_3) that returns to the earth along with rainwater. This acid rain makes soils acidic.

Solutions to fertilizer problem:

- Reduce the use of inorganic fertilizers in farms; use them in combination with organic manure like compost, vermicompost etc.

- Use biofertilizers like blue green algae, Azolla (a water fern) etc. in fields.
- Do not use fertilizers on slopes at all.
- Use slow release fertilizers if possible (such fertilisers have been introduced in the market. They release the chemicals slowly into the soil).
- Alternate rows of crops with pulses—the bacteria in the roots of pulses fix atmospheric nitrogen, so there is no need to give extra nitrogen to the plants.
- Plant some permanent plants between cultivated fields and nearby water bodies.

Pesticides: These are chemicals that kill pests, i.e., undesirable organisms like harmful insects. In a broad sense of the term, insecticides (chemicals that kill insects), fungicides (chemicals that kill fungi), weedicides (chemicals that kill weeds) etc. are all pesticides. However, in general, pesticides are considered to be the same as insecticides.

To save plants from pests of different types, pesticides like DDT, dieldrin, aldrine etc. are used. DDT is a non-biodegradable compound. It does not decompose but accumulates in the soil. Different types of weedicides like 2,4-D; 2,3,5-T etc are sprayed in croplands to kill unwanted weeds.

Harmful effects of pesticides:

- Pesticides pollute soil.
- They are washed off from the crop plants during rain or irrigation and reach water bodies too.
- The remains of pesticides on the leaves or fruits make them poisonous. Those who eat unwashed fruits and vegetables have a chance of getting food porsoning.
- Pesticides also remain on the grains.
- Alongwith the crop residue, they enter into the bodies of farm animals and make them ill.
- Dioxin is a by-product releated during the synthesis of herbicides. This compound was used extensively in the Vietnam War to kill crops in large areas. It can be seen many kilometers around the site of application. Dioxin is extremely toxic to mammals, causing liver disorders, nerve problems and even cancer.

Methods to control pesticide pollution:

- Apply pesticides only when needed.
- Reduce the need of pesticides by using biological control mathods.
- Use biopesticides like neem seed oil, neem oil cakes etc. in place of chemical pesticides.
- Reduce non-farm use of pesticides in lawns, public lands, golf courses etc.

- In animal farms, control animal density.
- Use runoff from animal farms on croplands or forests.

These measures are usually not taken seriously by many. According to a survey made by the WHO (World Health Organization), more than 50,000 people are poisoned and more than 5,000 die in developing countries as a result of toxic chemicals used in agriculture. Samples of fruits, vegetables, eggs and milk collected from many cities in India contain pesticide residues much higher than human tolerance limits by WHO. Pesticide content in soft drinks like Coco Cola and Pepsi has become a big controversy in the recent days.

(e) Waterlogging

Excess irrigation leads to waterlogging. Clayey soils with fine colloidal particles and high water holding capacity also get waterlogged easily. Such soils are heavy, with high plasticity and stickiness. It swells when moist and shrinks when dry. This is very harmful for plant roots. Interspaces in waterlogged soils are very small. All or most of them are filled with water, so there is no free movement of air. The lack of air affects the plant roots badly and water absorption is greatly retarded. Thus, even if there is water all around, plant roots cannot absorb water from the soil due to lack of air. This is a case of physiological dryness for the plants. If the condition persists for long, the plants die. In short, waterlogged soils cannot support plant life.

In the absence of soil air, many microbes also die in waterlogged soils. These include microbes that are useful to plants in many ways too, like the nitrifying bacteria, etc. The bacteria that are useful to plants are in general aerobic. They cannot thrive in anaerobic conditions resulting from waterlogging in the soil.

Moreover, the salts in the soil dissolve in the irrigation water and rise to the surface. When the water evaporates, the salts are left on the soil surface.

Thus, waterlogging leads to land degradation. Data of World Bank Survey (1995) revealed that India loses 1.2 to 6 million tonnes of food grains every year due to waterlogging. Therefore, land degraded by waterlogging must be reclaimed by various methods like drilling sinkholes in the soil and then filling them with pebbles and sand, using the area for cultivation of plants tolerant to waterlogging like species of Iris, Primula, Salix etc.

(f) Salinity

In areas where rainfall is low, if the drainage is not proper, soluble salts accumulate in the top soil. It also takes place as a result of evaporation of water from soils having salt water like sea coasts etc. Soils that have a high concentration of soluble salts are called alkali soils irrespective of their reaction. CO_3s of Na and K are alkaline; they give a black or dark brown colour to the soil. Such a soil is called black alkali. Cl, SO_4 and NO_3 of Na, K and Mg are neutral in reaction. Soils rich in these are called truly saline soils or white alkali, as these salts produce a white layer on the surface.

Sometimes, salts accumulate in the lower layers of the soil. This does not produce saline soils, but when they are irrigated, or when groundwater rises and evaporates, a high concentration of salts is left on the soil. Patches of alkali soils are thus formed.

Extremely alkaline and saline soils are found in UP, Punjab and Rajasthan. They have no vegetation as absorption cannot take place even if there is plenty of water in the soil due to excess concentration of minerals. Such soils are physiologically dry. Roots may lose water and plants may wilt. Roots are dwarfed and transpiration rates are low. Germination is retarded greatly; so plants take to vivipary. Plants that can grow on soils with high salt concentration are called halophytes.

Excessive salts are very bad for plants. Saline soils are a national waste. To make such soils fit for agriculture, salts should be removed.

Methods used to reclaim saline soils:

1. Mechanical—Water tables are lowered to 5 or 6 feet below the soil surface by a network of deep cuts at right angles to the slope. The land is then irrigated by good water. Large amounts of salts leach down below the normal root zone. Two or 3 leachings should be done to remove the salts. In the beginning, salt-tolerant plants like rice, sugarcane, barley etc. can be grown. Surface incrustations of salts are removed by scrapers or a rapid water stream.
2. Chemical—Saline soils are reclaimed by the addition of gypsum, i.e., $CaSO_4$. It reacts with Cl's of Na and K to form insoluble CO_3's and SO_4's of Na and K. Soluble salts can be washed out by heavy irrigation and drainage.
3. Biological—There are many biological methods for the reclamation of saline soils.
   (i) Reforestation through micropropagation—Degraded land can be reclaimed by planting forests using trees of wide adaptability and high productivity. Such trees can be grown by clonal micropropagation through tissue culture. Superior genotypes of many species capable of growing on saline and other types of degraded soils have been selected for mass multiplication. Examples are *Butea monosperma, Terminalia bellerica* etc.
   (ii) Through genetic engineering—Plants resistant to salinity have been found. Research is being done to find out the molecular basis of salt tolerance and to isolate genes responsible for this, so that salt tolerant plants can be developed through genetic engineering.
   (iii) Use of mycorrhizae in reforestation—On degraded land, only a brief period is available for growth. On saline soils, for example, concentration of salts will be comparatively less in the rainy season. If saplings do not establish by then, plants die. Mycorrizae can

improve seedling survival and growth by increasing the uptake of essential nutrients, particularly P and water by lengthening root life and providing protection against pathogens. A list of such fungi has been prepared. They can be inoculated in sapling roots.

(iv) Use of microbes for improving fertility—N_2 fixing bacteria like *Rhizobium* can be used to induce nodule formation in many legumes and non-legumes. These can be used to improve the fertility of degraded land.

Thus reclamation of saline land can be done by various methods.

(g) Case Studies

1. *The price of feeding livestock*: Natural resources are being used wastefully all over the world. Consuming beef is a good example. Eating beef ultimately leads to degrading the environment and contributing to hunger and poor health, consuming huge quantities of precious water etc. More than one-third of the world's grain harvest is used to feed livestock. About 70 to 80% of grain produced in the United States is fed to livestock. A lot of rainforests in the Amazon and elsewhere are cleared for raising cattle, not for local consumption, but for fast food restaurants elsewhere. Subsidies in farming in the US and many other western countries lead to encouraging unhealthy foods in comparison to cheaper and healthier foods. If the cost of water is added, the price of any food item will multiply manifold.
2. *Organic farming by farmers in one district of Maharashtra*: Farmers in one district of Maharashtra are taking to organic farming in a big way. In January 2005, a group of farmers who had stopped using chemical fertilizers a year ago gathered under sprawling canopies of tamarind and neem trees to share their experiences with government officials. The officials were briskly making notes, for a change!
3. *Increasing salinisation of land in Australia*: Seven to 8% of the dryland area in Australia has been identified as salinised at present. This may go up to a maximum of 50% in the next 50 years, according to a case study done by Mike Read and others.

2.2.5 Energy Resources

All forms of life need different types of resources to live. Energy is one of them. Life on earth depsends on the sun. Ninety-nine per cent of the energy that living organisms use comes directly or indirectly from the sun. Thus solar energy is the most important source of energy for us.

Green plants are the primary producers, i.e., they trap solar energy and use it in making food for themselves and all other organisms on earth. (This process by which plants make food using solar energy is called photosynthesis, i.e. synthesis or manufacture in the presence of light.) In addition to man's primary need of food, most of his other needs are also met with ultimately by plants.

It is the oxygen given out mainly by plants during photosynthesis that we breathe in. Similarly, the carbon dioxide that we breathe out is used by plants for photosynthesis. This shows that the whole cycle of exchange between oxygen and carbon dioxide, so essential for life to exist on earth, is running smoothly because of the solar energy.

Solar energy also helps directly or indirectly in the circulation of material in this world. For example, it is the heat of the sun that evaporates water in the water bodies and soil into water vapour which forms clouds in the sky. Rain formed from the clouds falls into the water bodies and on the soil. This is the main part of the water cycle or the hydrological cycle.

It is because of the water cycle that rivers do not dry up. The electricity that we get in our homes comes mainly from hydroelectric projects, which use the force of the running water of the rivers. Electricity is most essential for industries and to run agricultural machinery. In fact, we cannot imagine life without electricity.

Electricity is also produced using thermal power. For the generation of heat, coal is the fuel used. Coal is nothing but trees that lived many years ago and are now buried under the ground, which have been subjected to high pressure and temperature. Thus, the energy that we get while burning coal is actually the solar energy trapped by these trees thousands of years ago in photosynthesis.

Petrol and diesel used for transport and kerosene used for lighting by the rural poor are obtained from petroleum, which is another fossil fuel. It occurs as crude oil in underground oil fields in different parts of the world. Petroleum is formed by bacterial decomposition of marine microscopic plant and animal fossils. The energy in these organisms that lived in the past was of course solar energy converted into food energy. Transport of people, raw materials and manufactured goods from place to place plays a most significant role in today's every day life.

Thus almost all the energy the living organisms use comes ultimately from the sun. Other forms of energy are atomic energy, wind power, tidal energy etc.

(a) Growing energy needs

People need energy for many things—agriculture, domestic appliances, industry, transport etc. All these are much more in developed countries. Agriculture is almost fully mechanized there; industries and transport facilities are more there; people use a variety of electrical appliances there. In short, the energy needs of the developed countries far exceed the energy needs of the developing countries.

Some of the most important minerals such as potroleum and coal are used as fuel for power both in developing as well as developed countries. These and natural gases are fossil fuels. These form a major chunk of the fuel energy used all over the world. These resources evolved millions of years ago as plant and animal matter decomposed and was converted under high temperature and pressure under the earth's surface into hydrocarbon compounds that we now call fossils fuels. Since the beginning of the machine age, industrial societies have become increasingly dependent on fossil fuels. This is a cause of concern for two main reasons:

1. Fossil fuel resource is limited in amount. These were produced by solar energy in the distant past and when they are gone, there will be no more. Fossils are still being formed, but at a very, very negligible rate compared to the rate at which we are consuming them. We first began consuming fossil fuels at an appreciable rate only about 150 years ago. How long will they last? On a global scale, we will still have some coal for a few centuries more, but natural gas and oil will be in short supply in a few decades.

   In the United States, the situation is worse than the global average because they are depleting their resources at a faster rate than in other fossil fuel rich countries. In fact, the entire stock of fossil fuel available has been held in storage under the earth's surface for more than a hundred million years, and now it is being completely exploited in only a few centuries.

2. Unexpected environmental consequences result from the extensive use of fossil fuels for everything from cooking in our kitchens to powering our automobiles. When we burn coal, natural gas or oil to obtain energy, gaseous compounds are formed and dumped into the atmosphere. This is causing problems. For many years, it was felt that the emitted gases were not significant as the earth's atmosphere is vast. But now with increasing world population and industrialization, this is no longer true. Atmospheric pollution is causing health problems and even death. Carbon dioxide emissions are threatening to produce climatic changes over the entire globe. It is thus high time we find solutions to these problems of resource depletion and environmental pollution

Coal, petroleum and natural gas give conventional forms of energy. Wood which is used mainly in the developing countries, also provides a conventional form of energy. In these countries, large quantities of wood from forests are used as fuel. This leads to deforestation, which causes soil erosion, floods, desertification and many other grave consequences. Burning of firewood leads to atmospheric pollution. Thus, all conventional forms of energy cause pollution.

The energy that is sold in the market is called commercial energy. This is what the developed countries use most. In the developing countries, most people use non-commercial energy like firewood, dung and crop wastes that they collect on their own. These together can be called biomass. Thus, while the quantity of biomass used as energy source is just 3% in the developed countries, it is as much as 35% in the developing countries. Similarly, there are differences in other forms of energy used too. The percentage of natural gas and nuclear energy used by developed countries is much higher.

The developed countries are sure to face shortage of energy resources in a few decades; but the developing countries are already facing it. Cutting wood for fuel has led to deforestation in vast areas in these countries.

The USA uses maximum energy resources in the world. The Americans use 25% of the world's commercial energy even though they account for only 4.7% of the world's population. They also waste a large quantity of energy. At the same time, though 16% of the world's people are Indians, they use only 3% of the world's commercial energy.

Yet, a critical analysis will show that energy needs are growing everywhere. The developing countries are striving hard to become developed, and for this, they will have to spend more energy. At the same time, the developed countries are also making material advances fast. Thus, energy needs are increasing at a fast pace all over the world. It has been estimated that the world's energy needs have increased four times during the period from 1950 to 1990. The world's demand for electricity has doubled in the past 22 years. Added to this is the fact that electrification is far from complete in all settlements of the developing world. A majority of the rural areas in the LDC's do not have electricity at present. Even where there are electric connections, supply is irregular due to real shortage in generation or loss in transmission. In such a situation, wasting electric power is a crime.

It is because of non-availability of conventional energy that man is now trying to tap non-conventional energy resources more and more for his various needs.

(b) Renewable and Non-renewable Energy Resources

Energy resources are of 2 types—non-renewable and renewable. The renewable energy sources pollute the environment very little, while the non-renewable forms pollute the environment on a large scale.

There are basic differences between the use of commercial energy in the developed or More Developed Countries (MDC's) and the developing or Less Developed Countries (LDC's). While the MDC's use more of non-renewable energy, the LDC's use more of renewable energy.

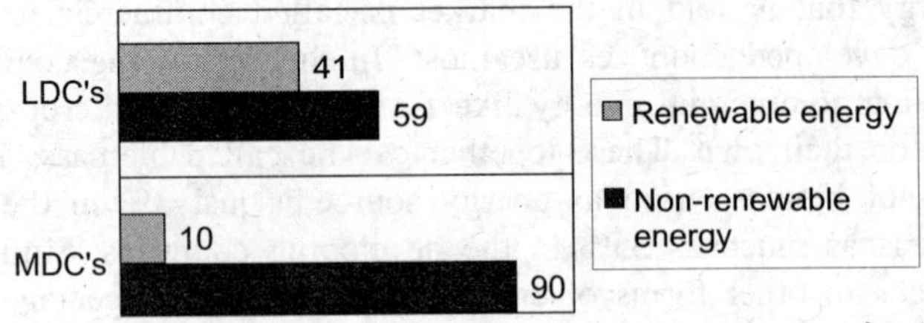

Figure 2.1 Commercial energy use

Table 2.2 Use of non-renewable energy

	Coal	*Oil*	*Natural gas*	*Nuclear power*	*Total*
MDC's	25	37	23	5	90
LDC's	25	26	7	1	59

Table 2.3 Use of renewable energy

	Biomass	*Hydropower, Geothermal, Solar*	*Total*
MDC's	33	7	10
LDC's	35	6	41

From this data reproduced from Tyler Miller Jr., it is clear that the MDC's are using more non-renewable sources of energy available in the world. But this does not mean that we can relax. We should also try to reduce our use of non-renewable energy, especially coal and oil, as they are highly polluting.

1. *Renewable energy sources*

If a resource can be renewed in a short time, such as a few hours to a few decades through natural processes, it is called a renewable resource. Examples of renewable energy resources are solar energy, hydropower, wind energy, geothermal energy, tidal energy, energy from burning biomass (firewood, dried dung, municipal solid waste, agrowaste etc.) and the like.

The non-renewable energy sources have been in use for centuries now. These are therefore also called conventional energy sources. The renewable sources, also called non-conventional or alternative energy sources are becoming more and more popular these days as the scare of the underground deposits of coal, petroleum and natural gas getting exhausted has become a major concern. On the whole, we should turn to alternative energy sources as soon as possible and as much as possible. There is no fear that they will be exhausted; they are environment friendly also.

2. *Non-renewable Energy Sources*

Energy resources that are available only in limited amounts in nature, as it takes a very long time, tens of thousands of years or more, to be renewed, are called non-renewable energy resources. So if continuously used, their supply will end one day. Examples of non-renewable energy resources are coal, mineral oil, natural gas and nuclear power. Of these, coal, oil and natural gas are organic in origin and can be considered fossils. So they are also called fossil fuels.

Coal: Coal is one of the oldest forms of energy. The location of industries was earlier decided by nearness to coal deposits. However, the discovery of oil and invention of electricity have reduced the importance of coal. Its low heat output, untidy appearance and great volume, which increases transport costs, are negative factors. Still, it is used widely in many countries as a domestic fuel, source of industrial power and raw material in some chemical industries.

Coal is a very important solid fuel needed for many industries. It is also the commonest domestic fuel. There are different types of coal containing different amounts of carbon. The more the carbon content, the higher is the quality of the coal.

Table 2.4

S.No.	*Type of coal*	*% of carbon*	*Qualities as a fuel*
1	Peat	50-60%	Lowest grade coal, producing less heat and more smoke and ash
2	Lignite or Brown Coal	60-70%	Burns easily with some smoke
3	Bituminous coal (Black hard coal)	70-90%	Burns with lesser smoke, used as domestic fuel
4	Anthracite	90-95%	The best quality coal. Very hard, shining, producing intense heat and minimum smoke

Anthracite comprises only 5% of the available coal. It is used mainly for making steam in industries and for home heating in cold countries. Due to scarcity and higher price, anthracite is used only when other types of coal are less suitable.

Bituminous coal is the most abundant. It is of 3 types. Steam coal is the superior grade used in steam ships, locomotives and engines. Coking coal softens on burning to form coke. Coke is used in blast furnaces to extract iron from iron ore. Gas coal is the lowest rank of bituminous coal. When burnt, it gives off coal gas and other by-products.

Lignite is a low category coal. It is widely used in domestic heating, to generate thermal electricity and in chemical industry.

Peat is formed from dead plant remains that settle at the bottom of ponds. It is taken out, dried, cut and burnt in homes. It has only local importance.

Petroleum: Petroleum is also called mineral oil. Oilfields are situated near coastal regions. An oilfield contains a layer of petroleum and natural gases above a layer of salt water. On drilling a hole into the oil well, which may be a few hundred to a few thousands of feet deep, crude oil and natural gases gush out. The crude oil is then carried by pipelines to oil refineries where it is refined into petrol, diesel, kerosene etc. A large number of by-products are also obtained from the crude oil.

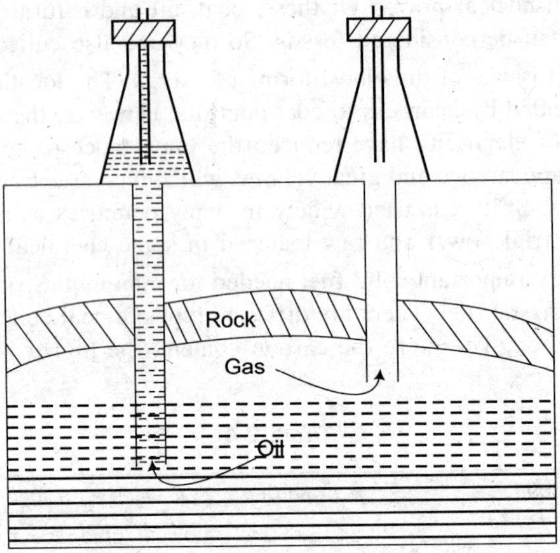

Figure 2.2 An oilfield

Uses of petroleum:

- It is an indispensable motor fuel. Without petrol and diesel, the modern transport system will simply break down.
- It is a vital source of power. Vast amounts of mineral oil are burnt daily in heaters, boilers and furnaces to provide power in factories and to generate thermal electricity.
- By-products are many, like grease used as lubricant for machines, paraffin and wax used to make candles etc.

Natural Gas: Natural gas occurs along with petroleum. Some wells yield oil, but not much gas, while some other wells yield only gas. Natural gas is mainly methane and ethane. The gas has to be properly purified before using as a source of power. It is used in industries and also for domestic purposes, such as heating and lighting. It is often cheaper than electricity. In America, Europe, Russia and New Zealand, natural gas is pumped through pipes for industrial and household purposes.

Nuclear Energy: Strictly speaking, only very minute quantities of Uranium are needed to make nuclear power, and so in effect, the supply is limitless. However, as uranium deposits underground are limited, nuclear energy is included in non-renewable resources.

Nuclear energy was discovered in 1938-39 in Germany. To produce nuclear energy, the nuclei of uranium atoms are split with the help of neutrons. When the nuclei split, mass is converted into energy. The amount of this energy is very large. The heat generated during fission (splitting of nuclei of atoms) is used to heat water. One kg of uranium can generate energy equivalent to that obtained from 3,000 tonnes of coal.

Initially nuclear energy was used for producing nuclear weapons for the world war. The first nuclear reactor was assembled at the University of Chicago in 1942. After the war, efforts were made to utilize nuclear reactors for the generation of commercial electricity, and propelling submarines and other ships of the navy.

Electricity for civilians was produced first by a reactor at Pennsylvania in 1957. This was just the beginning. Many countries set up nuclear reactors. In India, Dr. Homi Bhabha is the father of nuclear power development. Bhabha Atomic Research Centre (BARC) is doing research in this line. India has 10 nuclear reactors and produces 2% of our electricity. India has uranium deposits in Jadugoda in Jharkhand. Thorium deposits are found in Southern Kerala. Our power plants are located in Maharashtra, Rajasthan, Tamil Nadu, Uttar Pradesh and Gujarat.

Nuclear energy is an inexpensive and environment friendly source of energy. However, it also has some serious disadvantages:

1. Even when maximum care is taken, some radioactive gases may escape out through microscopic cracks in the reactor.
2. Serious accidents can occur with nuclear reactors as happened in Chernobyl.
3. The biggest problem is the safe disposal of nuclear waste.

Due to these drawbacks, many countries have stopped tapping this source. For example, Sweden has decided to become nuclear free by 2010.

(c) Use of alternative energy resources

Solar energy: The energy from the sun is a renewable resource as the sun will continue to give energy to the earth for more than about 6 billion years. So we can consider solar energy as inexhaustible. The energy that the earth receives from the sun in the form of electromagnetic radiation is solar energy. It includes infrared, ultraviolet etc., in addition to visible light.

To put solar energy to use, different types of solar cells have been designed. These are used for different purposes ranging from heating water to cooking to converting solar energy directly into electrical energy. Various technologies have been developed for these.

Solar energy is used most commonly for heating water. Such systems have two parts—a solar collector and a solar tank. The collector, kept on the terrace, heats the water; the heated water flows into a well insulated storage tank. Collectors are of many types. A flat-plate collector is an insulated rectangular box. It has a transparent cover that faces the sun. Inside the box is a metallic plate that is painted black. Many tubes are attached to the metallic plate. Water flows through these plates. As the collector gets hot, the water is heated too. Pipes carrying this hot water to the tank and from there to wherever it has to be used are well insulated too. Such solar water heaters have become very common. These can be seen in many hostels, restaurants etc.

Solar energy is non-polluting and safe. However variations in solar intensity over day and night, summer and winter and also changing cloud cover complicate the use of solar energy. There must be a storage system for cloudy and rainy days.

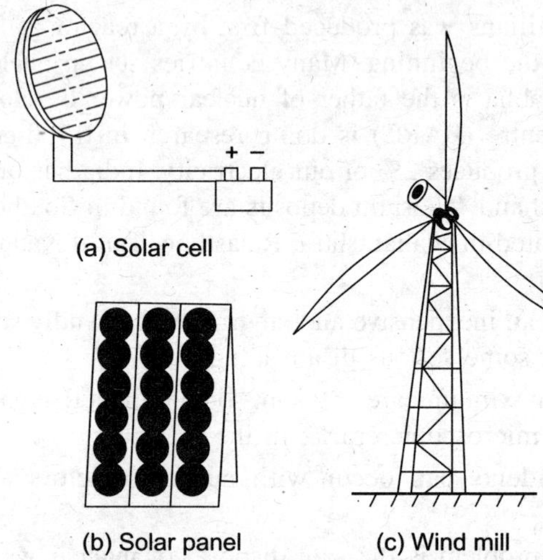

Figure 2.3 (a) Solar cell (b) Solar panel (c) Wind mill

Solar cells, also called solar photovoltaic cells, have become common in some countries. A solar cell has two layers of silicon. Each layer contains an electric charge. When sunlight strikes the cells, the charges move between the two silicon layers and electricity is produced. Calculators, watches etc. running on solar cells have become very common.

In India, in more than 300 villages in Rajasthan, Gujarat, UP and Haryana, solar energy is used for street lighting. Solar furnaces at Pune, Rohtak, Hissar etc. are functioning very well. As such, we dry our clothes in the sun. This is also in fact the usage of solar energy. In developed countries, imagine how much electrical energy is used in the washing machines with dryers.

Wind energy: In ancient times, sailors used wind blowing on the masts to steer their boats. The large ships of the nineteenth century could extract as much as 10,000 hp (horsepower) energy from wind. Wind mills in thousands have been operating in different countries from the nineteenth century onwards. Wind energy is used to run mills, pumps etc. and also to generate electricity.

In Denmark, 13% of the country's electricity is produced using wind energy. The Government of Denmark is encouraging people to establish wind co-operatives. In these co-operatives established by the Government, people can invest their money. Thousands of Danish families have shares in such wind co-operatives. Wind mills produce noise and obstruct view, so the government has made it compulsory for the investors who benefit from wind energy to live within 3 km of the site. This ensures that no movements of opposition arise from any quarter. On the other hand, the non-polluting nature of the energy is making it more popular.

In India, Gujarat is the state where wind energy is being used extensively. Tamil Nadu produces 850 mW of electricity from wind energy. India is the third largest producer of wind energy in the world at present.

Wind energy has two disadvantages. Firstly, as wind velocities are not constant, energy produced will fluctuate. Secondly, large areas are needed to set up wind mills. Yet advantages outweigh the disadvantages.

Geothermal energy: The naturally occurring heat from the interior of the earth flowing to the cooler surface is called geothermal energy. Where heat flows are greater, volcanoes, hot springs etc. occur. Underground hot water has been used to produce electricity. Steam from geysers is used to drive turbines. Deep wells can be dug into the underground hot water reservoirs and the hot water brought to the surface. Technologies have been developed for this. This has become necessary as natural geysers are very rare.

No widespread commercially beneficial technologies have been developed thus far that can use geothermal energy of hot rocks, molten magma etc. However, people have used geothermal energy from hot springs for bathing, aqua culture, green-housing etc. In India, about 350 geothermal springs have been located. Good sites for power generation exist in Ladakh, MP, UP and HP.

Tidal energy: Tidal energy is extracted from the potential and kinetic energy of the earth-sun-moon system. Ocean tides result from this interaction. Water level fluctuates twice daily, alternately filling and emptying along the shore line. These currents can be used to drive water turbines connected to generators to produce electricity. Two systems are in practice. In the one-way system, the incoming tide is blocked and the water collected is used to turn turbines when it goes back. In the two-way system, both incoming and outgoing tides are used to generate electricity. The technology is very similar to hydropower. Tidal power stations are operating, in Russia, France, Canada etc. In India, tidal power potential has been identified in the Gulf of Kutch, Gulf of Cambay, and the Sunderbans.

Disadvantages of tidal energy: The times of the tides keep changing. Storms may bring irregularities. Power stations disturb aquatic life and block navigation.

Hydroelectric power: Hydel power or hydroelectric power is the most important source of electricity. In large projects, water stored in dams built across large rivers is made to flow through huge pipes. This spins turbines to produce electricity. This is called hydropower. Fifty per cent of the electricity in the developing countries like India is produced by this method. In Norway, 95% of the electricity is hydropower. The hydroelectric plants have long durability. The source of energy is renewable, cost of operation is low, and it is a non-polluting energy. These are the advantages of hydroelectric power. However, large dams are not built anymore in any country as their disadvantages are many. Therefore only small projects are made. These use low dams on small streams, with or without reservoirs.

Energy from Hydrogen, the gaseous fuel: Hydrogen fuelled automobiles are set to hit the market on an experimental basis within months. BMW, the automobile manufacturers of Germany, have designed such cars. These cars will release water vapour instead of the poisonous carbon monoxide, nitrous oxides etc. into the air! If such cars do become popular gradually, much of our air pollution due to automobiles will be reduced.

Energy for biofuel from Jatropha: Jatropha is a plant that is being cultivated widely these days. Extracts from this plant can be distilled into a substitute for petrol. It is cheaper than petrol and is a non-polluting biofuel.

Energy from biomass: This is of different types. Agro-wastes like waste wood, green plant matter, straw, paddy husk, saw dust etc. can be converted into fuel. In 1989, a rice straw fired thermal plant was set up in Punjab. Paddy husk can be made into smokeless solid fuel briquettes. These can be used for cooking at home, hotels, kilns and boilers. Such briquettes are made at the alternate energy centre at Roorkee. Rice bran is used as a source of oil, which is later converted into methyl ester for use as fuel. Saw dust is converted into low calorific producer gas for thermal power generation. At the Central Mechanical Engineering and Research Institute at Durgapur, since 1988, solid garbage (leafy waste) is used to produce electricity. The garbage is digested anaerobically to produce biogas, which in turn is used to generate electricity.

Biogas plants: It is true that India lives in her villages. For the villages, the best form of non-polluting alternative energy is biogas. Biogas plants are very useful in converting solid waste into energy. Biogas is a mixture of methane (CH_4), hydrogen (H_2), carbon dioxide (CO_2) and nitrogen (N_2). It is produced by degradation of biological substances. Sixty to 70 % of biogas is CH_4; 25-30 % is CO_2 and the rest is composed of H_2 and N_2. About 600 litres of biogas can be produced per kg of dry organic matter.

In India, in many states, small and cheap biogas plants have been set up in villages. Biogas is commonly called gobar gas in India as the main raw material used is dung. Biogas is the best solution for energy problems in rural areas. The technology involved is environment friendly. Cow dung is available in our country in large quantities. Biogas made from cow dung can be used in place of firewood and kerosene in our villages. In the process, tonnes of wood can be saved every year. A byproduct of the process called slurry can be used as manure.

There are different designs for biogas plants, but in all modern ones the digesters (fermenters) are almost entirely underground. In a biogas plant, dung and water are mixed in the ratio 1:1. Dung is fed at a constant rate to the fermenter. Other than dung, the input may include recycled slurry, night soil, kitchen waste etc. These help in the growth of anaerobic bacteria. The pH is maintained at 7.

The cow dung mixing tank is placed on an earthen platform. It is fed with a continuous supply of cow dung mixture. From this tank, cow dung moves through a cow dung inlet pipe into the fermenter tank. There is a platform at the base of the tank.

The fermenter tank and other units are made with brick and cement or concrete. The lid is usually of tin. The biogas produced in the underground fermenter tank is kept under pressure below the lid held in place by a heavy iron rod. Biogas collects above the slurry level. From here, it is directed into gas outlet pipes. The flow of gas is controlled by the gas cock. The gas outlet pipe is connected to main pipes carrying biogas to gas stoves in homes. Biogas burns without smoke and is an excellent fuel.

The used slurry is directed to drying beds where it is converted into organic manure. Biogas can also be used as fuel in a generator to produce electricity. Using 1m^3 of biogas, one sixty watt bulb can give light for six to seven hours, or three meals for six persons can be cooked, or one HP motor can be run for two hours.

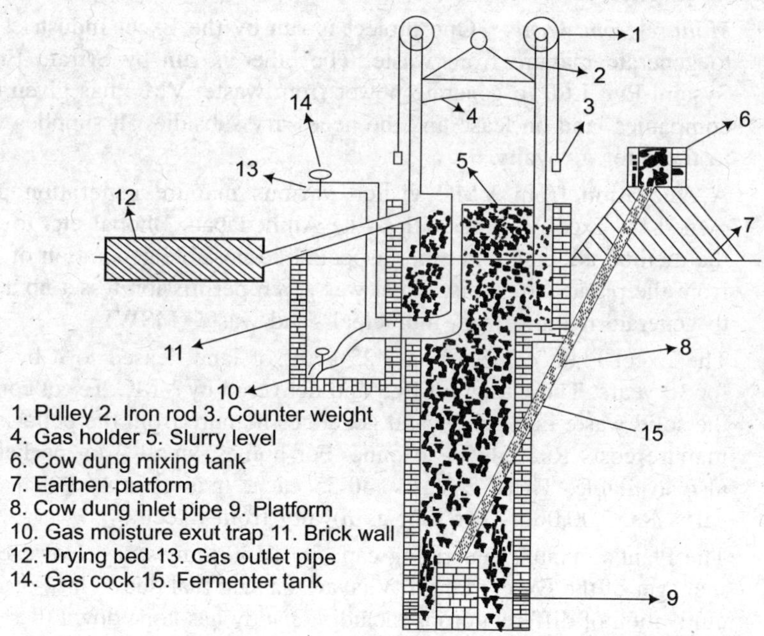

Figure 2.4 Design of a Biogas Plant developed at Indian Agricultural Research Institute (IARI)

Benefits of biogas plants: Biogas plants are beneficial in different ways.

1. Prevent pollution—a lot of smoke is emitted on burning *uplas* (goyttas). Moreover, making *uplas* is difficult in the rainy season.
2. Increase fertility—wastes from biogas plants are used as manure.
3. Deforestation is prevented—no need to fell small trees and shrubs for firewood.
4. Improves family health—by avoiding smoke and dust, soot and coal.
5. Improves economic condition—villagers can lead a more prosperous life as biogas is cheap, convenient and healthy.

In India, biogas produced in villages is being used as fuel in homes. We have a few million families using biogas as a fuel. In China also biogas is very popular. In many European countries like Germany, France, England, Denmark etc, biogas is used for producing electricity on a very large scale. In India, some cities have begun such large scale projects. There is need to set up more such plants.

Case Studies:

Within the city limits of Vijayawada in Andhra Pradesh, 500 tonnes of solid wastes are generated every day. Of this, the Vijayawada Municipal Corporation (VMC) is disposing of 350 tonnes of waste every day with the help of 2 innovative projects. The Corporation has involved the private sector in the process to put into practice the concept of 'Wealth from Waste'.

1. *Manure from waste*: One project is run by the Excel Industries Ltd. to generate manure from waste. The other is run by Sriram Energy System Pvt. Ltd. to generate power from waste. VMC has given the 2 companies land on lease and the necessary subsidies. It supplies waste to them for a royalty.

   A delegation from VMC visited various manure generation plants owned by Excel in other cities like Ahmedabad, Bhopal etc. to study the method adopted and other aspects like possible generation of gases from the plant. After this, Excel was given permission to set up a plant to generate manure from municipal solid waste (MSW).

   The Excel Plant came up on 8.25 acres of land, leased to it by VMC for 10 years. The waste is lifted and delivered by VMC. Excel converts the solid waste into pellets and generates manure from the pellets. This manure costs Rs. 2,400/- a tonne. For homes, small 2 kg packets are also available. These cost Rs. 30-35 each. It is estimated that VMC earns Rs. 7,00,000/- per year as royalty from Excel.

   The Plant's manure generating capacity is 50 tonnes/day. However, the manager of the Excel Plant in Vijayawada said that due to lack of water, cultivation of different crops including paddy has gone down these days and so the demand for manure has come down too. This may be just a temporary phase as the Govt. has plans to deal with water shortage on a large scale.

2. *Electricity from waste*: In 2002, VMC entered into an agreement with Sreeram Energy Systems Ltd., Hyderabad. The firm agreed to generate 6 MW of power per day from solid waste. The firm set up Rs. 45.86 crores power generation unit at Singhnagar. VMC leased 10 acres of land to the company for 10 years. The company agreed to pay a royalty of Rs. 50/- per tonne and an annual lease amount of Rs. 1,28,000/- and also water charges @ Rs. 20/- per 1.000 kilolitres to VMC. It is the responsibility of VMC to supply 600 kilolitres of water and the promised MSW per day to the plant.

   At present, the company gets 350 to 400 tonnes of solid waste per day from VMC. The company makes pellets out of the MSW and the power generated is sold to the AP Govt. at Rs. 3.37 per unit. The plant started power generation in 2003. In 2005, the production was 1,45,000 to 1,50,000 units per day.

3. *Biomethanation Plant*: Seeing the success of such non-conventional energy sources, VMC has now proposed setting up a 20 tonne capacity biomethanation plant to generate 135 kw of power, in addition to 1,500 cubic metres of biogas, using solid wastes generated from vegetable markets and slaughter houses. Manure would be a by-product of this plant. VMC, as a joint sponsor along with UNDP, is planning to share 25% of the cost and also give an acre of land for the purpose. VMC is expecting to earn Rs. 23,00,000 per annum from this plant, also likely to come up near Singhnagar.

VMC has ensured large-scale plantation in the disposal yards of all projects to mitigate pollution. Checks are also conducted regularly by the Pollution Control Board. However, a complaint has come up recently that Sriram Energy is not using the total MSW it takes and this unused waste is emanating stench. If the company does not solve this problem, its lease will be cancelled and then the firm will have to shift to a far-off place as per the Supreme Court guidelines. It is therefore expected that the company will take heed.

2.2.6 Land Resources

Land is another very important natural resource. The forests that maintain the whole balance of nature and conserve valuable wildlife, the croplands on which we grow all our crops, the pastures on which our cattle graze, the land on which rural and urban settlements are made and industries are set up etc. are all land resources. Thus, it is essential to maintain the quality of land all over the world. Areas under shallow water are called wetlands. These occur between the dry land and water bodies. The protection of wetlands is also essential to maintain the equilibrium of the ecosystems of the earth.

One-fifth of the earth's surface is land. Of this, 51% is not usable as it consists of ice or snow, or deserts or mountains. Forests, croplands, grazing land and arid land are included in the rest 41%.

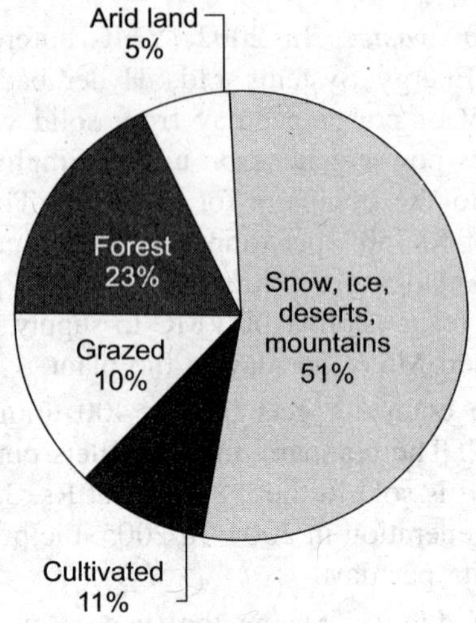

Figure 2.5 Land use percentages (world figures)

In India, however, these percentages are very different. First of all, cultivated land is much more here; it is the maximum in the world. This is because ours is an agricultural country and grazing lands (pastures) are very less here. Moreover, forest land is also less here compared to many countries in the world.

Wastelands include the alkaline, saline, arid, rocky and sandy deserts and the mountains and the highly uneven land. Part of the wasteland is used as residential land.

Land is also needed for transport, industries, commerce, recreation etc. As the total land is fixed, towns and cities will have to grow upwards now. Sidewise growth will hamper land being used for other purposes.

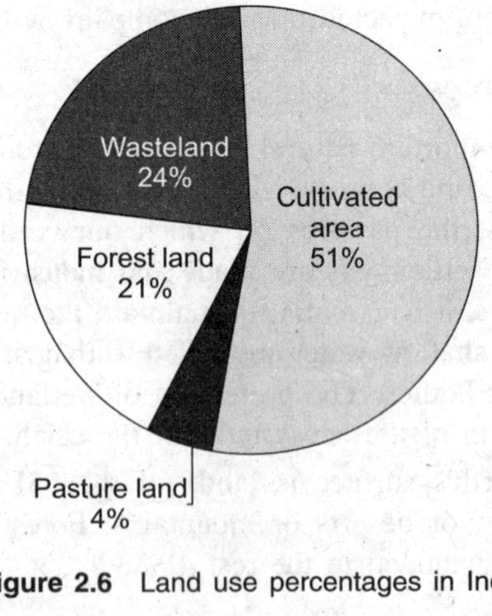

Figure 2.6 Land use percentages in India

Land use planning is not done properly in India. For example, fertile agricultural land in the Yamuna basin is being used for the spread of industry. Because of the high population, per capita land availability is very less in India. Man-land ratio is very low, 0.48 ha/person. The Government must take steps to survey the whole country properly, formulate the best land use plans and implement them strictly.

As 76% of Indians live in villages, it is all the more important to have proper plans because this vast population depends on land for their livelihood. If forests are depleted or pastures are overgrazed or toxic industrial wastes are dumped on it, valuable land will be converted into wasteland. On the other hand, if utilized carefully and intelligently, land becomes a renewable natural resource. Maintaining some natural areas as protected areas is also essential to conserve our vast biodiversity.

(a) Land degradation

There has been a great increase in agricultural land in India since 1950. This was necessitated by the ever increasing population. In addition to more and more land being converted into croplands, the existing croplands became over-utilized too. Increased irrigation has resulted in waterlogging, bringing the salts in the deeper layers of soil over the ground. Use of agrochemicals (fertilizers and pesticides) has polluted the land. These and many other factors have been responsible for the fast pace at which land is being degraded in many parts of the world. It has been estimated that every year about 5 to 7 million hectares of land is getting degraded in the world. Thus, the area of usable land is dwindling year after year.

Reasons for land degradation:

1. ***Deforestation:*** Deforestation is the main cause of land degradation. Unmindful of the consequences, large areas of forest land are being degraded all over the world. Since agriculture began about 10,000 years ago, word's forest areas have decreased from 34% to 26% of the world land area. Forests are not only disappearing, but also being fragmented due to roads, rails, human settlements, dams etc.

In India, the figures available are not much reliable. This is due to two reasons. First of all, many non-forest areas like scrublands are included as forest land. Secondly, contradictory reports from different sources and agencies are circulated and one cannot know which one is true. Moreover the area of plantation forests should not be added to that of natural forests. In India, the forest area is 23% of the land area, but the forest cover is only 19%. This difference is because plantation forests are taken to be forests, which they are not. What we see today in India is the sight of natural forests depleting and plantation forest increasing in area.

Dense forest cover that is undisturbed is called pristine forest cover. When the area of the pristine forest cover decreases, the population of wild animals like tiger also falls. Forest cover depletion is not only due to tree felling, but also due to human activities like converting forest land into croplands or fruit orchards.

In Himachal Pradesh, large areas of natural forests have been cleared for apple cultivation. This is causing land degradation on a large scale due to many factors:

- Apple starts yielding in 5 to 7 years. During this period, potato is cultivated as inter-crop. Potato plants cannot bind soil at all thus soil erosion increases greatly.
- Monoculture of apple leads to death of many wild plants, animals and microorganisms.
- To pack apples cultivated in 1 hectare of cleared forest land, wood from 7 hectares is needed for the packing cases. This leads to further deforestation.
- Huge amounts of fungicides sprayed on the apple trees to save the apples from the scab disease pollutes streams and land equally.

2. ***Urbanization:*** As urbanization goes on unchecked in the world, more and more land is degraded due to many reasons. As the population of cities increases, there is a corresponding increase in many factors. The total concreted (pucca) surface increases greatly. The quantity of wastes also increases in amount. Moreover, there is a rise in the number of:

- slums;
- buildings;
- roads;
- streets;
- sewage drains;
- vehicles;
- factories etc.

All these lead to deterioration in the quality of land. Soil becomes hard and loses the capacity to hold and retain water. Pollution kills many useful soil microbes.

3. ***Industrialization:*** Industrial development is good to boost the economy of a country, but it should not be at the cost of environment. The main problem created is pollution. The effluents, sometimes even hazardous wastes dumped outside the industries or even in the dumping yard are injurious to land. Substantial land area has been degraded due to dumping of industrial wastes.

4. ***Mining:*** During mining operations, large quantities of wastes are produced. For example, while mining metallic ores, large quantities of rocks, earthen materials and other wastes are produced. While mining for coal, surface materials from the earth's crust are removed to expose the coal seams. These are heaped at other places. These may contain toxic substances too. The wastes gradually become mountains that occupy the valuable land surface. The mined areas also, if not reclaimed, become degraded land.

5. ***Water logging:*** Excess irrigation leads to waterlogging. Waterlogged soils cannot support plant life.

6. ***Desertification:*** Desertification is a process by which the productive potential of a land reduces by 10% or more. It is caused mainly by human activities.

7. ***Acid Rain:*** Industries like thermal power plants, crude oil refineries and automobiles release large amounts of sulphur dioxide into the air. It dissolves in the water films on the suspended particles in the air forming sulphuric acid. This corrosive acid falls down to the earth along with rainwater. This is called acid rain. Nitrogen oxides released by some industries form nitric acid; this also causes acid rain. Acid rain makes soils acidic, thus degrading the quality of the soil.

8. ***Nutrient disorder:*** When a land is cultivated intensely and the crops absorb the nutrients they need from the soil, the land becomes degraded due to mineral deficiency. The opposite of this also happens in croplands. Too much fertilizer is added to the soil. Both these are included in nutrient disorder and lead to land degradation.

9. ***Extension of cultivation to marginal land:*** In many developing countries, when the poor have no land to cultivate, they try to cultivate the crops on marginal lands like slopes, sandy soils etc. Such lands are less fertile and get degraded fast.

10. ***Soil erosion:*** This is a major cause of land degradation. The loss of top soil or disturbance of the soil structure is called soil erosion. Since all the nutrients needed by plants are present only in the top soil, its loss is great resulting in land degradation.

(b) Man induced landslides

Tumbling down of huge landmasses is called a landslide. It may be natural or man-made. Landslides may be a natural phenomenon caused by heavy rain and the resulting softening of the soil, earthquakes etc. They can be anthropogenic (man-made) also. For example, mining on hills makes the ground shaky, leading to landslides. Deforestation on slopes results in severe soil erosion. In the absence of trees and other plants, which fix the soil in its place by their roots, huge masses of land tumble down.

If landmasses from hill slopes fall into rivers, the flow of water is blocked. This may result in flooding. If the landmasses fall from slopes on to passing vehicles on roads, loss of life and damage to the vehicles result. Roads also get blocked, and till the time the blocks are removed, all transport comes to a standstill.

There are serious instances of landslides caused by deforestation on the slopes of hills. When trees are cut down from mountains and mountainous slopes, the roots of tree no longer hold the soil particles in place. If these hills or mountains are made of comparatively soft and weak rocks, the slopes are unstable and they fall down in a mass, causing landslides.

Natural landslides are often caused by earthquakes. In 1970, there was a severe earthquake in Peru. Due to the shocks of the earthquake, ice caps on the peak of a high mountain collapsed. This mountain was located near a town. When the ice cap collapsed, it tumbled down. The falling ice mass dislodged rocks from the sides of the mountain which also started falling down at high speed. Tonnes of rock masses fell down along with ice blocks at a speed of 320 kms/hr. As the landslide continued, more and more debris and water joined the falling mass. Thus, the volume of the debris became very much. It travelled 15 km down the mountain sides within a few minutes. Many houses were destroyed completely and 25,000 people were killed.

Another major landslide was caused by an earthquake in 1989 in Tajik, part of the former USSR. A lot of snow had melted and filled the pore spaces of mountain slopes. Because of this, the slopes had become unstable. So when the earthquake struck, huge landslides resulted. The mud dissolved in the water to form muddy rivers. These rivers were 20 metres deep and flowed down with great speed. Whole villages were wiped off. The earthquake that struck Tajik was not very severe. It measured only 5.5 on the Richter scale. But it had occurred in the rainy season in a mountain, therefore it had caused a landslide. Earthquakes occur in the mountains in the dry season also, but people do not notice this. The slopes are stable in the dry seasons and hence there are no landslides.

Deep underground mining also leads to landslides. On many occasions, rocks are blasted for the construction of roads, dams and reservoirs. When major dams are built, enormous quantities of water are stored. This water causes the building up of great hydrostatic pressure underground, which may result in landslide and flow of mud along the reservoir walls.

One of the reasons why construction of a high dam at Tehri in the River Bhageerathi has always been opposed by environmentalists is the real scare of landslides. As the whole region falls under earthquake-prone zone, environmentalists argue that the pressure resulting from storing such a huge quantity of water in the dam will cause earthquakes with fissures in the ground, resulting in huge landslides from the sides of the mountain.

(c) Soil erosion

Soil is the top cover of the earth in which plants can grow. 4 inches of soil is necessary for growing any economically important plant—for food, clothing or timber etc. We depend completely on plants directly or indirectly for all our basic needs, and plants depend on soil. This shows how much important soil is for us. Destruction of the top soil has dangerous consequences. One inch of top soil takes 400 to 1000 years to form. Yet, a single dust storm can blow it away or a single flood can wash it away.

Removal of the upper part of the soil is called soil erosion. In soil erosion, the soil particles of the upper layers are removed from their original places and deposited in some other place.

Soil erosion is one of the most serious problems faced by the world today. Every year, about 2,500 million tonnes of soil are lost in the world. Due to unchecked soil erosion, large areas in Africa, China and India and parts of America have been converted into deserts. In the USA alone, 3 million acres of land has been damaged by soil erosion. In India, thousands of tonnes of top soil containing nutrients worth thousands of crores of rupees is washed away every year. Soil erosion results in siltation of lakes and reservoirs on which we have invested thousands of crores. The Bhakra Nangal dam was designed to last for 400 years, but due to siltation caused by soil erosion, it may not last even for 150 years. Silt is choking our harbours. Eroded soil is deposited on river beds, raising them. This leads to floods, killing people and damaging property worth crores.

Soil erosion by wind is also severe in many parts of the world. In India, it affects about 50 million hectares of land, mostly in Rajasthan. This area is spreading due to deforestation, overgrazing and desertification. The fine dust is always suspended over the desert, so moisture laden clouds passing over Rajasthan do not precipitate into rain.

Soil erosion can be broadly classified into 2, based on the rate of loss. These are:

1. *Normal (Geological) erosion*: This is a natural, slow process. Here, equilibrium between loss and build up is never lost. Soil is lost only as slowly as it is formed.
2. *Accelerated erosion*: This is very rapid. It is much faster than soil formation.

A. *Erosion by water*: It is of 5 types:

(a) *Sheet erosion*: Removal of soil in small but uniform amounts all over like a sheet is called sheet erosion. The removed soil can be seen as silt in some other place.

(b) *Rill erosion*: Here, the running water moves rapidly and cuts small streams like channels on the land.

(c) *Gully erosion*: Many rills converge along slopes and join to form broad channels called gullies. The surface becomes uneven.

(d) *Riparian or stream bank erosion*: This happens on river banks. The surface of fast moving rivers cuts the bank laterally. When a lot of soil is cut off in this way from beneath, the soil above tumbles down into the river with a big splash.

(e) *Slip erosion (landslide)*: In the rainy season, rocks or masses of soil slip down. This is called slip erosion or landslide.

B. *Erosion by wind*: It is also of different types:

(a) *Saltation*: Salty soil carried by wind in small quantities is called saltation.

(b) *Suspension*: In suspension, fine dust is lifted in the air and remains suspended there.

(c) *Surface creep*: This is pushing of heavy soil particles by strong wind.

Classification of land into Capability Zones: For proper management, land is divided into 8 capability zones. Class 1 and 2 are level lands with no erosion. These are recommended for intensive agriculture. In classes 3 to 8, restricted agriculture or no agriculture is permitted. These must be maintained as grazing lands or made into natural forests or national parks.

Prevention of soil erosion: Soil conservation by different means must be done to prevent soil erosion. The method adopted should be suitable to the area concerned, and the topography (form and surface behaviour) of the land concerned.

Methods to prevent soil erosion by water: These can broadly classified into 2—biological methods and mechanical methods.

1. *Biological methods*: These are of 4 types:

A. *Agronomic practices*: The economic agriculture is called agronomy. Agronomic methods of preventing soil erosion are of 4 types.

- *Increasing soil fertility*—Plants do not grow well on infertile soils. Such a soil with poor vegetation is more exposed to water and wind erosion. So it is essential that soil is made fertile. The fertility depends on the mineral and humus content. In a fertile soil, plants grow luxuriantly and leave more residual material, which in turn, adds to the humus. To increase soil fertility, many practices are followed. The 3 most important ones are:

  Mulching: In this method, cereals like wheat, maize, oats, barley etc. and cotton, tobacco, potato etc. are cultivated. The basal parts of the plants called stubble are left in the fields after harvest. These form a protective layer, holding the soil together and reducing evaporation from soil and rate and amount of runoff. All these prevent soil erosion. Later, the stubble is ploughed under the soil. It is soon converted into humus and provides nutrients to the next crop. The organic matter increases soil aeration and stimulates the activity of microorganisms. The stubble is thus used as mulch and the process is called mulching.

  Adding fertilizers: Plants absorb many minerals from the soil. So the soil becomes poor in nutrients. If the next crop is planted in such a soil, the plant cover becomes poor and water and wind erosion result. To avoid this, soil fertility should be maintained. For this, farmers add different inorganic commercial fertilizers like urea, ammonium sulphate, potassium nitrate, superphosphate, ammonium diphosphate etc. and organic fertlizers like cow dung, green manure, compost, vermicompost etc. to the soil.

Ploughing: Other than absorption by plants, minerals are also lost by leaching. Leaching is a process by which soluble minerals dissolve in rainwater and move deep down into the soil. The bad effects of leaching can be undone to some extent by ploughing. By deep ploughing, some subsoil minerals are brought to the surface. Ploughing also breaks big lumps of soil and increases aeration and the water absorbing capacity of the soil.

Crop rotation: Crop rotation is also an important agronomic practice. It is a very old practice of growing different crops in regular order on the same area during successive years. During crop rotation, in the 1st year, maize or cotton, potato etc. are planted and in the 2nd year, oats or other grain crops are planted. In the 3rd year, pulses like any gram, beans, ground nut etc. are planted. Crop rotation decreases soil loss and preserves the productivity of the land, controls erosion, weeds and diseases. The same crop year after year depletes the soil of minerals. This can be overcome by planting pulses as they have nitrogen fixing bacteria in their root nodules. So when they are ploughed under, valuable Nitrogen is added to the soil.

Contour planting or contour farming: This is one of the oldest methods of cultivation used in areas of low rainfall. The fields are prepared with alternate ridges and furrows. Ridges at the same level are called contours. Water is caught and stored in the furrows. This reduces runoff and erosion.

Strip cropping: Strip cropping involves planting of crops in rows or strips to check flow of water. This is of 2 types:

(a) *Contour strip cropping*: If entire slopes are ploughed and planted as one unit, rainwater will flow down and collect in lowlands. If the slope is long, flow of rainwater becomes faster and it cuts the sides more and carries more soil away. So long slopes are broken into many strips laid across the slope. Strips of dense growing and soil covering species are alternated with other crops, which are not dense, and are cultivated leaving bare soil between rows. In this method, the force of running water is reduced. More water filters into the soil. Planting is done at 90° to the direction of the slope.

(b) *Wind strip cropping*: In this method also, strips are made, but in parallel rows at 90° to the direction of wind.

B. *Dry Farming*: This is the second biological method after agronomic practices. This is useful for crops grown in low and moderate rainfall areas. In such areas, ordinary farming will be a loss, so such areas are put to different uses.

- *Developing grasslands*: Grasslands are very effective in preventing soil erosion. India is not very rich in natural pastures (grazing grounds), so grasslands have to be grown. Wherever possible, vacant lands should be developed into grasslands. Open places in educational institutions, public offices and private concerns, hospitals and residential areas in the cities, wastelands around villages and towns, banks of canals etc. should all be utilized. Bare soil between trees in orchards and open forests should be covered with grasses. Grasses provide food to our cattle, sheep, goats and horses. They also help to absorb rainwater and increase the stability and fertility of the soil.
- *Land fallowing*: This method is expensive. Here, the fields are ploughed and then allowed to lie idle for one season. This method builds up a good supply of valuable minerals, especially nitrogen in the soil.

C. *Agrostological Methods*: All practices here involve the use of grasses in different ways. Grasses like the 'Doob grass' (*Cynodon dactylon*) is commonly used as erosion resisting plants. This species has a very high conservation value of 90 to 97%. Conservation value is the percentage of soil that would have been eroded in the absence of the species. Thus, if *Cynodon dactylon* plants were not covering an area, 90 to 97% of the top soil of that area would have been washed away in a heavy rain. There are 3 main agrostological methods of soil conservation.

- *Lay Farming*: Here, grass is grown in rotation with field crops. This helps to build up the soil structure, improves its fertility and prevents soil erosion.
- *Retiring land to grass*: Where a lot of top soil has eroded, only grasses are grown.
- *Strip growing of grass*: Here, grasses are grown in strips between crops. The grasses act as stabilizers.

D. *Control of Grazing*: The control of grazing is also included as a biological method of soil conservation. As cattle and sheep eat the green parts of the plants, the resulting loss of leaves leads to starvation of not only the shoots, but also the roots that hold down the soil. It is therefore necessary to prevent excess grazing. Fodder trees should be grown in semi-arid overgrazed areas. Trees like ku-babul supply high quality food to the animals. Villages should be asked to grow such plants. Also, goats and sheep can be kept in enclosures and fed on tree fodder. These animals waste large quantities of energy roaming about the pastures. This energy can be converted into milk and wool respectively if they are kept in healthy well ventilated enclosures.

2. *Mechanical Methods*: These are used to supplement biological methods.

A. *Terraced cultivation or Contour terracing*: Here, the slopes are divided into a series of flat fields called terraces. In each terrace, a broad bottomed channel is made, so that there are ridges on both sides of the channel. These channels break up long slopes into a number of small watersheds. This method is called broad based ridge terracing. Another method is channel terracing. Here, channels are made at suitable intervals and the dug up soil is deposited as a wide, low ridge along the lower side of the channel. These serve as bunds along the contours. This helps to slow down the flow of water and leads to the sides of the fields where it flows down slowly without eroding soil.

B. *Basin listing*: Here small basins are constructed along the contours to retain water. This reduces the velocity of the water flow and thereby erosion.

C. *Gully control*: Fast flowing water cuts channels in hills. Such channels are called gullies. To prevent the formation of gullies, some measures can be taken such as:

- Building dams or bunds across the gullies to catch silt
- Planting grass, trees or shrubs in gullies
- Blocking the flow of water by boards at regular intervals across gullies
- Constructing drains or diversions through which excess water can run off etc.

D. *Stream bank protection*: To check the cutting and caving of stream or river banks, stream bank protection must be done. For this, the following measures can be taken:

- Vegetation can be grown alongside river banks
- Drains can be constructed
- Concrete or stone pitching can be done

Methods to prevent soil erosion by wind: This is also called control of wind erosion. Many methods are used for this:

- Covering the soil with vegetation
- Keeping the soil rough rather than smooth
- Strip cropping across the direction of wind
- Making furrows across slopes to hold water
- Planting rows of trees across the wind. (These are called wind breaks.)

These methods help to check the blowing away of valuable soil and its drifting on to crops, roads and farms. Furrows hold water badly needed by crops. Making furrows thus help in water conservation along with preventing soil erosion by keeping the soil moist for longer periods. This produces more plant growth and provides a better vegetation cover which protects the soil.

Gigantic efforts are being made in Rajasthan to arrest the spread of the desert and to make the land productive. The Central Arid Zone Research Institute located at Jodhpur is doing a lot of research work in prevention of wind erosion. Control of wind erosion is an expensive process. So to avoid increase in costs, it is essential to implement projects with full vigour and complete them in time.

The Rajasthan Canal has been a boon to the people of Rajasthan. It is fed by both the Bhakra Nangal Dam and the Beas Dam. Work done by Rajendra Singh, the Magsaysay Award winner, is an example of how a dedicated man forming an NGO with some followers can work wonders. He made a river, which had gone dry for the past 15 years, flow again! He had adopted rain water harvesting methods to make this seemingly impossible task possible.

(d) Desertification

Desertification is a process by which the productive potential of a land reduces by 10% or more. It is caused mainly by human activities. If land near deserts is left to itself, sand storms blowing at high speeds will cover neighbouring vegetation and residential areas with sand. Effects of wind erosion are clearly seen in Rajasthan. West of the Aravalli Mountain Ranges, there is a vast sandy waste.

Overgrazing is one main reason for desertification. Forest margins are very commonly overgrazed. The surface soil is trampled under the hoofs of heavy animals. This makes the soil hard so that rainwater does not soak into it readily. The ground remains drier than under conditions of undisturbed nature. As the tops of the plants are chopped off, re-seeding reduces. After many years, the soil becomes weak and useless and is converted into desert.

Deforestation is another reason. As more and more trees are cut, the forest becomes less dense and open. The place of the trees is taken over gradually by shrubs. As exposure to sun and wind increases and evaporation from soil increases, the deforested land is gradually turned into desert, with only some xerophytic shrubs, herbs and mosses scattered here and there.

The spread of the desert robs the land of fertility and moisture. Pastures are also degraded if more animals graze on it than the area can support.

Methods to manage deserts and prevent further desertification: Deserts are dry zones due to actual deficit of rainfall. Here, the soil is sandy. Pore spaces between sand particles are large and so water holding capacity is very low. Soil erosion takes place by wind. To make the soil cultivable, therefore, some practices must be done.

- Making furrows for holding water—this helps in keeping the soil moist. A moist soil is not loose; it can hold water and is not blown away by sand.
- Plant sand binders (plant species which can bind the sand) first.
- Later, as the soil improves, do strip cropping (planting in rows) across the direction of the wind.
- Plant trees in rows, across the direction of the wind. Such trees are called wind breaks.
- Prevent overgrazing by animals and deforestation near the desert as both these will lead to desertification of normal land.

2.3 ROLE OF AN INDIVIDUAL IN CONSERVATION OF NATURAL RESOURCES

Following is a list of what individuals can do and to conserve natural resources, as depletion of these resources will have serious consequences. (Add relevant points from the list on what students can do to create environmental awareness.)

1. Plant as many trees as possible in as many places as possible. Saplings can be planted not only in own gardens, but also on many public places like roadsides, near railway lines, canal banks, farm boundaries, around villages, school and college yards, public buildings, places of worship, cemeteries, market squares etc.

   In many of these cases, permission from authorities may be needed. But this should not be considered an obstacle. Individuals can always approach the authorities. It will not be difficult to convince the authorities as all know that trees prevent air pollution. What is needed is initiative and determination.

   There are many institutions and agencies that supply saplings at nominal rates or even free of cost. They can be approached for saplings. Saplings can be planted during the Van Mahotsav or on any day in the rainy season. Care must be taken to look after them properly after planting.
2. Paper is made from trees. So wasting paper is a crime. Write on both sides of paper.
3. Use recycled paper if possible. This is made by recycling used paper. Popularise the use of recycled paper among your friends. It can be used not only for writing, but also for making greeting cards, flags, curios etc.
4. Do not crumple paper before throwing it; tear it into pieces. This is because crumpling breaks fibres and recycled paper made out of this will be of inferior quality.
5. Try to use envelopes more than once by carefully tearing one side and pasting a new label on the address written.
6. Do not use paper napkins. Instead, use cloth ones.
7. It is meaningless to say 'Don't cut trees' or 'Don't use teak for furniture'. Trees will have to be cut for different purposes; teak will be used for doors, windows and furniture as it is hard, durable and resistant to white ants. Therefore the aim is to make use of natural resources, but at the same time to regenerate them. Only in the case of non-renewable resources, we need to be extra cautious.
8. For example, petrol is a non-renewable resource. Therefore, use petrol sparingly. If the distance is not much, use a bicycle in place of a

scooter/motor cycle/car. If the distance is walkable, just walk. (In any case, walking is good for health!)

9. Car pools are ideal when people have to go to the same place or nearby places at the same time. Taking out individual cars on such occasions is wastage of petrol.
10. Enquire whether organic farming is being done for the vegetables and fruits you buy. If many people make such enquiries, word will spread and finally reach the growers.
11. Do not waste food.
12. Do not waste water. Do not keep the tap on while brushing, shaving etc.
13. If any tap leaks in your house, get it repaired immediately.
14. If you see a leaking tap in your neighbourhood, report it to the authority.
15. Learn how to do rainwater harvesting and try to get one system in your house. If you are living in a flat complex, persuade the landlord to get one installed. He will most probably agree, collecting a certain amount from all the residents.
16. Do not waste electricity. Switch the lights and fans off when you are not there.
17. If the electric meter in your house is not working, report the same to the authorities.
18. Use solar energy as much as possible. Solar water heaters, solar calculators etc. are available in the market.
19. Do not let land go waste. Even if you have a small plot behind your house, have a flower garden or kitchen garden there. In addition to enhancing beauty and utility, plants prevent soil erosion.
20. If you see soil being washed off and underlying pebbles being shown above in public places, inform authorities.
21. As far as possible, maintain lawns instead of concreting courtyards. This helps water to seep into the ground.
22. Spread environmental awareness by organising various activities in your locality.
23. Write articles on the importance of natural resources and publish in local newspapers.
24. Appreciate the beauty of plants and birds and butterflies – that will make you a nature lover and you will never feel like disturbing nature or any organism in nature any longer.

2.4 EQUITABLE USE OF RESOURCES FOR SUSTAINABLE LIFESTYLES

In today's world, there is no equitable use of resources. As per latest figures available, in 2005, the wealthiest 20% of the world accounted for 76.6% of total private consumption; the poorest fifth just 1.5%:

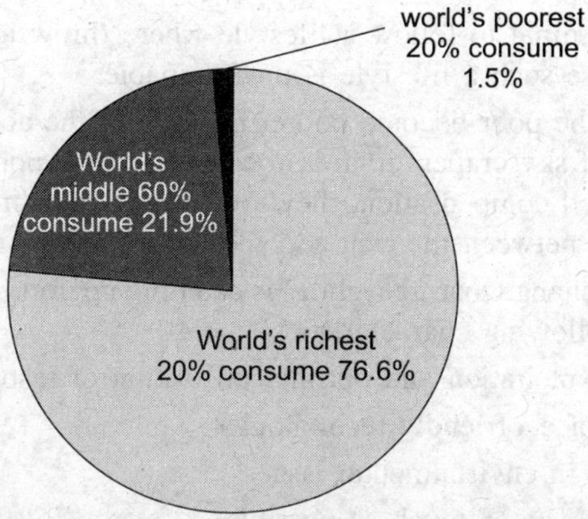

Figure 2.7 Share of world's private consumption, 2005

Today's consumption is upsetting the environmental resources. The gaps among consumption-poverty-inequality-environment relations are widening. If these trends continue unabated, today's problems of consumption and human development will worsen. If this should not happen, then redistribution of resources must take place from high-income to low-income consumers, and a shift must come into effect from polluting goods to cleaner goods and production technologies. Consumption of resources for vulgar display of wealth must give way to utilization of resources for meeting basic needs. The real issue is not consumption itself but its patterns and effects.

Inequalities in consumption are very prominent. Globally, 20% of the world's people in the highest-income countries account for 86% of total private consumption expenditures; the poorest 20% a minuscule 1.3%. In other words,

- The richest fifth consume 45% of all meat and fish; the poorest fifth 5%
- The richest fifth consume 58% of total energy; the poorest fifth less than 4%
- The richest fifth have 74% of all telephone lines; the poorest fifth 1.5%
- The richest fifth consume 84% of all paper; the poorest fifth 1.1%
- The richest fifth own 87% of the world's vehicles; the poorest fifth less than 1%

Rapid growth in consumption in the past 50 years is putting strains on the environment never seen before.

(Data from Human Development Report 1998; Overview, UNDP (United Nations Development Programme)

If developing nations follow the same path as today's developed countries, their consumption patterns will also be damaging to nature.

Many aspects of nature can be classified as sustainable and non-sustainable. Resources, economy, development, lifestyle etc. can all be sustainable or otherwise. It is criminal to follow a lifestyle where throw away society ethics is followed, because such a lifestyle is unsustainable.

Many a times, the poor become poorer because of the activities of the rich, like construction of skyscrapers in areas occupied by the poor. Unsuccessful in getting the promised compensation, they are compelled to move to still poorer localities. The gap between the rich and the poor enlarges further.

It is a welcome change that 'recycling' is catching up everywhere. Sustainable lifestyle has the following characteristics:

- Rational exploitation and optimal utilization of resources
- Adoption of ecofriendly technologies
- Awareness in environmental issues
- Avoiding waste as much as possible
- It leads to sustainable economy
- Maintaining quality of life along with ecological balance
- Assuring continued yield of resources to the present and future generations.

This means that the final goals of sustainable lifestyles of the people in a society are two: 1. To provide a strategy (action plan) that will improve the quality of human life, and, 2. To provide a strategy that ensures the vitality and diversity of the earth.

Developing countries like India accounts for a very large population; in fact every 6th world citizen is an Indian. The developed countries are pointing fingers at such countries (mainly India and China) and accusing them of depleting natural resources much faster than they can regenerate by natural processes. On the other hand, the industrialized west is utilizing a major portion of the earth's resources, in spite of having low population levels.

Disparities between the rich and the poor countries are becoming more and more glaring. In each category, there is an advantaged group and a disadvantaged group. The burden of polluted air, water and land and of dwindling natural resources always falls on the poor. Often, it is the poor who are compelled to encroach into a forest just to survive; or keep their surroundings dirty as they cannot afford to keep them clean.

The challenge of bringing in an era of equity is very large indeed. To make this happen, there must be sincere efforts at the national level and also at the global level. Responsibility for the present position of the planet has to be shared between the rich and the poor countries. Therefore, for finding out solutions and implementing them, the rich and the poor countries must work together. For this, trade talks and flow of money is not enough. Social and political ties between governments and civil societies must be strengthened.

Disparities between the rich and the poor exist within countries also. In fact, the disparities are more in India than in America. Some waste resources and live lavishly while some have almost nothing to eat. Resources belong to all, irrespective of class. But when rich contractors are given permission to fell trees, it is the poor 'vanvasi' who suffers. In addition, the poor forest dwelling tribal people are accused of destroying the forests. Same is the case with poor fishermen. The ocean does not belong to anybody. But when mechanized trawlers catch fish, the poor fishermen suffer losses and even go hungry. The poor fishermen have as much right to the fish in the sea as the rich and powerful trawler owners have.

In short, for a sustainable lifestyle, it is essential to have an equitable distribution of natural resources in place.

QUESTIONS

A. Very short answer type:

(a) Multiple choice – Tick the right answer:

1. Coal is a
   (a) Renewable source of energy
   (b) Inexhaustible source of energy
   (c) Non-renewable source of energy
   (d) None of the above
2. Which of the following is a non-polluting source of energy?
   (a) Coal (b) Petroleum
   (c) Natural gas (d) Solar energy
3. For the formation of one inch of top soil, it takes
   (a) 10 to 20 years (b) 100 to 200 years
   (c) 300 to 500 years (d) 400 to 1000 years
4. DDT is a
   (a) Fungicide (b) Pesticide
   (c) Fertilizer (d) Disinfectant
5. Deforestation is caused by
   (a) Mining (b) Overgrazing
   (c) Dam building (d) All of the above
6. Major dams may cause
   (a) Earthquakes (b) Hurricanes
   (c) Storms (d) Cyclones
7. The percentage of water accounted by oceans and seas is
   (a) 90% (b) 87%
   (c) 97% (d) 99%
8. Maximum usage of water is in the field of
   (a) Industries (b) Homes
   (c) Agriculture (d) None of the above
9. Medha Patkar is famous for
   (a) Chipko Movement (b) Narmada Bachao Andolan
   (c) Green Revolution (d) Silent Valley Project

10. Which of the following is a non-metallic mineral?
    (a) Gypsum (b) Iron
    (c) Zinc (d) Copper
11. Examples of renewable resources are
    (a) Wind energy (b) Solar energy
    (c) Tidal energy (d) all of the above
12. Biogas is made up mostly of
    (a) Hydrogen (b) Carbon dioxide
    (c) Ethane (d) Methane

(b) Define:

1. Resource ecology
2. Natural resources
3. Atmosphere
4. Hydrosphere
5. Lithosphere
6. Biosphere
7. Inexhaustible resources
8. Exhaustible resources
9. Renewable resources
10. Non-renewable resources
11. Afforestation
12. Reforestation
13. Forest conservation
14. Drought
15. Minerals
16. Ores of metals
17. Mining
18. Green revolution
19. Pesticides
20. Industrialisation
21. Soil
22. Soil erosion
23. Soil (land) reclamation
24. Sheet erosion
25. Monoculture
26. Developing grasslands
27. Contours
28. Sustainable lifestyle

(c) Explain briefly:

1. Atmosphere
2. Hydrosphere
3. Lithosphere
4. Biosphere
5. Van Mahotsav
6. Joint Forest Movement (JFM)
7. Timber extraction
8. Flash floods
9. Reasons for drought
10. The three types of minerals resources
11. Minerals and mineral ores
12. Prospecting
13. Exploitation
14. Gold mining
15. Mulching
16. Types of agriculture
17. Traditional agriculture
18. Food resources in India
19. Acid rain
20. Solutions to fertilizer problem
21. Methods to control pesticide pollution
22. Uses of petroleum
23. Natural gas
24. Disadvantages of nuclear energy
25. Disadvantages of wind energy
26. Geothermal energy

27. Tidal energy
28. Hydroelectric power
29. Energy from biomass
30. Benefits of biogas plants
31. Land degradation due to mining
32. Land degradation due to waterlogging
33. Desertification and land degradation
34. Land degradation due to acid rain
35. Role of nutrient disorder in land degradation
36. Types of soil erosion by water
37. Wind erosion
38. Overgrazing
39. Dry farming
40. Terraced cultivation
41. Contour planting
42. Stream bank protection

(d) Name the following:

1. Three main reasons for deforestation
2. Five renewable energy resources

(e) Differentiate between the following pairs:

1. Contour strip cropping and wind strip cropping
2. Geological (normal) erosion and accelerated erosion

B. Short answer type:

Write short notes on:

1. Renewable and non-renewable resources
2. Natural resources and associated problems
3. Human activities leading to deforestation
4. Plan of action for forest management
5. Advantages of major dams
6. Uses of water
7. Over-exploitation of water
8. Methods to prevent desertification
9. Methods of flood control and flood management
10. Reasons for today's water shortage
11. Waterlogging
12. Salinity
13. Environmental effects of extracting and using minerals
14. Stages in mining operations
15. Coal
16. Petroleum
17. Sustainable agriculture
18. Methods to reduce droughts and their effects
19. Changes caused by agriculture and overgrazing to environment
20. Green Revolution

21. Fertiliser pollution
22. Harmful effects of pesticides
23. Renewable and non-renewable energy resources
24. Solar energy
25. Wind energy
26. Land resources
27. Land degradation due to deforestation
28. Land degradation due to urbanization and industrialization
29. Consequences of soil erosion
30. Increasing soil fertility
31. Crop rotation
32. Control of grazing
33. Wind erosion
34. Biogas plants

C. Long answer type:

(a) Answer the following in detail:

1. Give an account of the uses of forest resources.
2. What are the effects of dams on forests and tribal people?
3. Give an account of different types of mineral resources
4. Explain in detail the causes of floods.
5. What do you understand by 'Equitable use of resources for sustainable lifestyles'? Explain in detail.

(b) Write long notes on any two:

1. Harmful effects of major dams
2. Conflicts over water
3. Causes of deforestation
4. Salinity and methods to reclaim saline soils

(c) Almost all energy that man uses comes from the sun. Discuss.

(d) Write long notes on any two:

1. World food problems
2. Nuclear energy
3. Biogas plants
4. Reasons for land degradation

(e) Describe in detail the methods for the prevention of soil erosion.

(f) Write notes on any two:

1. Uses of minerals
2. Reasons for world food problems
3. Effects of modern agriculture
4. Man-induced landslides

Unit 3

Ecosystems

3.1 INTRODUCTION

Plant and animal communities of any region influence each other and interact with one another. They together form a biome or a biotic community. In a biotic community, plants, animals, microbes and the non-living environment are constantly interacting and exchanging materials. Trees provide shelter to birds. Birds help in pollination and dissemination of fruits and seeds. Plants purify the atmosphere. Animals provide carbon dioxide to the plants for respiration. Microorganisms decay dead plants and animals thereby returning the minerals to the soil for plants to absorb. These minerals enter animals through plants. These are just a few examples. In short, the living and non-living components in nature together form a stable system. In this system different components, like the well oiled parts of a machine, are co-coordinated to form a single unit. This is an ecosystem. A natural unit of living communities and the non-living environment interacting and exchanging materials between themselves is called an ecological system or an ecosystem.

Ecosystem is the basic functional unit of ecology. The study of ecosystems in fact covers the major part of the study of environmental science. This is because now-a-days, most of the ecological studies are done at the level of ecosystem. This is called the bioenergetic approach to ecology.

The term 'ecosystem' was proposed by A. G. Tansley in 1935 in England. He defined it as a system resulting from the interaction of all living and non-living factors of the environment. Different terms were used earlier to refer to the ecosystem, like biocoenosis, geobiocoenosis, biogeocoenosis, microcosm, biosystem, ecocosm etc. However, the term ecosystem is the most preferred and has become the most commonly used term. 'Eco' means environment, and system means an interacting, interdependent complex.

3.1.1 Size of Ecosystems

There is no fixed size for ecosystems. They may be very small or large or very large. An ecosystem may be as large as a grassland or a river or a desert or an ocean. It may be as small as water in a ditch, a small lump of soil or a single fallen leaf. A forest is an ecosystem; a single tree is also an ecosystem; a single fallen leaf is an ecosystem too.

A small roadside ditch is an ecosystem. There may be some bacteria or blue green or green algae in it. The water in the ditch has a certain temperature and pH. It is influenced by light, wind, atmospheric humidity etc. It gets minerals from the soil underneath it. It contains organic and inorganic compounds formed by the decomposition of the living organisms in it. Thus, a roadside puddle is an ecosystem.

A tree is an ecosystem. It produces food by photosynthesis. It gives shelter to so many birds, sometimes monkeys and snakes too. Many bugs, ticks, fleas etc. live on the birds' bodies. There may be parasitic fungi and bacteria on or inside the bodies of these birds. The tree gives shade to the plants that grow below it. It reduces the light intensity and heat and increases the humidity available to these plants. The plants growing under the tree are protected from strong winds. The light they get is of a different quality as the tree reflects off most of the green rays in the incident light from its leaves. Many red and blue rays are used by the leaves of the tree for photosynthesis. Raindrops do not fall forcefully on the plants underneath. The soil under the tree contains more humus formed by the decomposition of the fallen leaves. It is more fertile. As the fallen leaves form a soil cover, there is less evaporation from the soil. There is no erosion under the tree as the soil is held in position by the roots of the tree. There are many insects and worms in the soil. There may be burrows of rats, snakes or rabbits under the tree. There are trillions of microorganisms in the soil. All these perform different chemical reactions like nitrification, denitrification, ammonification, nitrogen fixation etc. Some microbes are parasitic on other animals or plants. Altogether, there are so many actions, reactions and interactions in and around a single tree. It is one complete ecosystem. (Thus when one tree is cut, it is not one single tree, but an entire ecosystem is destroyed. Understand this truth, and tell the same to all your friends.)

A single fallen leaf is also an ecosystem. It is being decomposed by thousands of microorganisms. Environmental factors are of different types. Climatic factors like light, temperature, wind, rain; edaphic factors like soil water, soil temperature, soil pH, soil acidity and alkalinity and all other factors related to soil. Topographic factors like exposure to sun and wind and rain, biotic factors like competition among the microbes, interrelationships among them etc. are all working on the dead leaf fallen on the soil. Thus it is also an ecosystem.

Forests, grasslands, deserts, croplands, ponds, lakes, streams, rivers, seas, oceans, estuaries, marshes etc. are all large ecosystems. The largest ecosystem is the whole earth. This is called the biosphere. This is the biggest ecosystem possible on earth.

It is for convenience in studying that we take up different ecosystems differently as forests, grasslands, deserts, rivers etc. However, in fact, all of them are linked with each other. There are no functional boundaries between them. For example, molecules of different gases and minerals circulate throughout the earth, in and out of living beings. Ecosystems of different places are all thus connected to each other. They are connected through time also. Ecosystems present at any one place at a given time are indeed related to those present at these spots years ago and also to those that will appear there in future through soil, water, gases etc. Energy is flowing through all these ecosystems. Energy is transformed, circulated and accumulated in different parts of the ecosystems. The source of energy responsible for maintaining this giant system is the sun.

Green plants absorb sunlight and use the energy contained in it in making food. This process is called photosynthesis. The food produced is used by the green plants as well as the animals, which do not have the ability to produce food. When plants and animals die, their dead bodies are decomposed by microorganisms. The complex molecules in the dead bodies are broken down into simpler molecules and are finally released into the environment to be used by living beings again. This circulation of materials taking place continuously in nature is responsible for the maintenance of the ecosystems.

3.1.2 Types of Ecosystems

Ecosystems may be natural or artificial. Natural ecosystems like undisturbed forests, deserts etc. function under natural conditions without human interference. In artificial or man-managed ecosystems, there is constant human interference. For example, in a cropland ecosystem, the man adds manures and fertilizers, provides irrigation, sprays pesticides periodically etc. In a zoo, he provides food to the animals and looks after their health and safety.

Based on the habitat, ecosystems can be classified into terrestrial (on land) and aquatic (in water). Examples of terrestrial ecosystems are forests, deserts, grasslands, croplands etc. Aquatic ecosystems may be freshwater or marine. Freshwater ecosystems may be lotic (in running water like rivers, streams, springs etc.) or lentic (in standing water like lakes, ponds, marshes, puddles etc.) Marine ecosystems are the seas, oceans, estuaries, backwaters etc.

3.2 STRUCTURE AND FUNCTIONS OF ECOSYSTEMS

To study any ecosystem, its structure and functions must be understood.

Structure includes three aspects:

1. The composition of the living beings (the biotic communities) of the ecosystem, like the species, their numbers, life history, spatial distribution etc. The living beings may be producers, consumers or decomposers.
2. The quality and distribution of the non-living materials like water, inorganic and organic nutrients etc.
3. The range of climatic conditions like temperature, light etc.

The functional aspects of ecosystems include the following:

1. Food chains
2. Food webs
3. Ecological pyramids
4. Biogeochemical cycles
5. Energy flow etc.

3.2.1 Components of Ecosystem

The two main components of an ecosystem are the abiotic component and the biotic component.

(a) Abiotic component: This is the non-living component of the ecosystem. The abiotic component influences the biotic component and is also influenced by it.

The abiotic component comprises of three parts—inorganic, organic and physical.

1. *Inorganic*: The inorganic component includes water, different minerals, different gases etc.

   These form the nutrients or raw materials for the green plants. Chlorophyll – the green pigment responsible for photosynthesis—is also an important inorganic chemical. The amount of inorganic substances like phosphorus, sulphur, carbon, nitrogen, hydrogen etc. present in the environment of an ecosystem at any given time is called the standing state or standing quality.

2. *Organic*: The organic component includes compounds like carbohydrates, fats and proteins belonging to the dead animals and plants and the excreta of the animals in the ecosystem. The intermediates and end-products of the decomposition of these substances are also part of the organic component. Decomposition by bacteria and fungi ultimately converts these organic compounds into inorganic form, which are reutilized by the plants. Bacteria and fungi thus link the biotic and the abiotic components of the ecosystem.

3. *Physical*: The physical component includes climatic factors like light, temperature, wind etc. In water bodies, in addition to these, there is also the pH of water, wave action etc. The most important physical component is the solar energy that flows through the entire biotic community.

(b) Biotic component: The biotic component includes living plants, animals and microorganisms. They all need food to live. Ultimately, all energy comes from the sun. Organisms in an ecosystem can be divided into producers, consumers and decomposers. All these organisms are placed in different energy levels (food levels). Each energy level in an ecosystem is called a trophic level. Different trophic levels differ from one another in the relative distance, which separates from the primary source of energy, i.e. the sun.

Based on the nutritional level, the biotic component has 2 components.

1. *The autotrophic component*: This component includes the autotrophs. Photosynthetic plants that can manufacture their own food are called autotrophs. This includes all plants that contain the green pigment chlorophyll and also the photosynthetic bacteria containing bacteriochlorophyll. Chemosynthetic bacteria are also autotrophic. All these autotrophs fix energy and build up complex substances from simple substances.

   In an ecosystem, the autotrophic plants are the producers or primary producers (P or PP) and form the first trophic level (T_1)

2. *The heterotrophic component*: Heterotrophs are organisms that do not have chlorophyll and therefore cannot prepare their own food. Organisms that cannot make their own food and get their food from the green plants directly or indirectly are the consumers (C). Animals, fungi and a majority of bacteria are consumers. They take in complex food materials and break them down into simple forms of food. Conversion of one compound into another by rearrangement also takes place inside the bodies of consumers.

Consumers are of two types—macroconsumers and microconsumers.

(A) *Macroconsumers:* Macroconsumers are large sized consumers. They include the herbivores, carnivores and top carnivores.

1. *Herbivores*: Animals and parasitic plants that feed directly on the green plants are called herbivores. They are consumers of the first order (C_1) and belong to the 2nd trophic level (T_2). They consume only food manufactured by the producers.

   As herbivores are eaten by organisms of the next trophic level, they are sometimes called secondary producers. But it is better to avoid this usage as herbivores do not produce any food. Their bodies are eaten wholly by the animals of the next level.

   Examples of herbivorous insects are grasshoppers, bees, butterflies etc. Examples of herbivorous animals are cows, buffaloes, goats, rabbits, meadow mice, deer, elephants, camels, monkeys etc. Parrots, koels, bulbuls etc. are herbivorous birds.

2. *Carnivores*: Animals eating herbivores are called carnivores. They are consumers of the 2nd order (C_2) and constitute the 3rd trophic level (T_3).

   Insects like praying mantis, beetles and crickets; animals like small fishes, frogs, lizards, snakes etc; and birds like sparrows, crows, woodpeckers, mynahs etc. are carnivorous.

3. *Top Carnivores*: Top carnivores are animals that are not killed and eaten by other animals. They are tertiary consumers (C_3) and constitute the 4th trophic level (T_4). Top carnivores eat carnivores and herbivores, while carnivores eat only herbivores. This is another

difference between carnivores and top carnivores. Examples of top carnivores are animals like tigers, lions, sharks, crocodiles etc. and birds like eagles, hawks, kites, owls, kingfishers, gulls etc. Animals of T_3 and T_4 are also called predators.

The trophic levels of some animals are not clear. Best example is man. He uses both plants and all types of animals as food and is an omnivore.

(B) *Microconsumers:* Microconsumers are decomposers. These are the saprophytic fungi and bacteria that live on or near the soil level. They obtain their food from dead organisms and animal excreta. On the land, fallen leaves and twigs and in water, dead algae and parts of other water plants that fall to the bottom of the water body form the detritus. The detritus is eaten by animals like protozoa, nematodes, snails etc. and these are called detritus eaters.

The function of microbes like bacteria, fungi etc. and the detritus eaters is similar; they are all decomposers.

Decomposers (D) are the top consumers (C_4) and occupy the 5th trophic level (T_5): They have a very important role in the functioning of the ecosystem. Minerals necessary for plants and animals will remain trapped in the dead bodies if there are no decomposers. This will lead to mineral deficiency of the soil, which will lead ultimately to the death and extinction of all plants and animals. This calamity is prevented by the decomposers. Reutilisation of minerals is the basis of an active ecosystem. Some decomposers fix atmospheric nitrogen also, thereby increasing soil fertility.

Moreover, in the absence of decomposers, also called scavengers, land and ocean will be filled with dead bodies. In such a situation, there will be no place for the living!

3.3 CYCLING OF MATERIALS IN THE ECOSYSTEM (BIOGEOCHEMICAL CYCLES)

In ancient Hindu scriptures, the concept of 'Panchatatva' is very clear. These are 'Prithvi', i.e., the earth (mineral nutrients); Jalam, i.e., water; Agni, i.e., fire (energy); 'Vaayu', i.e., gases and 'Aakaasam', i.e., the sky or space. This reflects the circulation of materials.

Living organisms obtain many nutrients from the environment. These are gases like oxygen, carbon dioxide etc., and minerals like calcium, potassium, iron, nitrogen, phosphorus, sulphur etc. and water. These materials pass through the different trophic levels and are ultimately returned to the environment in many ways and are reutilized by the plants. Every nutrient atom enters the biotic component of the ecosystem from the abiotic component and finally returns to the environment. In this way, atoms are used over and over again, i.e., they are repeatedly recycled. This circulation of materials between the biotic and the abiotic components of the ecosystem is called the biogeochemical cycle.

There are many types of biogeochemical cycles in nature. The more important of these are:

3.3.1 Water Cycle (Hydrologic Cycle)

About 73% of the earth's surface is covered by water. Part of it evaporates and forms clouds. Rain returns this water to the earth. Plants absorb water from the soil and lose most of it to the atmosphere through transpiration; part is passed onto the next trophic level. Animals get water from plants and the environment. They also return water to the environment through evaporation. Thus the water cycle goes on continuously.

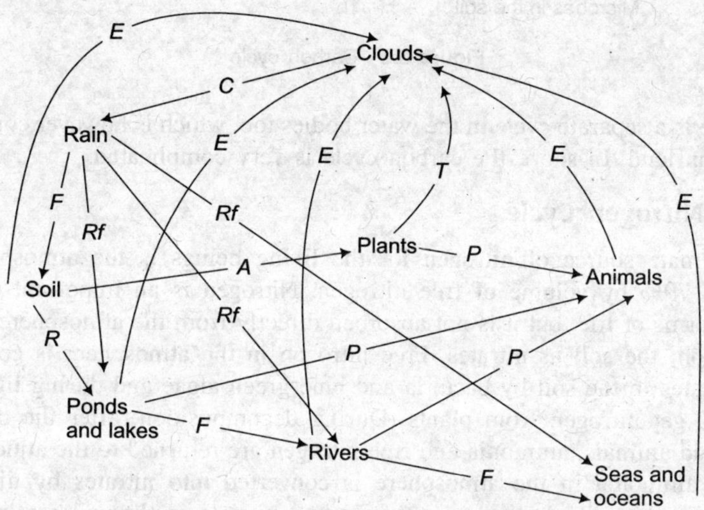

Figure 3.1 Water cycle

3.3.2 Carbon Cycle

This is a gaseous cycle. 0.03% by volume of CO_2 in the atmosphere is the source of carbon for the biotic community. Carbon is also obtained from water where it is present in the soluble form. Carbon is the most important element of all organic compounds. Coal deposits, limestone rocks, volcanoes, lime deposits in the seas and oceans etc. are also sources of CO_2 in the atmosphere. Other sources are respiration, death and decay of plants, animals and microbes, weathering of limestone rocks, burning of coal and firewood etc. CO_2 from factories, automobiles etc. is in addition to all this. Thus there are many sources of CO_2 in the atmosphere, but it is removed from the atmosphere only by the plats. They absorb it from the atmosphere and convert it into organic food during photosynthesis.

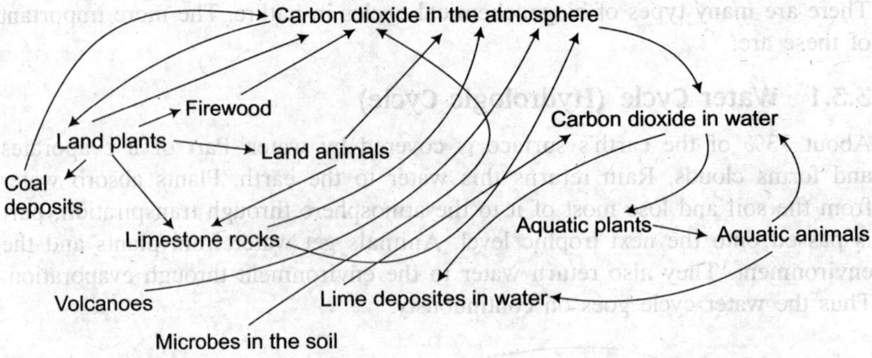

Figure 3.2 Carbon cycle

There is a separate cycle in the water bodies too, which is however connected to that on land. In short, the carbon cycle is very complicated.

3.3.3 Nitrogen Cycle

The ultimate source of nitrogen for the living beings is the atmosphere. It contains 79% by volume of free nitrogen. Nitrogen is an important element for all forms of life, but it is not absorbed directly from the atmosphere. Plants get it from the soil as nitrates. Free nitrogen in the atmosphere is converted into nitrates in the soil by bacteria and blue green algae and during lightning. Animals get nitrogen from plants. During decomposition, after the death of plants and animals, ammonia and free nitrogen are returned to the atmosphere. Part of ammonia in the atmosphere is converted into nitrates by nitrifying bacteria in the soil. At the same time, some nitrates in the soil are lost back into the atmosphere by the activity of denitrifying bacteria. Nitrogen cycle is thus a gaseous cycle.

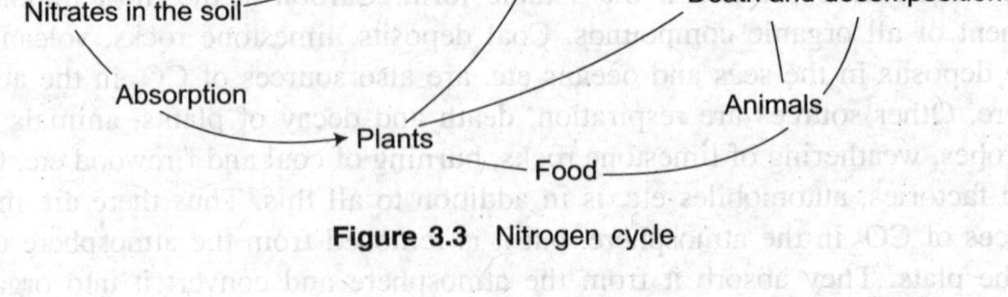

Figure 3.3 Nitrogen cycle

3.3.4 Sulphur Cycle

Sulphur is seen mostly as sulphates in the soil. A small amount of hydrogen sulphide in the gaseous form is also present in the soil. Addition of fertilizers leads to increase of sulphates in the soil. Plants take in sulphate from the soil and pass on part of it to animals. On death and decay, plants and animals return sulphates to the soil. Burning of fuel, automobiles and some industries release sulphur dioxide into the atmosphere. Part of this forms acid rain, which falls to the earth and the water bodies. Some sulphates from the soil are washed into the water bodies. Aquatic plants and animals get their supply of sulphur from the water itself in the form of sulphates.

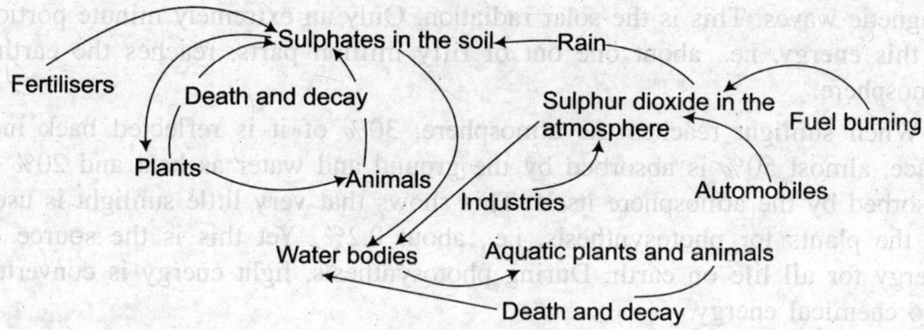

Figure 3.4 Sulphur cycle

3.3.5 Phosphorus Cycle

Plants get phosphorus in the form of phosphates from the soil. Sources of phosphorus in the soil are phosphate rocks and volcanoes. Plants take in phosphates from the soil; animals get it from plants. When plants and animals die and decompose, the phosphates are returned to the soil. Some phosphate is lost from the cycle by sedimentation in seas and oceans.

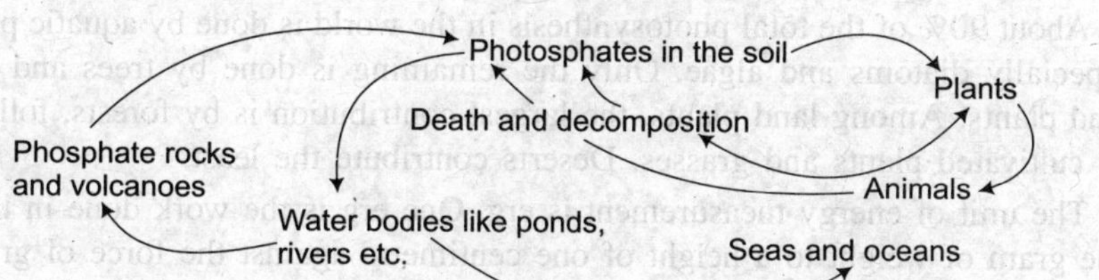

Figure 3.5 Phosphorus cycle

In this way, all minerals are continuously circulating through the biotic and abiotic components of the ecosystem.

3.4 ENERGY FLOW IN THE ECOSYSTEM

If there are no plants, the light energy of the sun will be converted into heat energy which will heat up the earth. However, plants convert light energy into the chemical energy of the food during photosynthesis, and it is this food that is used by the animals. In short, if there are no plants, there will be no life on earth. The life of all animals depend on the light energy of the sun (solar energy) trapped by the green plants.

The sun is extremely hot. In the sun, nuclear rearrangements are taking place continuously, converting hydrogen to helium. This process releases huge amounts of energy which radiates out in all directions in the form of electromagnetic waves. This is the solar radiation. Only an extremely minute portion of this energy, i.e., about one out of fifty million parts, reaches the earth's atmosphere.

When sunlight reaches the atmosphere, 30% of it is reflected back into space, almost 50% is absorbed by the ground and water as heat and 20% is absorbed by the atmosphere itself. This shows that very little sunlight is used by the plants for photosynthesis, i.e., about 0.2%. Yet this is the source of energy for all life on earth. During photosynthesis, light energy is converted into chemical energy.

Solar energy travelling in the form of radiation is called radiant energy. Of this, rays having wavelength between 400mμ to 760mμ is the visible range or light energy. Light can be easily converted to heat. In the chloroplasts of green plants, light passes through the electron transport chain of the photosystems, and is fixed ultimately in carbohydrates. These are the immediate food sources.

The first law of thermodynamics proposed by Joule states that 'energy cannot be created or destroyed, but may be converted from one form into another'. This law is applicable in all cases.

About 90% of the total photosynthesis in the world is done by aquatic plants, especially diatoms and algae. Only the remaining is done by trees and other land plants. Among land plants, the largest contribution is by forests, followed by cultivated plants and grasses. Deserts contribute the least.

The unit of energy measurement is erg. One erg is the work done in lifting one gram of weight to a height of one centimetre against the force of gravity, which is 981 ergs. 10^7ergs = 1 joule. All forms of energy can be completely converted into heat energy. Heat is measured in calories (cal). 1 cal = 4.2 joules. 1000 cals = 1 kilocalorie (Kcal or Cal). In environmental studies, energy is therefore measured in calories or kilocalories; sometimes in kilojoules.

A large fraction of the solar energy received by the earth's atmosphere does not reach the ground, as it is scattered by suspended particles in the atmosphere. The constant rate at which solar energy reaches the earth's outer atmosphere is called solar constant. Twenty-five percent of the incoming solar radiation is reflected by clouds and 9% by suspended dust. Nine per cent is held by ozone, water vapour and other gases and 10% is absorbed by the clouds. Thus, about 47% of the radiation reaches the earth's surface.

The quantity of solar radiation that is received in any place thus depends on the clarity of the atmosphere in that place. Latitude is another important factor. Thus, regions near the equator receive maximum solar radiation, while it reduces gradually towards the poles.

In ecological energetics, studies are carried out mainly in three fields:

1. The quantity of energy reaching an ecosystem/unit area/unit time (for example, 1 sq. metre/one hour or one day etc.).
2. The quantity of energy absorbed by green plants and converted into chemical energy.
3. The quantity and path of energy flow from green plants to different trophic levels in a known area for a known time period.

The available energy decreases from producers to consumers of different levels. Green plants are the primary producers. The total energy fixed in food by green plants is called gross production. Some of this is used for respiration and other activities; the remaining is called net production. Some of this is consumed by herbivores, some is decomposed and some is not utilized and accumulates as sediments.

Of the total energy that herbivores get from the autotrophs, again, some is used in respiration, some is used by carnivores, some decomposes and some is not utilized.

Some herbivores in the ecosystem are eaten by carnivores. So the herbivores are called secondary producers. The carnivores get energy from herbivores. Of this, some is used in respiration, some is decomposed and some is not utilized.

From this, it is clear that energy flows only in one direction in the ecosystem. This is called the unidirectional energy flow. As food passes from one trophic level to the next in an ecosystem, the energy lost as heat at various levels of transfer does not return to the living organisms. The energy that goes to herbivores cannot pass back to the autotrophs. The energy of the carnivores cannot go back to herbivores or autotrophs. Thus, the energy flow in an ecosystem is always unidirectional. As it moves through different trophic levels, it is no longer available to the previous level. So if the primary source is cut off, the whole system will collapse. The one way flow or the single channel flow of energy is a universal phenomenon, and it is true to the basic law of thermodynamics. To maintain the ecosystem, therefore, a constant supply of external energy is needed. This is the sun.

Materials like carbon, nitrogen, oxygen, water etc. are repeatedly used and reused through the cycling of materials. All these pass from the living to the non-living and back to the living in an unending cycle. But energy flows unidirectionally.

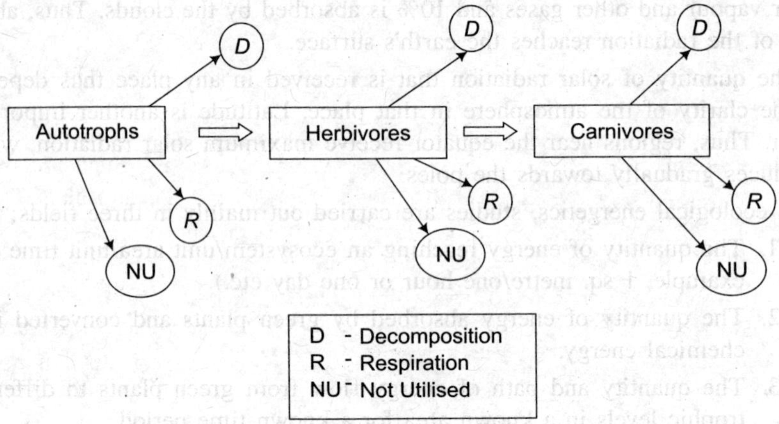

Figure 3.6 Energy flow diagram of a lake

Energy does not flow freely, but through organic materials from one trophic level to the next. For this, inorganic material has to be converted into the organic form. This is done in primary production. An ecosystem remains balanced as long as the energy utilization by animals is matched by the energy trapped by plants. Otherwise, food levels become short and a famine results. Consumers die if no food is available from outside the ecosystem.

From the energy flow in the ecosystem, it is also clear that a progressive decrease occurs at each trophic level. This is because of dissipation of energy as heat in different metabolic activities and also unused energy. This means that if a food chain is shorter, the available food energy will be more. The longer the food chain, the food energy available for the organisms of the last trophic level will be very less.

Materials like carbon, nitrogen, oxygen, water etc. are repeatedly used through the cycling of materials. All these pass on from living to non-living and back in a cyclic form. But energy flows in a unidirectional manner, in a single channel. This is one of the most important characteristics of energy flow in any ecosystem.

In any ecosystem, it is of great ecological significance to know the following:

1. The efficiency of primary and secondary production at different levels
2. The total input of energy
3. Efficiency of assimilation (photosynthesis)
4. Loss through excretion
5. Loss through respiration
6. Gross and net production at each level

A thorough understanding of the ecosystem is essential for better management of resources for human welfare. Every citizen should be educated in non-technical terms about the cycling of materials and the flow of energy in an ecosystem, of which he himself is a part.

In the history of mankind, we see many instances where man himself has upset the ecosystem, which has resulted in great harm to man himself. Killing of some animals at a higher rate than their capacity to regenerate leads to an increase in the population of plants eaten by these animals. Such and other disturbances in the ecosystem will affect the cycling of materials and the energy flow.

Similarly, indiscriminate felling of trees upsets the entire balance of water and other mineral cycles as well as plant and animal populations. Rajasthan was under forest vegetation a few centuries ago, But man removed plants faster than he could regenerate. As the primary production reduced, the fauna also gradually declined. Now, most of the Rajasthan is desert. Efforts are being made at present on a large scale to stop the marching of the desert (further spreading of desert) by water conservation and planting trees.

Maintenance of life on earth is dependent on one energy source — the solar energy. This is fixed only by green plants. Therefore, if forests are destroyed, the net energy fixation capacity of the earth will fall. On the other hand, human population is increasing drastically. To support this ever increasing human population, the population of green plants or the efficiency of photosynthesis or both should be increased. Another possibility is to consume water plants directly, especially marine flora as fish resources are already being over-exploited.

3.5 ECOLOGICAL SUCCESSION

In nature, different kinds of organisms grow in association with each other. In a particular area, many plants belonging to different species live together in small or large areas. Such an aggregate of pants having mutual relations among themselves is called a plant community.

All communities have an origin; they develop and finally mature. In the beginning, an area is bare, without any vegetation. Seeds and spores arrive here by migration from other communities through wind, water or animals. Many seeds and spores die, but some germinate. At the seedling stage also, many die. Ultimately, some seedlings grow into mature plants. They reproduce and multiply. Soon, competition starts. Again, many plants die. During all these developments, there is a continuous interaction between the plants and the environment. As a result, once lighted areas become shaded, humus accumulates, dry areas become moist while wet areas become drier. When the conditions change so much that the initial plants find them unsuitable, new species that find these conditions favourable establish in the area. The pioneers thin out. The newcomers also change the environment. The process continues. As the plant communities change, the animal communities also change as animals depend on plants for survival. The diversity of plants and animals increases during this process. The final composition of the vegetation is determined by the climate of the area. So the final stage in this series of changes is called climax community. This process of orderly evolution of communities from birth to maturity, leading to the establishment of a relatively stable community in total harmony with the environmental conditions is called ecological succession. It takes many centuries for a succession to reach its climax.

The bare area on which succession begins may be primary or secondary. If no plants grew in the area before, it is a primary area. Such areas are sterile and conditions are not favourable for life in the beginning, like newly formed islands, newly deposited volcanic ash on rock, silt deposited at river mouths etc. A succession that begins here is called a primary succession. If plant life had existed earlier in the area, nutrients are already present in the soil and conditions are comparatively favourable. Examples are cut out or burnt out forests, abandoned croplands, new ponds formed after heavy rains or floods etc. Successions beginning on such secondary areas are called secondary successions.

In most successions, vegetation interacts with and modifies the environment and this results in its own replacement by newcomers. This is called autogenic succession. In some cases, however, replacement of a community is caused largely by external conditions and not by the existing vegetation. Such successions are called allogenic.

The developmental stages in the process of plant succession are called seral stages and the entire sequence of communities on a given area is called a sere. A succession beginning in water or watery habitats is called a hydrosere, while a succession beginning in dry conditions is called a xerosere. It may be a lithosere, i.e., succession beginning on rocks or psammosere, i.e., succession beginning on sand. Successions beginning in salt water are called haloseres, while those starting in areas with enough moisture are called mesoseres.

3.5.1 Hydrosere

A hydrosere originates normally in a newly formed pond or lake. The water is deep in the middle and progressively shallow towards the bank. Hydrosere starts with phytoplankton. Associated zooplankton appear soon. The evolution of the pioneer community proceeds progressively, and passing through various intermediate stages ends in a forest, which is the climax.

Though succession takes plant in both pant and animal communities, changes are more obvious in plants. So the stages are named as per the dominant plant communities of that particular stage.

(a) *Phytoplankton stage*: This is the pioneer stage. In the beginning, in the newly formed water body, there are no nutrients in the substratum or water. Algal spores are brought by wind or with soil particles and deposited in the water. These pioneers are the phytoplankton consisting of unicellular and colonial algae and bacteria. pH is about 5 in the bottom mud. Pioneers multiply fast. Zooplankton appears soon. Dead bodies of these plankton fall on the bottom mud, adding organic matter to the soil. Some flies also appear.

(b) *Submerged stage:* Floating algae add large quantities of organic matter. Nutrients are also brought with dust blown by the wind. A soft mud is formed at the bottom. The pond thus becomes shallow and light penetration becomes easier. The habit is now suitable for submerged plants like *Hydrilla, Vallisneria* etc. They grow fast, and when they die, and accumulate at the bottom, the pond becomes shallower still. Moreover, they catch the soil particles falling from the edges of the pond and deposit it at the bottom. The bulky remains of the submerged plants are converted to humus and this binds the bottom mud. The pond is now too shallow for the submerged plants. Crustaceans are the prominent animals of submerged stage.

(c) *Floating stage*: When the submerged plants gradually move into deeper parts of the pond, floating plants make their appearance at the margin, which is 5 to 8 feet deep. Most of the floating plants are rooted, like lotus and water lily. They have extensive rhizomes and long leaf stalks, which raise the leaves to the surface of the water. Free floating plants like *Lemna, Pistia* etc. also grow at the margins. The leaves of the floating plants cut off the light for the submerged plants so that those still present move away to deeper water where floating plants cannot live. The process of mud building continues. When the pond becomes too shallow for the floating plants, reeds that prefer swamps appear. Gill bearing snails, small fish, frogs, beetles etc. make their appearance in this stage.

(d) *Reed swamp stage*: Water is only 2 to 3 feet now at the margins. So swamp plants like *Typha, Juncus* etc. invade the area. These plants are rooted in the mud, but their upper parts are well above the water surface. They have well developed rhizomes and propagate fast. They cut off light to the floating plants. Catching eroded silt and bottom building go on. The habitat is now unsuitable for the reed swamp plants also, and so the next stage sets in. Dragon flies, water scorpions, giant water bugs, worms and fish are seen in this stage. Fish eating birds fly over the water.

(e) *Marsh meadow stage*: The water stands only 2 to 3 inches now at the margins. The soil may even be dry in summer. Plants like *Carex, Cyperus* etc. find this habitat favourable. The grasses form a thick mat. The swamp plants have a high rate of transpiration and so the loss of water from soil is rapid. Mud is now exposed to air. Sulphides and ammonia are oxidized to sulphates and nitrates. The soil has become mesic now. The marsh plants now migrate inward, and the next stage sets in. Snails are the common animals here. Fish are excluded.

(f) *Woodland stage:* Soil is now dry for most part of the year. Terrestrial plants like different shrubs now invade the area. The pioneers are plants that can tolerate some waterlogging in the soil. The shrubs transpire vigorously and make the soil drier. They also cast shade and catch wind-borne soil. Plant debris accumulates and is decomposed to humus. The soil is now rich in microorganisms. At this stage, trees appear. By this time, most of the terrestrial animals also appear.

(g) *Climax forest*: The species of trees vary depending on the climate of the place. In tropical climates with heavy rainfall, rainforests develop. In temperate regions, mixed forests develop. In regions of moderate rainfall, deciduous forests develop. Under the canopy of the trees, air becomes very humid. Their shade makes conditions unfavourable for seedlings of the pioneer sun loving trees. Thus, after a few generations, pioneers are replaced by shade tolerant species. Ferns, mosses and lichens love humidity and begin to thrive on the forest floor. Climbers also make their appearance. Ultimately, the forest becomes very complex. Associated animals, birds, insects are also seen in the climax forest.

Thus, the area once under deep water finally becomes a dense forest. This will happen only if the climate is moist. In dry climates, the climax may be a grassland or even a desert.

3.5.2 Xerosere

A xerosere originating on a rock is called lithosere. The substratum is dry and has no organic matter. It has only minerals in an unweathered state. Crustose lichens are the pioneers here. A climax forest is formed through different intermediate communities.

(a) *Pioneer stage or crustose lichen stage*: A rock is extremely xeric. It has no water, or nutrients. Exposure to sun and wind is great. Temperatures are extremes. In such a habitat, only crustose lichens can grow. Symbiotic association of a specific alga and a specific fungus growing together in symbiosis as one plant is called a lichen. Thin and papery lichens are called crustose lichens, like *Graphis, Licidia* etc. These can tolerate extreme conditions and grow very slowly. Whenever there is rain, they absorb water like a sponge. Migration takes place by lichen fragments or spores. Crustose lichens produce some acids, which weather the rock. The dead lichens add organic matter to the thin layer of soil formed on top of the rock by corrosion brought about by the lichenic acid. Mites are the associated animals in this stage.

(b) *Foliose lichen stage*: This layer of soil on top of the rock is suitable for the growth of foliose lichens like *Parmelia, Physcia* etc. These have thin leaf-like plant bodies. They absorb and retain more water and accumulate dust particles. The foliose lichens grow in depressions in the rock and other parts where there is soil. They overshadow the crustose lichens, which now move away to more exposed portions of the rock. Foliose lichens also secrete lichenic acid. More soil is formed, which is made more fertile by the dead remains of the lichens. Enough soil is now formed which can sustain xerophytic mosses. Some ants and spiders live in the crevices.

(c) *Moss stage*: Xerophytic mosses start growing first in the cracks and crevices of the rock, e.g., are *Polytrichum, Bryum* etc. Their rhizoids compete with the lichens and their erect stems overshadow them. The foliose lichens gradually move away or disappear. Death and decay of both lichens and mosses make the soil a thicker layer and makes it rich in organic compounds. The water holding capacity of the soil increases. The soil is now fit for the growth of herbs. The number of mites and spiders increases in the moss stage.

(d) *Herbaceous stage*: At this stage, there is a thick mat of humus and mineral particles and enough moisture in the soil. These particles provide foothold to the roots of the herbs. Small herbs like *Tridax, Poa* etc. and ferns invade the area. The first herbs are annuals, i.e., plants living for one year or one season. Later, biennials (plants living for two years) and perennials (plants living for many years) arrive. The roots of these plants accelerate soil formation. Large quantities of humus are added to the soil. Evaporation from the soil decreases and humidity increases. Soon, the soil becomes fit for shrubs. With ants, mites and spiders, there are other animals now like different insects and nematodes.

(e) *Shrub stage*: Common shrubs to establish are *Rhus, Capparis, Zizyphus* etc. As they are large sized and have longer life cycles, they exert greater influence on the environment. Overshadowing of herbs, formation of soil, addition of humus and lessening of evaporation continues. The wind movement is retarded and humidity increases. Soil is now mature. Species diversity of plants and animals increases. Slugs, snails, millipedes, centipedes, bugs, mice, snakes etc. appear. All these conditions are favourable for growth of trees.

(f) *Forest stage or climax forest*: The first species to appear are xeric. Their growth is stunted and they are far spaced. Their saplings cannot tolerate shade. All the changes in the environment continue to happen. As trees become taller and their growth denser, tolerant species (species tolerant to shade) grow more and more and the intolerant ones disappear. Similarly, xeric trees gradually give way to mesic trees. New herbs and shrubs adapted to humid air and rich soil develop on the forest floor. Other than all animals mentioned earlier, squirrels, rabbits, foxes, different birds, snakes etc. also appear now. Thus, a mesophytic forest is formed.

A study of hydrosere and xerosere shows that ecological succession is convergent. Whatever the initial habitat, extremes gradually give way to an environment suitable for plants and animals living in normal conditions. Thus, watery habitats become drier and dry habitats become moister. Beginning as a hydrosere or a xerosere, the climax is always a mesophytic forest. The deciding factor is the climate. If the initial conditions are already mesic, succession takes place with time, and again, the climax is a mesophytic forest.

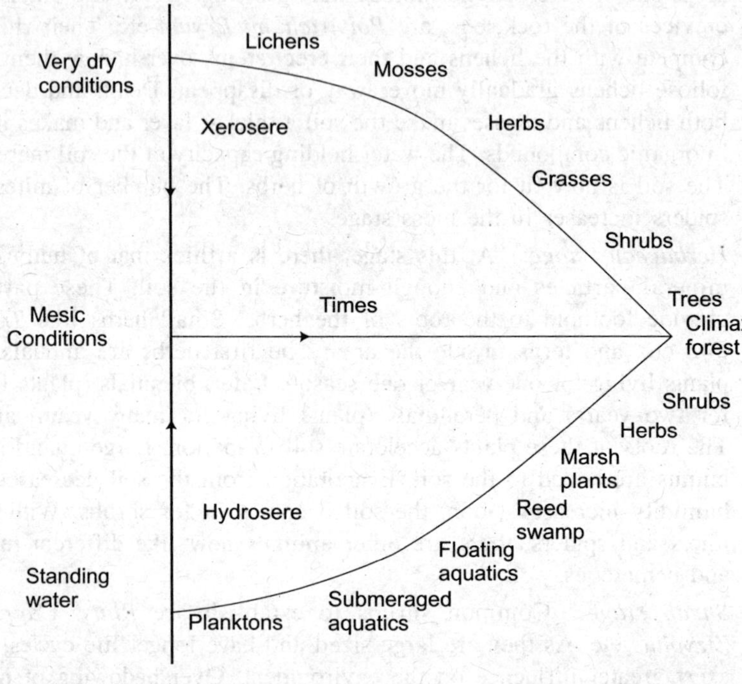

Figure 3.7. Diagrammatic representation of the changes during xerosere and hydrosere. Both end in mesic conditions.

3.6 FOOD CHAINS

The sequence or order of trophic levels through which food energy passes from the primary producers to the top carnivores is called the food chain.

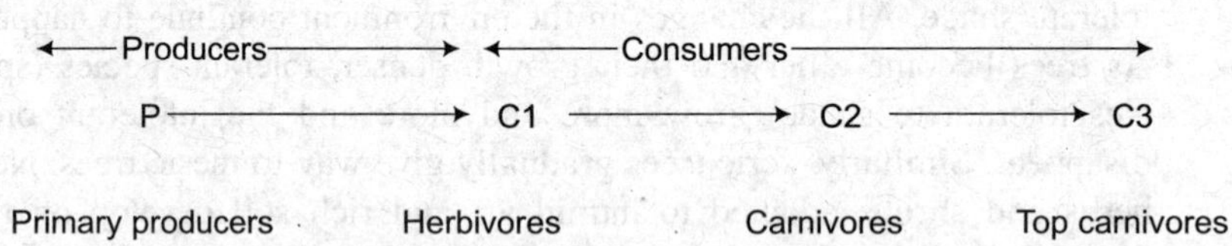

Figure 3.8 Food chain

Decomposers get food from all trophic levels. Omnivores eat everything and are finally decomposed.

Scriptures of all old civilizations have references to food chains. The most ancient of these is the *Mahabharata*, which mentions the importance of food chains and the interdependence of organisms for food. The Greeks also knew this. In the modern times, it was Antonie Van Leeuwenhoek who elaborated the principles of food chains at the end of the 17th century.

This classification of the living organisms of any ecosystem is based on their function and not species. Species that are very much different from each other may occupy the same trophic level. For example, in a pond, water lily and lotus that are flowering plants, non-flowering plants like water ferns and algae, bacteria, all occupy T_1.

In grasslands, the food chain starts with the grasses, goes through grass hoppers, frogs, snakes, hawks etc. In a pond, it starts with phytoplanktons and goes through water ferns, small fish, big fish, birds, large animals etc.

Food chains are of two types.

1. *Grazing food chain*: A grazing food chain starts from living plants and goes through herbivores to carnivores. Following are some of the examples:

Phytoplankton .... Zooplankton .... Fish

Grass .... Rabbit .... Fox

This type of food chain depends on photosynthesis. Most of the ecosystems in nature follow this type of food chain.

2. *Detritus food chain*: If a food chain starts with dead organic matter into microbes and then into larger organisms, it is called a detritus food chain. Such ecosystems do not depend directly on solar energy. For example, leaves in marshy vegetation fall into water. These dead leaves are eaten by insect larvae, different types of worms, snails etc. These are the detritus eaters. They are eaten by small fish, which form the food for big fish. Detritus food chains also start from forest floors.

Dead leaves in water .... Worms and Snails .... Small fish .... Big fish

Dead leaves on the forest floor .... Insects .... Small birds .... Snakes

3.7 FOOD WEBS

Food chains do not operate independently of each other. They are interconnected at some points so that an interlocking pattern is formed. This is called a food web.

Consumers always have alternative arrangements in nature. Snakes may eat mice or small birds; hawks may eat snakes, mice or rabbits. Besides these, vultures, foxes and human beings etc. also make part of a grassland ecosystem. Thus food webs get interlocked at many points making the whole system very complicated.

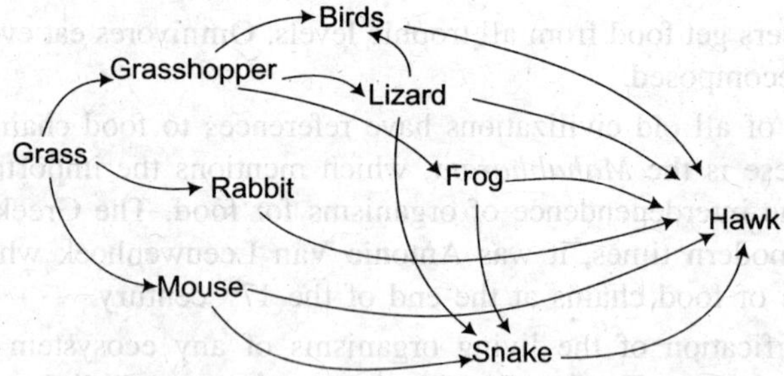

Figure 3.9 Food web in a grassland ecosystem

In a pond, primary producers are the microscopic phytoplanktons and the macrophytes (large plants). Consumers of different levels are the zooplanktons, different types of insects, snails, crabs, small and big fish etc.

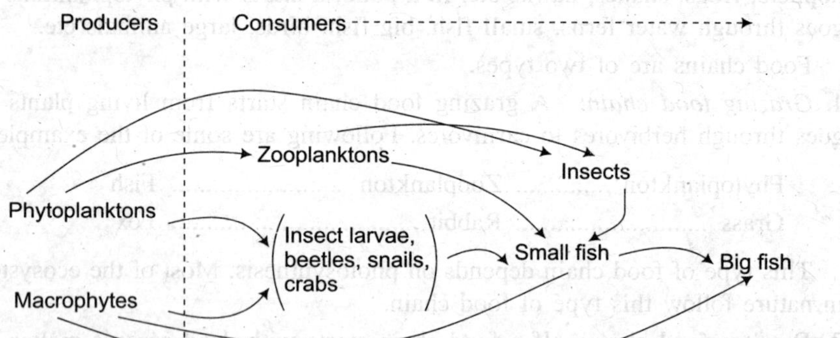

Figure 3.10 Food web in a pond

Food webs are very important in maintaining the stability of ecosystems in nature. For example, in the grassland ecosystem, if the population of rabbits decreases due to some reason like a disease, population of alternative herbivore, i.e., mice increases. This may decrease the population of carnivores that prefer rabbits. Thus, alternatives help to keep the ecosystems stable. This is necessary for the survival of all living beings in nature. If herbivores are absent, the producers will multiply so much causing overcrowding and resulting in competition. Same is the case of herbivores. Some of them should be consumed by the carnivores.

As we move further away from the primary source of food, i.e., the green plants, the energy available to the animals of successive food levels goes on decreasing. As a result, the population of animals that can be sustained also becomes smaller as we move up the food chain. As a herbivore, more energy is available to man than as a carnivore. If man starts taking more animal food, the available energy can support only fewer people. Then the human population will fall. Thus in nature, each species is under a check.

The more different the organisms in a system, the more complex is the food web. It depends on the length of the chain and the alternatives at different points. When alternatives are more, interlocking is also more. In seas and oceans, food webs are very complex.

Other than the producers and consumers, the ecosystem includes two more physiological groups—the parasites and the symbionts. Parasites are organisms that depend on other living organisms totally or partially for food. Symbionts are pairs of organisms that live in association with each other and benefit from each other. Examples of parasites are the disease causing bacteria and fungi, which live in or on plant and animal bodies. Examples of symbionts are bacteria that live inside the nodules of roots of pulses etc. When these are also included, food webs become more complex.

Ecological niche: Each organism, whether plant or animal or microorganism, plays a definite role and occupies a definite position in the structure and operation of an ecosystem. This status or role of an organism within an ecosystem is called the ecological niche. It includes all the physical, chemical and biological factors that an organism requires to live. It also describes the food source, the organisms for which it serves as food, and its effects on other organisms and the environment.

Biomass and standing crop: The total amount of living organisms present in an ecosystem is called biomass. It may be expressed as living (wet) weight, dry weight, carbon content, nitrogen content, or calorie content per unit area (square metre, acre, hectare etc.) This varies at different times of the day and in different seasons. The total amount of living material in an ecosystem at any given time is called the standing crop.

3.8 ECOLOGICAL PYRAMIDS

Producers are most closely related to the primary source of energy, the sun, as they get the energy directly from it. So they occupy the 1st trophic level. Herbivores eat green plants and are therefore included in the next trophic level. Carnivores constitute the 3rd trophic level and so on. The different trophic levels in any ecosystem can be represented graphically and these graphic representations are called ecological pyramids.

Each ecosystem has its own characteristic ecological pyramid. Ecological pyramids are basically of 3 types—pyramids of number, pyramids of biomass and pyramids of energy. At the base of the pyramid is the producer, the green plants. The apex is occupied by the top carnivores (as the decomposers get food from all trophic levels). Herbivores and carnivores are placed in between.

(a) *Pyramids of numbers*: In pyramids of numbers, the number of individuals in the different trophic levels at any given time is shown.

The shape of the pyramid depends upon the size of the individuals of the primary producers. If primary producers are small in size compared to the consumers, the pyramid is erect. The pyramid of numbers of freshwater ecosystems like that of a pond is erect.

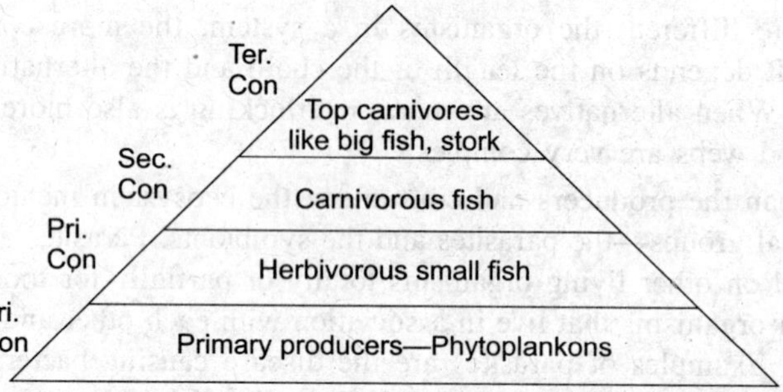

Figure 3.11 Erect pyramid of numbers of the freshwater ecosystem of a pond

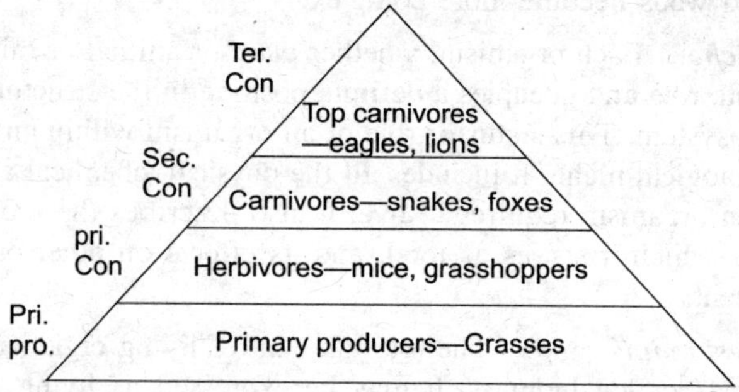

Figure 3.12 Erect pyramid of numbers of grassland ecosystem

The primary producers in a pond are the phytoplanktons (microscopic algae). Though these are extremely small, their numbers are very large. So they form the broad base of the pyramid. Above this are the herbivorous fish followed by carnivorous fish. The top carnivores at the top may be large fish or birds like storks. Here, the number is maximum at the base, decreasing gradually towards the tip of the pyramid. Pyramid of numbers of grasslands and croplands is also erect.

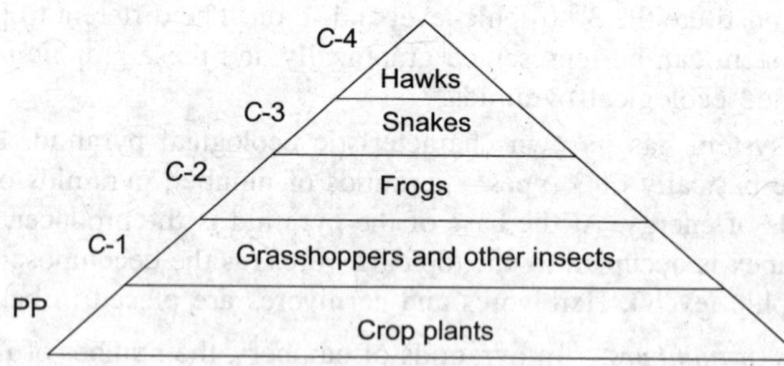

Figure 3.13 Erect pyramid of number of a cropland ecosystem

If the primary producers are large in size, and the size goes on decreasing at the consumer levels, the number of organisms goes on increasing in the successive trophic levels. This results in an inverted pyramid. A very good example is a tree ecosystem. A single tree is an ecosystem too, as different types of birds live on the tree. Moreover, monkeys, snakes, squirrels etc. may depend on the tree for various purposes.

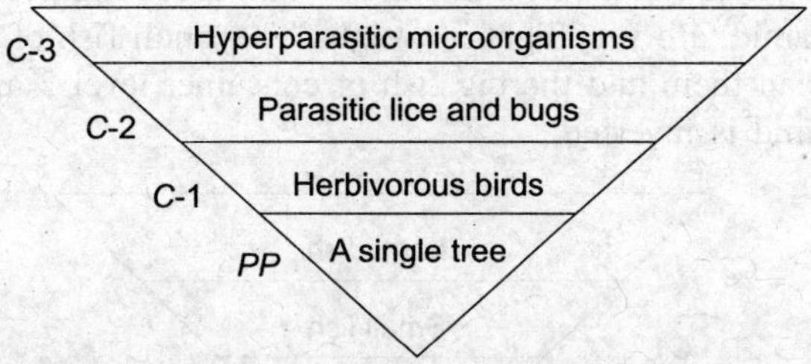

Figure 3.14 Inverted pyramid of numbers of a tree ecosystem

Here, a single tree is the producer. Many herbivorous birds live on the tree. These are consumers of the 1st order, C_1. On the body of each bird, many parasites like lice, bugs etc i.e., consumers of the 2nd order, C_2 live. On each louse or bug, many bacteria, fungi etc. live. These hyperparasites constitute C_3. Here, the number of organisms gradually increases from the producer to the consumers.

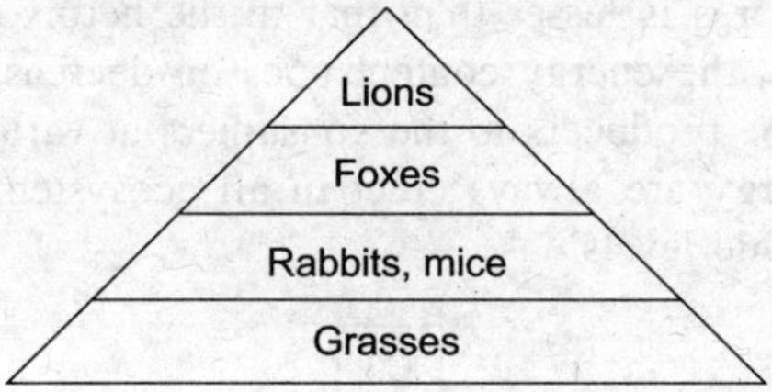

Figure 3.15 Erect pyramid of biomass of a grasland

(b) *Pyramids of biomass*: Pyramids of biomass indicate the relationships between the different trophic levels of an ecosystem on the basis of living weight—fresh or dry.

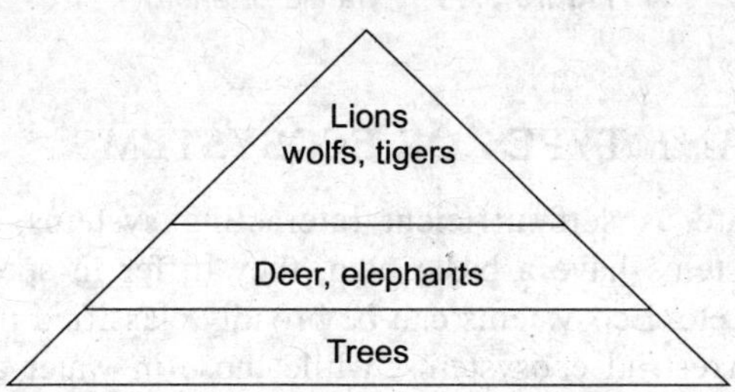

Figure 3.16 Erect pyramid of biomass of a forest

If the biomass of the producers is more, the biomass goes on decreasing at successive consumer levels. In such a case, the pyramid is erect. The best example is a grassland ecosystem. Pyramids of forests are also erect.

On the other hand, pyramid of biomass of a pond ecosystem is inverted. This is because phytoplanktons like diatoms and green algae, which form the base of the pyramid, are very light in weight. The small fish of consumer level 1 are heavier than them and the big fish of consumer level 2 are the heaviest. Thus, the pyramid is inverted.

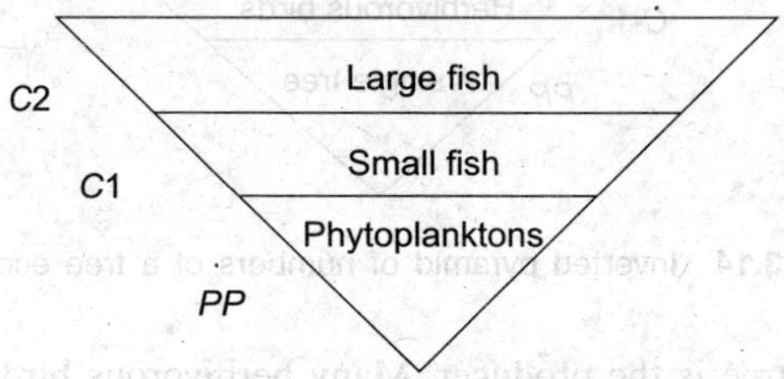

Figure 3.17 Inverted pyramid of biomass in a pond

(c) *Pyramids of energy*: Pyramids of energy show the total energy contained in the different trophic levels in ecosystems. The total energy in the producers is the highest. Herbivores contain lesser energy than the producers. The total energy in the carnivores is lesser than that in the herbivores. Thus, as we go up the trophic levels, the energy content goes on decreasing. This is because energy flows from the producers to the consumers at various levels. Therefore the pyramids of energy are always erect in all ecosystems. Decomposers get energy from all trophic levels.

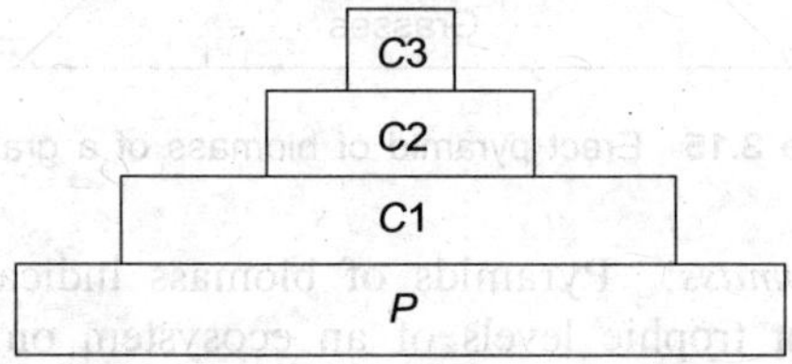

Figure 3.18 Pyramid of energy

3.9 IMPORTANT TYPES OF ECOSYSTEMS

Ecosystems operate as self-sufficient interacting systems in the biosphere. Though all ecosystems have a basic plan, they differ in species composition, rate of production etc. Ecosystems can be broadly classified into two. Those on land are called terrestrial ecosystems, while those in water are called aquatic ecosystems.

3.9.1 Terrestrial Ecosystems

Main types of terrestrial ecosystems are forests, deserts, grasslands, croplands, savannas, tundras etc.

(a) Forest ecosystem

Forests are of different types. They may be tropical, temperate, alpine etc. based on the geographical zones in which they are seen. They may be coniferous, deciduous, evergreen, thorn, mangrove etc. based on which types of plants grow there. They may be rainforests, dry thorny forests etc. based on the climate of the region. There are many more types of forests. In any country, one-third of the land should be under forests. But in India, only 8 to 10% of the land is under forests.

Abiotic components: Light, humidity, humus in the soil etc. are important abiotic components. Light conditions are different in different strata. The composition of the organic and inorganic materials in the soil differs greatly depending on the tree species living in the forest.

Biotic components: Producers in forests are trees. Species differ in different types of forests. Ground vegetation is formed of various shrubs, ferns, herbs and terrestrial orchids. In tropical deciduous forests, teak (Tectona), palash (Butea), shisham (Dalbergia) etc. are seen. In temperate coniferous forests, conifers of different types are prominent. Examples are pine (Pinus), deodar (Cedrus) etc. Ground vegetation is very less in temperate forests.

Primary consumers: Primary consumers are the herbivores like ants, flies, beetles, leaf hoppers bugs etc., animals like elephants, monkeys, deer, pigs, squirrels etc., and birds like koel, bulbul, parrots etc.

Secondary consumers: These are carnivores like snakes, lizards, fox, bear etc.

Tertiary consumers: Lion, tiger, wolf etc. are tertiary consumers of the forest.

Decomposers: Detritus eaters are millipedes, mites etc. Different types of bacteria and fungi are very active. The rate of decomposition in tropical and subtropical forests is greater than that in temperate forests.

(b) Grassland ecosystem

About 20% of the land area is occupied by grasslands. Grasslands are formed where rainfall is low and soil is not deep and fertile. Grassland ecosystems are very common in the temperate countries, but they are not very common in the tropical countries.

Abiotic components: These are water, minerals and organic materials in the soil, gases in the air, light, temperature etc.

Biotic components

Producers: Producers are the grasses of different types. Some herbs and shrubs are also present.

Primary consumers: cattle, deer, rabbits, mice etc. are primary consumers. Different herbivorous insects, termites and millipedes feeding on grass, leaves and some birds are also primary consumers.

Secondary consumers: Fox, jackal, snakes, frogs, lizards and many birds are secondary consumers. Hawks, kites, eagles etc. may eat primary or secondary consumers, so they may be secondary or tertiary consumers.

Decomposers: Different species of fungi, bacteria and protozoa decompose the dead organic matter.

(c) Desert ecosystem

About 17% of the land area is desert. Here, annual rainfall is less than 25 cm. Deserts are hot; winds blow at high speed. Soil is sandy.

Abiotic comoponents: Organic matter in the soil is very less as vegetation is scanty. Water is also very little in the soil. Temperatures are high; wind velocity and exposure are also very high.

Biotic components

Producers: The primary producers are mainly shrubs and also some herbs. Shrubs have widespread branched root systems with variously branched stems and branches. Thorny plants and succulent (water storing and fleshy) cacti are common. Lichens and xerophytic mosses also occur.

Consumers: Reptiles and insects which can tolerate high temperatures are the consumers. Different types of birds, snakes, rodents etc. are also seen. Camel is the commonest herbivore.

Decomposers: There are very few decomposers in the sandy soil of the deserts. This is because there is very little organic matter in the soil. Some thermophilic (heat loving) fungi and bacteria live in the soil.

There are some more terrestrial ecosystems like those of croplands, savannas, tundras etc. Cropland ecosystems are the largest artificial ecosystems. Most of the environmental conditions in croplands are controlled by us. Savannas are grasslands with some shrubs and trees scattered in them. Tundra is an Arctic plain with permanently frozen subsoil. Some types of lichens, mosses and dwarfed vegetation live in tundras.

3.9.2 Aquatic Ecosystems

Aquatic ecosystems may be freshwater (those of rivers, lakes, ponds etc.) or marine (those of seas, oceans, backwaters etc.) A pond ecosystem can be taken as a very good example of an aquatic ecosystem.

(a) *Pond Ecosystem*: This is a self-sufficient, self-regulating system. The living and the non-living exist in harmony in a pond ecosystem. The nature of the pond influences the living organisms in it; the living beings in the pond also have an influence on the pond.

Abiotic components: These include all the non-living components present in the pond. These are physical, inorganic and organic.

A. *Physical:* Important physical components in a pond are light, temperature, pH of water, wind etc. If the pond is very deep, the bottom may be dark as there is no light penetration. But in shallow ponds, there is some light even at the bottom. Based on water penetration and the associated plants and animals, a pond can be divided into 3 zones:

(a) *Littoral zone*—This is the shallow water zone where there is enough light and free floating plants grow. This zone is rich in phytoplankton as well as zooplankton.

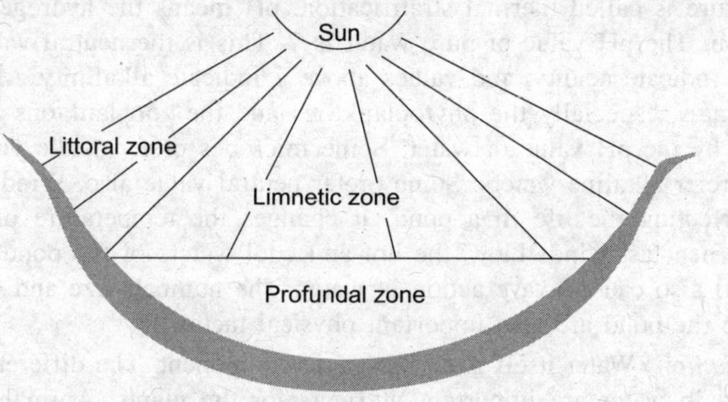

Figure 3.19 Zonation in a deep pond

(b) *Limnetic zone*—Here light is less but is enough for photosynthesis. Many rooted floating and submerged flowering plants and algae, crustaceans, insects and insect larvae grow in this zone.

(c) *Profundal zone*—This is the deep water where there is no effective light penetration. Saprophytic (living on dead matter) bacteria and fungi and different types of worms, snails, crabs etc. live in the profundal zone.

The particles suspended in water reduce light penetration too as they absorb some light. The quality of light also changes depending on the depth of the pond. White light is made of 7 colours—violet, indigo, blue, green, yellow, orange and red. Deeper waters get light of some colours more than other colours. Depending on all this, the types of plants that grow at different depths in a pond differ. The types of animals change too, as they depend on plants for food.

Temperature has a great influence on plants, animals and microorganisms.

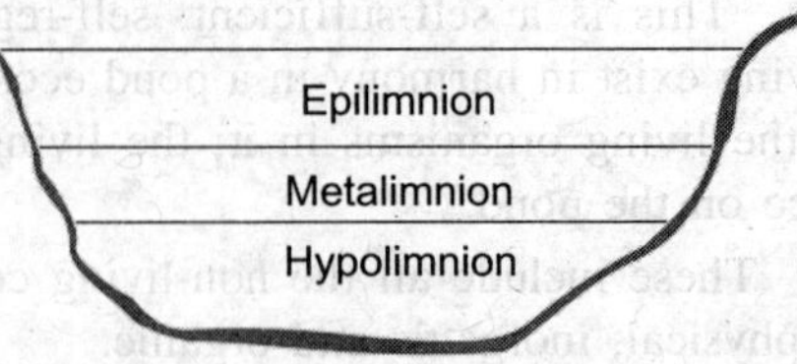

Figure 3.20 Thermal stratification in a pond

Based on the variations in temperature, the water in a pond can be divided into three zones or strata. This is called thermal stratification. The three strata are known as hypolimnion, epilimnion and metalimnion. The deeper part of the pond with cooler water is called hypolimnion. The upper part with warm water is called epilimnion. The middle part where these two layers meet is the metalimnion or the thermocline. The presence of these layers based on temperature is called thermal stratification. pH means the hydrogen ion concentration. The pH value of pure water is 7. This is the neutral value. Values below 7 indicate acidity, and values above 7 indicate alkalinity. Many plants and animals, especially the phytoplanktons and the zöoplanktons are greatly affected by the pH value of water. Some microbes prefer acidic waters while others prefer alkaline waters. Some prefer neutral value also. Wind is another factor affecting the life in a pond. It changes the temperature of the pond water. When fast winds blow, the hot and cool waters of the pond get mixed up. Wind also causes wave action in water. The number, size and duration of waves in the pond are also important physical factors.

B. *Inorganic:* Water itself is an inorganic component. The different minerals dissolved in water are important nutrients for the plants. Animals get these minerals from plants. Nitrates, phosphates are always present in pond water. Many gases like carbon dioxide, oxygen etc. remain dissolved in water. The solubility of CO_2 is more than that of O_2 in water. However, the exchange of gases between the plants and water is not much, as the air stored inside the plants form a reservoir of gases for them. The oxygen dissolved in water is used by the aquatic animals for respiration.

Chlorophyll, the green pigment present inside chloroplasts in all green cells of plants is essential for photosynthesis. This is an inorganic component too. The amount of inorganic matter at any given time is called the standing state or the standing quality.

C. *Organic:* The organic component in a pond includes organic compounds like carbohydrates, fats and proteins belonging to the dead plants and animals in the pond. The intermediates of their decomposition are also organic. Continued decomposition of this dead matter by bacteria and fungi ultimately converts them into basic inorganic form. This is reutilized by the plants. Bacteria and fungi thus form a connecting link between the abiotic and the biotic components of the ecosystem.

Biotic components: The biotic component includes the producers, consumers of different levels and the decomposers.

A. *Producers*: The primary producers of a pond are the autotrophic green plants and the photosynthetic bacteria. The large sized plants or the macrophytes may be flowering plants or non-flowering plants. These may be submerged fully in water or with leaves floating and other parts submerged or free floating. Of the flowering plants, *Hydrilla*, the insect catching plant Bladderwort or *Utricularia* etc. are submerged; Lotus (*Nelumbium*), water lily (*Nymphaea*) etc. have floating leaves but they are rooted in the mud at the bottom of the pond while duckweed (*Lemna*), water hyacinth (*Eichhornia*) etc. are free floating. Non-flowering plants are water ferns like *Salvinia*, *Azolla* etc. and green algae like *Chara*, *Nitella* etc. All these are macrophytes. The commonest microphytes functioning as primary producers in the pond ecosystem are the phytoplanktons. These include, most commonly, green algae like *Spirogyra*, *Zygnema*, *Scenedesmus, Ulothrix* etc. and the unicellular golden brown floating algae, diatoms.

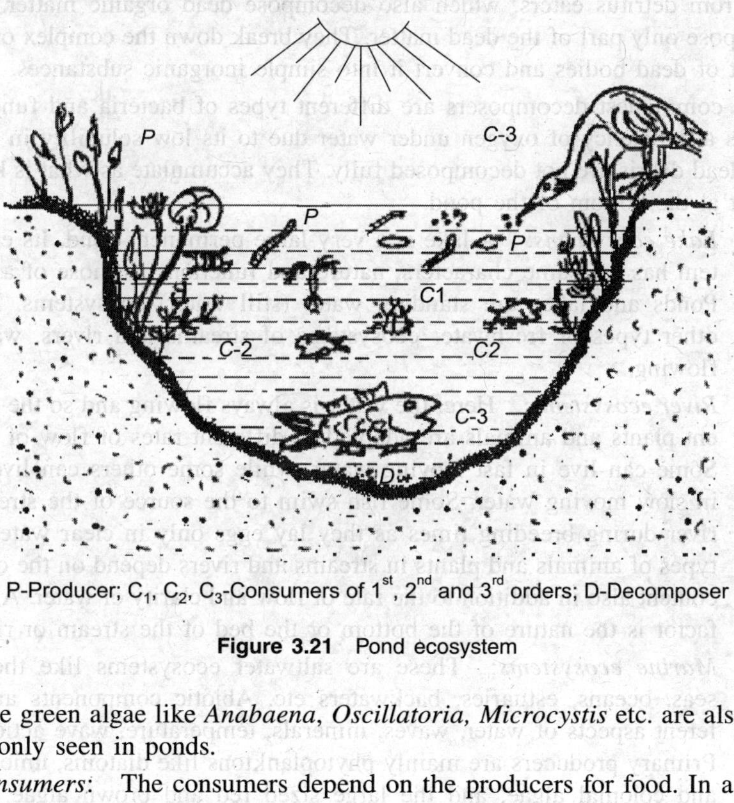

P-Producer; C_1, C_2, C_3-Consumers of 1st, 2nd and 3rd orders; D-Decomposer

Figure 3.21 Pond ecosystem

Blue green algae like *Anabaena, Oscillatoria, Microcystis* etc. are also very commonly seen in ponds.

B. *Consumers*: The consumers depend on the producers for food. In a pond, consumers are of 3 types.

Primary consumers: These are herbivores. The zooplanktons that feed directly on the phytoplankton are 1st level or primary consumers. Examples of zooplanktons are the rotifers. Some protozoans like *Cyclops*, *Euglena* etc. are also primary consumers. The insects and animals that live at the bottom of the pond and eat plants and plant remains are called detrivores. Some other insects, some small fish etc. eat phytoplankton and parts of other plants living in the upper waters of the pond. These, along with the detrivores, form the benthos. Benthic consumers are insects and insect larvae, beetles, crustaceans, molluscs etc.

Large animals like cows, buffaloes etc. and some birds visit the pond to eat the macrophytes at the edge of the pond. These are also primary consumers.

1. *Secondary consumers*: Insects and small fish that feed on primary consumers like zooplankton are the secondary consumers.
2. *Tertiary consumers*: Large fish and some birds that eat small fish are the tertiary consumers in a pond.

Thus, in a pond, fish occupy all levels of consumers.

C. *Decomposers:* These are sometimes called microconsumers to differentiate them from detritus eaters, which also decompose dead organic matter. They decompose only part of the dead matter. They break down the complex organic content of dead bodies and convert it into simple inorganic substances.

The commonest decomposers are different types of bacteria and fungi. As there is a deficiency of oxygen under water due to its low solubility in water, some dead debris are not decomposed fully. They accumulate as what is known as peat at the bottom of the pond.

(a) *Lake ecosystems*: A lake is a very large permanent pond. Its ecosystem has the same characters, nature and functions as those of a pond. Ponds and lakes are standing water (still water) ecosystems. In the other types of freshwater ecosystems of streams and rivers, water is flowing.

(b) *River ecosystems*: Here, the water is always flowing and so the different plants and animals are adapted to different rates of flow of water. Some can live in fast moving water, while some others can live only in slow moving water. Some fish swim to the source of the stream or river during breeding times as they lay eggs only in clear water. The types of animals and plants in streams and rivers depend on the oxygen content also in addition to the rate of flow and clarity of water. Another factor is the nature of the bottom or the bed of the stream or river.

(c) *Marine ecosystems*: These are saltwater ecosystems like those of seas, oceans, estuaries, backwaters etc. Abiotic components are different aspects of water, waves, minerals, temperature, wave action etc. Primary producers are mainly phytoplanktons like diatoms, unicellular and colonial algae, and the large sized red and brown algae. Some brown algae are very large, measuring a few metres in length. These are the seaweeds. All these producers show zonation, i.e., they live at different depths.

The primary consumers are the marine zooplankton. Some fish, crustaceans and mollusks are also primary consumers. Sea fish are different from river fish. The variety of fish in the seas and oceans is much more than in ponds or rivers. Other sea animals like crustaceans, starfish, jelly fish etc. also live in large populations in the marine environment. Fish occupy different trophic levels. While some small fish are herbivorous and occupy the 1st trophic level, some others occupy higher trophic levels. Thus, while fish like Herring, Mackerel etc. are secondary consumers, some other fish like Cod, Halibut etc. are tertiary consumers. Decomposers are the various types of fungi and bacteria.

The flora and fauna (plants and animals) growing near the sea coast in shallow water is different from those that live in deep sea. Many types of animals live in the coral reefs under water in the shallow seas. The plants and animals that live on the seashore differ based on whether the shore is sandy, rocky, muddy etc.

From this discussion on various types of ecosystems, it is clear that each part of an aquatic or any other ecosystem has a definite role to play. Each part has its own function. The function of the producers is to produce food. The function of the consumers is to use part of the chemical energy trapped by the producers in the form of food for their various activities. They also function as a check against the overcrowding of the earth by the producers. The consumers act on the dead producers and consumers and release the elements bound in their dead bodies. This helps in the recirculation of the elements throughout the system.

Mother Nature has planned out everything to perfection. It is for us to maintain this equilibrium by not disturbing the different ecosystems in our biosphere.

QUESTIONS

A. Very short type:

(a) Multiple choice questions – tick the right answer:

1. Pyramid of numbers in a pond ecosystem is always
   (a) Inverted (b) Erect
   (c) Irregular (d) None of these
2. Upper warm layer of water in a freshwater body is called
   (a) Hypolimnion (b) Metalimnion
   (c) Epilimnion (d) Profundal zone
3. Grasslands with trees here and there are called
   (a) Tundras (b) Estuaries
   (c) Plateaus (d) Savannas
4. Cycling of materials in the biosphere is called
   (a) Biochemical cycle (b) Biophysical cycle
   (c) Biogeochemical cycle (d) Geochemical cycle

5. The term 'Ecosystem' was coined by
   (a) Tansley (b) Odum
   (c) Warming (d) Blackmann
6. Pyramid of energy is always
   (a) Inverted (b) Erect
   (c) Rectangular (d) None of the above
7. Herbivores are
   (a) Primary producers (b) Consumers
   (c) Decomposers (d) All of the above
8. Butterflies and bees are
   (a) Carnivorous insects (b) Detritus eaters
   (c) Herbivorous insects (d) Nocturnal insects
9. The sequence of organisms in a food chain is:
   (a) Consumer → Producer → Decomposer
   (b) Decomposer → Consumer → Producer
   (c) Producer → Decomposer → Consumer
   (d) Producer → Consumer → Decomposer
10. Which of the following is a top carnivore?
   (a) Tiger (b) Crocodile
   (c) Hawk (d) all of the above
11. One method of Nitrogen fixation in nature is by:
   (a) Thunder (b) Rain
   (c) Lightning (d) Storm
12. Energy flow in any ecosystem is always:
   (a) Bidirectional (b) Unidirectional
   (c) Multidirectional (d) None of the above

(b) Define the following:

1. Psammosere
2. Ecosystem
3. Autotrophs
4. Heterotrophs
5. Ecological pyramids
6. Biogeochemical cycle
7. Hypolimnion
8. Epilimnion
9. Metalimnion
10. Symbionts
11. Biomass
12. Parasites

(c) Name the following:

1. Three terrestrial ecosystems
2. Three aquatic ecosystems
3. Three natural ecosystems
4. Three man-made (artificial) ecosystems
5. Two lotic ecosystems
6. Two lentic ecosystems
7. Two freshwater ecosystems
8. Two marine ecosystems

9. The two components of an ecosystem
10. Three herbivorous animals
11. Three carnivorous animals
12. Three top carnivorous animals
13. Three herbivorous birds
14. Three carnivorous birds
15. Two top carnivorous birds
16. Two herbivorous insects
17. Two carnivorous insects
18. Two ecosystems with erect ecological pyramids of numbers

(d) Write very brief notes on:

1. Ecological niche
2. Littoral zone
3. Limnetic zone
4. Profundal zone
5. Biomass
6. Detritus eaters
7. Predators

(e) Differentiate between the following pairs:

1. Primary and secondary producers
2. Primary succession and secondary succession
3. Carnivores and top carnivores
4. Herbivores and carnivores
5. Producers and consumers
6. Autotrophs and heterotrophs
7. Macroconsumers and microconsumers

B. Short answer type:

(a) Write short notes on:

1. Inorganic components of an ecosystem
2. Organic components of an ecosystem
3. Physical components of an ecosystem
4. Decomposers
5. Pyramids of numbers
6. Pyramids of biomass
7. Pyramid of energy
8. Unidirectional energy flow
9. Pyramid of biomass in a pond
10. Pyramid of numbers in a tree ecosystem
11. Water cycle
12. Carbon cycle
13. Nitrogen cycle
14. Phosphorous cycle
15. Sulphur cycle
16. Importance of green plants on maintaining life on earth
17. Sun as the main source of energy on earth
18. Forest ecosystem

19. Grassland ecosystem
20. Desert ecosystem
21. Grazing food chain and detritus food chain
22. Marine ecosystem
23. Climax forest
24. Lichens.

(b) Explain the role of decomposers in the circulation of minerals.

(c) Why is a fallen leaf or a roadside puddle an ecosystem? Explain.

C. Long answer type:

(a) Describe in detail the abiotic and biotic components of an ecosystem.

(b) Give an account of the various types of ecological pyramids.

(c) Write an essay on the cycling of materials in an ecosystem.

(d) What do you understand by unidirectional energy flow? Explain in detail.

(e) What is the difference between food chains and food webs? Explain in detail with the help of suitable examples.

(f) Give a detailed account of a pond ecosystem.

(g) Describe in detail a hydrosere or a xerosere.

Unit 4

Biodiversity and its Conservation

4.1 INTRODUCTION

Life originated on the earth millions of years ago. It is believed that some organic molecules like simple nucleic acids were the first to evolve. Gradually, the simplest forms of life developed from them. The organisms to appear first were unicellular. (A cell is a structural and functional unit of life; an organism made of one cell is called unicellular.) After thousands of years, multicellular organisms made their appearance. (A multicellular organism is made of many cells.) They evolved along so many lines that the whole pattern of evolution became a complicated web. The result of this is the biological diversity or biodiversity that we see today. Biodiversity can be defined as the totality of genes, the species, the ecosystems in a given region of the world. In a broad sense, it stands for all life on the earth.

It was in 1992 that environmental issues came to be accepted as of global importance. From 3rd to 14th June 1992, the Earth Summit or the UNCED (The United Nations Conference on Environment and Development) was held at Rio de Janeiro, capital of Brazil. It was a historic meet. Leaders from 170 countries participated in this summit. The definition formulated for biodiversity at this meet goes like this: 'Biodiversity is the variability of living organisms from all sources including terrestrial, marine and other aquatic ecosystems and the ecological complexes of which they are a part; this includes diversity within species, between species and of ecosystems.'

The number of species of known flowering plants alone is 2,00,000 at present. Add to this the number of species of non-flowering plants and those of animals—the number of species that has been scientifically described will be about 2 million. The total number is much higher.

4.1.1 Levels of Biodiversity

Biodiversity is visible at 3 levels in nature.

(a) *Genetic diversity*: In any organism, whether unicellular or multicellular, the cell contains some type of hereditary material. This is in the form of nucleic acids. Most of the organisms have Deoxyribonucleic acid (DNA) as their hereditary material. Each DNA molecule is made up of many genes. Thus, genes are the bearers of hereditary characters. The characters that pass on from one generation to the next of any organism are controlled by genes.

The number of genes varies greatly in different organisms. Even a bacteriophage has about 100 genes. Viruses are not organisms as they are not made up of cells. However, even though they are made of only nucleic acids and proteins, once they enter the living cells, they can multiply. Thus viruses show characters of both living and non-living things. Viruses cause many diseases in plants as well as animals. Bacteria, on the other hand, are unicellular organisms—they are unicellular plants. Bacteria are of different types. Many of them are useful to us, while some cause plant and animal diseases. There are viruses which attack bacteria, these are called bacteriophages.) *E. coli*, a common bacterium has 4,000 genes, while each cell of the common fruit fly (*Drosophila melanogaster*) has 13,000 genes. A cell of rice (*Oryza sativa*) plant contains 30 to 50 thousand genes, while a cell of a human being (*Homo sapiens*) contains 35 to 45 thousand genes.

The genes differ from one another chemically. Therefore, the enzymes they code are also different. Chemical reactions in the cells depend on the enzymes. And the characters of any organism are the ultimate result of all the chemical reactions that takes place within their cells. This is how genes control the characters of an organism. In addition to this, many new characters can be developed by the interaction of genes, the position effect of genes etc. Thus, the number of characters controlled by genes runs into tens of thousands, based on the number of genes. During reproduction, the genes pass on from one generation to the next. This is how genetic characters pass on to succeeding generations.

However, genes are not immune to changes. Under certain conditions they undergo changes, which are called mutations. The mutated gene will behave differently. When the mutated genes pass on to the next generation, changed characters will be seen.

In nature, many mutants (organisms with mutated genes) are seen. These along with the original wild type organisms are constantly affected by various environmental conditions. The result is natural selection. Those forms best suited to the environment are chosen by nature. These succeed in establishing themselves in nature. These may interbreed with the original forms and again give rise to new characters. The sum total of all the genes in all the wild species make what is known as the gene pool or the gene bank or the gene reserve.

The gene pool in nature is almost unlimited. The diversity in the genes in our crop plants and domesticated animals has developed over thousands of years. This gene reserve or gene pool is used by plant and animal breeders in their breeding experiments to give rise to better varieties of useful plants and animals. The gene pool is used by genetic engineers and biotechnologists also. They do research on producing genetically modified (GM) foods and other crops better suited for disease resistance, resistance to drought, salinity in the soil etc., developing better medicines and other products for our industries.

(b) *Species diversity*: When man studied more and more plants and animals on the earth, classification became necessary for convenience in studying. This led to the establishment of different systems of classification during different periods of scientific study. Accordingly, many units of classification were decided upon, like order, family, genus, species etc. Species is the smallest unit of classification. Individuals resembling each other in all important external characters, both vegetative and reproductive, constitute a species. From this it is clear that there are quite a large number of species of living beings on the earth. The number of species of plants and animals present in a region constitutes its species diversity. Some regions have greater species diversity than others. For example, a rainforest has much greater species than a desert. The species diversity is the least in a cropland, where only one crop is cultivated. Natural forests have great species diversity, while plantation forests have only one or a few species planted and maintained by man. Thus, though afforestation (planting of trees where there were no trees earlier) is a good practice, it cannot compensate fully for deforestation (felling of trees in a forest and the resultant destruction of the forest). This is because of the vast difference in species diversity between the two. Natural forests give us a large variety of products in terms of timber as well as other forest produce. They sustain whole communities dependent on them for their livelihood. Areas that are rich in species diversity are called hotspots. The world has 15 nations that have hotspots; India is one of them.

(c) *Community and ecosystem diversity*: Ecosystems are of different types. Natural ecosystems are seen in the wild, while man-made ecosystems are seen in inhabited areas. As ecosystems are in general large sized, when we speak about ecosystem diversity, we refer to countries or states or districts. Ecosystems undisturbed by human activities have more chances of having greater species diversity. Biodiversity is the basis of the stability of ecosystems. Reduction in the number of individuals or species reduces its stability. Therefore protection of a single species at the cost of other species is not advisable. The whole ecosystem of which it is a part must be conserved. Overuse and misuse of ecosystems will have long standing harmful effects on the environment.

Community and ecosystem diversity are of 3 types:

1. Alpha (α) diversity: This is the diversity shown by the organisms living in the same community. These organisms live in more or less the same habitat. If a community has more alpha diversity, it shows that the community is stable and the interrelationships between the different organisms are healthy.
2. *Beta* (β) *diversity*: This type of diversity is seen in gradients, i.e., in areas where two types of ecosystems meet. Here, the climate and other environmental factors show a gradual change from one ecosystem to the other.

3. *Gamma* (γ) *diversity*: This is also known as regional diversity. Gamma diversity is the total diversity shown within a large zone, which may be a specific geographical area or a landscape. It indicates the total species richness in all the habitats of the zone put together.

Ecosystem diversity indicates the variety of ecosystems present in a large area. Each ecosystem has its own food chains and food webs, trophic levels, nutrient cycles, energy flow mechanisms, ecological niches etc.

4.2 BIOGEOGRAPHICAL CLASSIFICATION OF INDIA

India is the seventh largest country in the world. In Asia, it is the second largest. Its total land area is 3,287,263 km^2. The mighty Himalayan ranges are located to the north, northwest and north-east of our country. The southern part of India is a peninsula with Arabian Sea on the west, Bay of Bengal on the east and the Indian Ocean on the south. Pakistan is situated on our north-west. China, Tibet and Nepal are the countries on our north-east. Myanmar (previously known as Burma) is located on our east.

India has climates varying from the equatorial to the alpine in character. The chief controlling factors of the climate are rainfall and temperature. Depending on the rainfall, India can be divided into 4 regions—wet zone with rainfall above 80″, intermediate zone with 40-80″ rainfall, dry zone with 20-40″ rainfall and arid zone with less than 20″ of rainfall. Temperatures vary greatly. The highest recorded was 50.6°C in Alwar, while the lowest recorded was – 45°C in Kashmir. There are 4 major seasons in India—winter from December to February, summer from March to June, rainy season from July to mid-September and autumn from mid-September to November. As per our traditional calendar, we have 6 seasons—Spring (Vasant), Summer (Greeshm), Monsoon (Varsha), Early Autumn (Sharad), Late Autumn (Hemant) and Winter (Shishir), each of two months.

We have a rich variety of climates and vegetation. There are 10 biogeographical regions in our country. These are:

4.2.1 Trans-Himalayan Zone

This is the cold region of Ladakh. It is a snow covered mountainous region. The Trans-Himalayas come under the alpine zone. 15,000′ is the snow line. Above this there is usually no vegetation. There is only snow. Animals of this zone include wild sheep and goat, yak, snow leopard, Tibetan asses and wolfs etc.

4.2.2 The Himalayan Ranges and the Valleys of Kashmir, Himachal Pradesh, Uttarakhand, Assam and Other North-Eastern States

This zone starting from the north-west of Kashmir extends eastwards and covers the north-west, central and eastern Himalayas. Vegetation-wise, this biogeographic zone is divided into the western Himalayas (from Kashmir to Kumaon) and the eastern Himalayas (Kumaon eastwards, up to Assam). Rainfall is more in the eastern part, while the climate is dry towards the west. Going up the mountain, climate changes as if travelling from the equator to the poles. Due to these latitudinal variations, there are definite belts of vegetation here. Transition from one zone to the next is rapid. Three climatic zones, each with its own characteristic vegetational belts, are recognized in the Himalayan ranges. These are the submontane (lower or basal) zone, the montane (temperate) zone and the alpine zone.

(a) *The submontane zone* extends from the foot of the Himalayas up to 5,000′ height. Climate is tropical and subtropical. In the lower belts, continuous belt of sal (*Shorea robusta*) forms a tropical deciduous forest. Many other trees like mango (*Mangifera indica*), amaltash (*Cassia fistula*) etc. are seen in the western Himalayas and kadam (*Anthocephalus cadamba*), *Terminalia* species etc. in the eastern Himalayas. Riverian (forests near rivers) and mangrove forests and savannas (tall grasses with trees here and there) are also seen in this zone.

(b) *The Montane zone* extends from 5,000′ to 12,000′. The lower belt here is made of coniferous forests, while the upper belt has deciduous forests. Common trees are different species of Pine (*Pinus*), deodar (*Cedrus deodar*) etc. in the western Himalayas and white oak (*Quercus incana*), champa (*Michelia champa*), jamun (*Eugenia jambolana*) etc. in the eastern Himalayas. The alpine zone lies above 12,000′, which is the timber (tree) line, i.e., no trees grow here. The lower alpine zone has some shrubs like *Juniper, Rhododendron* etc. in both west and east Himalayas.

(c) *The Alpine zone* has herbs with brightly coloured flowers in the western Himalayas. Some examples are *Ranunculus* with yellow flowers, *Polygonum* and *Pedicularis* with red flowers, *Epilobium* and *Geranium* with pink to red flowers, *Primula, Aster, Delphinium, Gentiana* etc. with blue flowers, crucifers with white flowers, *Potentiella* with scarlet flowers, *Clematis* with black flowers….the colour festival can be visualized in this beautiful part of India. Diversity of animals is more in the Himalayas than in the trans-Himalayas. It includes sambar, musk deer, wild bear, snow leopard, brown bear, pheasants etc. Sikkim dog, which used to live here earlier, has now become extinct.

4.2.3 Desert Zone

Rajasthan, Punjab, Haryana, Gujarat, Delhi, Rann of Kutch etc. are included in this zone. The Thar of Rajasthan is a famous desert. Rainfall is scanty, and winter temperatures are low, often nearing the freezing point. Summers are hot and dry with scorching dry winds and sometimes dust storms. Soil is sandy. Except in irrigated fields, land dries up very quickly after rain. Vegetation is therefore poor. Tree species are very few. They are seen only near the rivers, forming dry forests. Examples of trees are poplar (*Populus euphratica*), shisham (*Dalbergia sissoo*), *Cassia* etc. Some shrubs and many short-lived annuals are seen. Most common shrubs are the cacti, Indian ber (*Zizyphus jujuba*) etc. Common herbs are prickly poppy (*Argemone mexicana*), mako (*Solanum nigrum*) etc. South-west of Punjab merges into the hot dry desert of western Rajasthan, west of the Aravalli hills. In the south-eastern part of Aravalli, dry thorny forests of *Acacia*, *Zizyphus*, *Euphorbia* etc. are seen. The fauna of the desert zone includes wild ass, fox, desert cat, sand grouse, the great Indian bustard etc.

4.2.4 Semi-Desert Zone

This zone includes Madhya Pradesh, Chattisgarh, parts of Orissa and parts of Gujarat. Rainfall ranges from 25″ in the west and centre to 70″ in the east. Depending on the rainfall, forests of different types grow here. Sal (*Shorea robusta*) forms forests in eastern parts of M.P. and Orissa. Mixed deciduous forests of teak (*Tectona grandis*), shisham, *Cassia,* palash (*Butea frondosa*), amla (*Phyllanthes emblica*) etc. are seen in the central parts of M.P. and Chattisgarh. Thorn forests of acacia, *Zizyphus* etc. occupy the drier parts. The endemic (characteristic to a region and not found elsewhere) animal of the semi-desert zone is the Gir lion.

4.2.5 The Gangetic Plain

The Gangetic plain extends from Eastern Punjab to the Sunderbans in West Bengal. It passes through Haryana, Delhi, U.P., parts of Bihar and Orissa. The land is very fertile. The rainfall is less in U.P., and more in Bengal. Vegetation is mostly tropical moist and dry decidous forests. Northern regions have sal forests. Other trees are mango, peepal (*Ficus religiosa*), neem (*Azadirachta indica*) etc. The forest floor is covered with shrubs and herbs. At the eastern end are the mangrove tidal forests of the Sunderbans in West Bengal. These occupy the tidal areas close to the Gangetic delta. Marsh plants form swampy vegetation here. Fauna includes gazelle, elephant, rhinoceros, crocodile, black buck, barasingha, turtle etc.

4.2.6 The Western Ghats

This zone comprises the Malabar Coast and the Western Ghats. It extends from South Gujarat to Cape Comorin. Rainfall is very heavy. Vegetation is of 4 types—tropical moist evergreen forests, mixed deciduous forests, subtropical temperate evergreen forests and the mangrove forests on the sea shores. Tropical evergreen forests are multi-storied with tall trees like teak, mango, shisham etc. Bamboo (*Bambusa arundinaceae*) is common. Growth is luxuriant. Forest floor is densely covered with shrubs. Climbers and epiphytes (plants growing on trees) are also many. Mixed deciduous forests occur high on the mountains, up to 5,000′ altitude. Teak and shisham trees growing here are not very tall. Several bamboos grow here. Temperate evergreen forests are seen on the Nilgiris at 5,000′ and above. They are known locally as 'sholas'. Prominent trees are champa, jamun etc. Mangrove forests form large areas on the coast of Mumbai and its suburbs. Examples are *Rhizophora, Phoenix* etc: The Western Ghats have a very large number of endemic flora and fauna. Threatened species of animals are crane turtle and Travancore tortoise.

4.2.7 Deccan Peninsula

The Deccan peninsula includes the Deccan plateau in the south, the central and eastern plateaus, Chota Nagpur plateau and central highlands. Average altitude is 2,500′. Rainfall is less; it is 20 to 40″ as the area lies in the rain shadow region of the Western Ghats. Common plants are acacia, siris (*Albizzia siris*), sandalwood (*Santalum album*), teak etc. The fauna includes sloth bear, leopard, wild boar, barasingha, chital, wild buffalo, sambar, wild elephant etc.

4.2.8 Coasts

The coastline of our country is on both sides of the southern peninsula. Dry evergreen forests are seen in the eastern Coromandel Coast. Rainfall here is 40 to 50″. Besides the plants of the Deccan plateau, trees like *Strychnos*, jamun etc. are also seen here. In the drier parts, *Zizyphus*, date palm (*Phoenix*), *Casuarina* etc. grow. The spinous grass *Spinifex* is common in the sandy shores. Halophytes grow in the marshes.

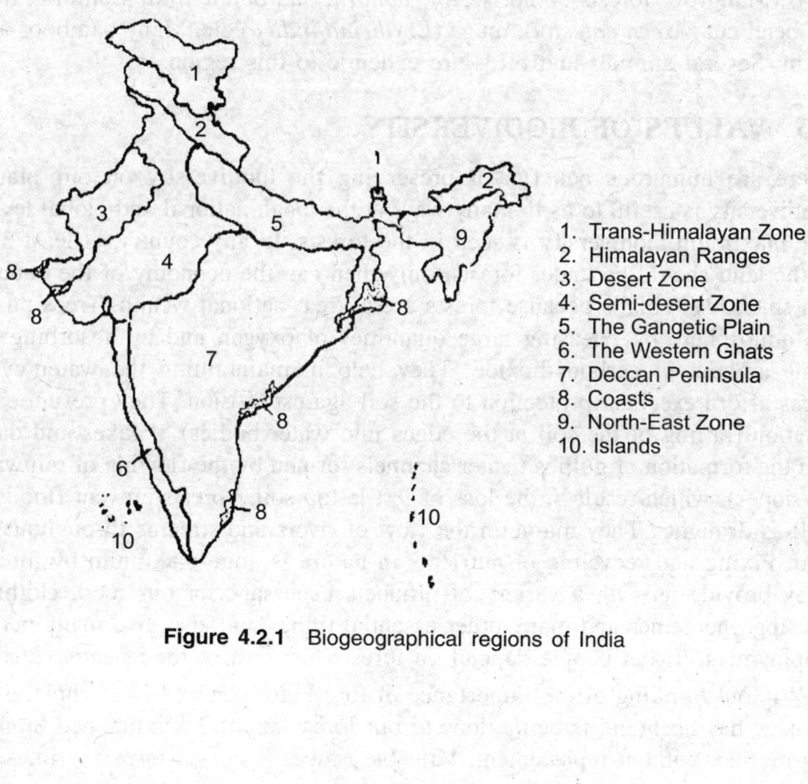

Figure 4.2.1 Biogeographical regions of India

Ratnagiri in Maharashtra and Pichavaram near Chennai have typical estuarine (lower tidal) vegetation. Animals characteristic to the region are inshore dolphins, estuarine turtles and crocodiles etc.

4.2.9 North-East Zone

This zone has a very heavy rainfall. Annual rainfall is more than 80″. Cheerapunji has more than 400″ annual rainfall. This heavy rain is responsible for dense tropical evergreen forests. In the valleys of Brahmaputra, outside the Himalayas, common plants are *Sterculia*, champa, mulberry (*Morus*), *Ficus* etc. *Dendrocalamus* is the common bamboo. Limbers and evergreen shrubs are very common. On the riverbanks, siris, silk cotton (*Bombax*), insectivorous Pitcher plant (*Nepenthes*) etc. grow. Sal forests occur in the Garo Hills. The fauna includes elephants, rhinoceros, swamp deer, pygmy hog, buffalo etc.

4.2.10 Islands

The islands include Andaman and Nicobar in the Bay of Bengal and also the Lakshadweep in the Arabian Sea. The flora of Andaman and Nicobar resembles that of Myanmar and Malayasia. The Andaman has a wide range of spreading coastal vegetation like mangroves, beech forests and in the interior, evergreen forests of tall trees. Some dry pockets are also seen in the interior parts. Mangrove forests comprise *Rhizophora*, and *Sonneratia*. Common trees are betel nut (*Areca catechu*), muga (*Erythrina indica*) etc. Many bamboos also occur. Several animals and birds are endemic to this region.

4.3 VALUES OF BIODIVERSITY

There are numerous benefits of preserving the biodiversity on our planet. Biodiversity is useful to us in many ways at the local, national and global levels. The maximum biodiversity is seen in the forests. In any country, at least 33% of the land should be under forests, only then can the economy of the country run smoothly. This is because forests are a great national wealth. Trees purify the atmosphere by releasing large quantities of oxygen and by absorbing the same amount of carbon dioxide. They help in maintaining the water cycle. Trees afford excellent protection to the soil against erosion. They prevent early siltation (falling of the soil at the edges into water bodies) of lakes and dams and the formation of gullies (water channels formed by the flowing of rainwater on slopes), which result in the loss of fertile top soil. Forests prevent floods as well as droughts. They maintain the flow of rivers and streams throughout the year. Fixing and recycling of nutrients in nature is done maximum by forests. They provide us with a variety of products necessary for our food, clothing, housing, medicines and many other essential things and also give many people employment. Tribal people depend on forests for most of their needs.

Without thinking of the importance of forests to mankind as a whole, great damage has been and is being done to our forest wealth by unplanned lumbering by man without replacement. Valuable timber is cut up to many times the permission given by governments. This has damaged the whole environment in many parts of the world. Floods and droughts and soil erosion has increased manifold. Valuable wildlife is gradually moving towards extinction. As forest covers are depleted and the number of polluting industries and automobiles increases, there is an increase in global warming that is extremely dangerous to the world environment. It also causes many changes in our atmosphere, and raises temperatures.

Moreover, wild plants and animals in forests constitute a never ending gene pool for our plant and animal breeders. From this vast gene reserve, they select useful genes for introducing into cultivated plants and domesticated animals.

Each species has a specific function in the ecosystem—it has an ecological niche. The ecological niche of an organism concerns basically 3 aspects: 1. The space where it lives; 2. What it does, i.e., how and where it gets energy from, how it behaves with other members of the system, how it responds to the environment, how it modifies the environment etc., and 3. How it is constrained by other species. Thus, the importance of any species in nature may not be known to us, but it is an essential part of nature. One of the main benefits of preserving biodiversity is thus maintaining the balance or stability of the ecosystems in nature.

The values of biodiversity can be categorized into 6 groups.

4.3.1 Consumptive Use Value

Local communities all over the world make direct use of different plants and animals available in their respective areas for various purposes. The biodiversity in the ecosystem provides the local people with most of the things they need on a daily basis. Rural people get the grains, pulses, oils, vegetables and fruits, milk, meat, eggs, fish etc. from their own villages or nearby villages. The rural folk depend on other plant and animal produce like medicinal herbs, fodder for their animals, building materials, firewood, honey etc. on the local ecosystems. Fishermen catch fish from ponds, lakes, rivers and seas and oceans. If diversity was not there in nature, people would have had to be content with the same produce over and over again.

4.3.2 Productive Use Value

It is because of the genetic diversity present in nature that scientists, farmers and biotechnologists are able to selectively improve plants and animals. All the species of plants cultivated for various uses at present had their origin in wild species. The process of cultivating wild species for human needs is called domestication of plants. This has played a vital role in human civilization. The origin of agriculture dates back to 7,000 to 13,000 years ago. It started in the fertile lands near Indus, Nile, Euphrates and Tigris. Cultivation started with plants propagating vegetatively by buds, tubers etc., but soon man came to know how seeds give rise to new plants. Cereals were brought into cultivation first. These were unipurpose plants. Later, multipurpose plants were domesticated.

Soon, plant breeding was devised as an important technique for plant improvement. Different types of selection were the first methods tried for plant improvement. Later, hybridization (crossing 2 types of plants) was found out. Tissue culture is a modern method. This is based on a phenomenon called totipotency, which is the capacity of any cell to give rise to a new plant. The more recent is Genetic Engineering. In genetic engineering, one or more genes from a desirable variety is introduced into or deleted from another variety, which needs to be improved. Genetic engineering is a part of biotechnology.

Plants into which one or more foreign genes have been introduced permanently are called transgenic plants or genetically modified (GM) plants. Such plants have been produced in tobacco (*Nicotiana tabacum*), tomato (*Lycopersicum esculentum*) etc. and many others. For best results in these fields, selection of the best variety among many is most important so that it can be still further improved by human effort. This is provided by the biodiversity in nature. For this, plant materials or genes of as many varieties in nature as possible are collected (in the form of seeds) and stored in live form. Tubers, rhizomes, bulbs etc. can also be preserved. Institutions where plant materials (genes) are stored are called gene banks. Where seeds are not produced, tissue culture techniques are used for storing germplasm (reproductive cells or the germ cells and the cells from which they are formed).

The discovery of new types of plants and animals is in no way complete. It is an ongoing process. Also, among the known species, only very few have been studied till now for their use as medicines, industrial potential etc. Loss of such organisms will also be a great loss to humanity.

4.3.3 Social Values

In the earlier days, population was less and the needs of humans were limited. In those days, nature was not tampered with. But as human population grew and human beings became more and more demanding, natural resources started getting depleted faster and faster. Coupled with their excessive usage, when replacement was not done properly, nature's resources started dwindling further. Biodiversity, one of the most valuable gifts of nature to man, was no exception. However, even now, we can see that traditional societies are preserving nature better than the modern society. Therefore, we need to learn a lesson from them.

The Vedic scriptures make explicit references as to how forests and other natural resources should be maintained along with best utilization. Trees like peepal (*Ficus religiosa*), neem, mango, vat (*Ficus bengalensis*) and many others; herbs like Thulsi (*Ocimum sanctum*); grasses like Nagarmotha (*Cyperus rotundus*), Doob (*Cynodon dactylon*); animals like cow, ox, tortoise etc; and birds like peacock, owl etc. have been given place of worship in the scriptures. Medicinal plants and trees (*vanaushadhis*) were personified as goddesses of forests. All this inspired and motivated the people of those times to conserve biodiversity.

Traditional agriculture systems in India always followed multiculture (cultivation of different crops) as against monoculture practised in some places today. Moreover, some governments in India are giving incentives to those farmers to start cash crops for sale in national or international markets. This has led to local shortage of food, unemployment, drought, flood etc. For the prosperity of local people, farm produce made locally must be available to them at reasonable prices. Otherwise, there will be unrest in society. Cultivating various crop plants and rearing domestic animals provide employment to lakhs of people in India. Thus conserving biodiversity has a great social value.

4.3.4 Ethical and Moral Values

Protecting all forms of life was one of the principles on which life was based in the Vedic period in India. The Vedas preach, "Aatmavat sarvabhootheshu": "Regard every being like your own self." The Smritis, Puranas and other ancient texts integrate the civil and social duties with religious faith. It was believed that all forms of life have a right to live on the earth. Man is only part of nature, just like the other organisms in nature. All ancient Indian scriptures propagated this idea.

The attitude of respect towards earth as mother is widespread among the Indian society. Moral and ethical norms had been designed to ensure whole-hearted participation of the masses in sustaining the biodiversity of nature. Many cultural, moral and ethical values are associated with nature even today in our country. One example is the sacred groves. These are small or large undisturbed forest areas in villages with tiny places of worship in the centre. In Kerala, sacred groves along with their associated ponds constituted a unique network of ecosystems. These are intertwined with the life and culture of the people of Kerala. They are dedicated mainly to goddesses/snakes. Most of them are private properties, owned by families. Some are owned by communities or temples. Due to neglect, the sacred groves with their immense biodiversity have been dwindling of late. Realizing the importance of such groves, local community members have formed committees to manage and protect sacred groves in some places. Pattuvaom Gram Panchayat was in news recently for preparing first Community Biodiversity Register in which the medicinal plants and their uses were listed. Such efforts from the people will go a long way in creating awareness regarding the ethical and moral values of biodiversity among the general public.

4.3.5 Aesthetic Value

Biodiversity adds to the aesthetic value of any place. Variety, they say, is the spice of life. How monotonous life would have been if there was no variety in nature. It is due to biodiversity that we have so many different varieties of plants and animals on the earth. We live amidst varying ecosystems of forests, rivers, lakes, ponds, grasslands, agricultural pastures, deserts, snow bound peaks...even urban areas with some remains of vegetation and water bodies. All these add to the aesthetic beauty of nature. India is a vast country with varying types of climates and the resulting immense variety of plants and animals. These are a great attraction for national as well as international tourists. The more the biodiversity in a country, the more will be the number of tourists that visit that country. In this age of eco-tourism, preserving biodiversity will therefore make a government earn more revenue. This will boost up the economy. Tourism has become a big industry in India. Conserving our biodiversity and thereby the places of aesthetic value in our country is therefore very beneficial.

4.3.6 Option Value

We never know which species of animals or plants will become useful to us in future. Only very little has been found out as yet. Before the discovery of penicillin, who would have thought that the lowly fungus *Penicillium* has such great therapeutic value? Nothing is impossible. Not only plants and animals, but many microorganisms also are of use to us in different ways. There is no guarantee that those which are not used at present will remain useless in future. Any day, any use of any organism may be found out. So it will be wise to preserve all forms of life present today in the best possible way. To get improved plants and animals, we go back to the wild types very often. So the wild dominant genes and genomes (a single set of genes is a genome) must be preserved. This is the option value of biodiversity.

4.4 BIODIVERSITY AT GLOBAL, NATIONAL AND LOCAL LEVELS

The number of known and described species of living organisms in the world today is about 1.8 million. According to some scientists, more than this number remains undiscovered. This shows how vast the extent of biodiversity on the earth is.

Life evolved on the earth more than 600 million years ago. Many species have evolved and many others have become extinct over these long years. This is proved by fossils. By studying fossil records, we can know how the species evolved, and when these forms dominated the earth. The fossil records also show that there have been periodic cycles of mass extinction in the history of the earth. Fossil history shows that over the last 150 million years, 1, 50, 000 species of birds have evolved and become extinct. Yet, as per the information we have on birds, 9,000 species of birds are living worldwide today. This is a good number of species, we might think. But the problem is that human induced extinctions are much more rapid than those by natural processes. For example, even in the 19th century, 2,000 million individuals of passenger pigeons were living. But due to hunting and habitat destruction, the species became extinct by 1914.

The developed nations have a lesser biodiversity than the developing ones. The economically developed nations have exploited their diversity for centuries.

The developing nations, including India, should therefore take a lesson from this. India has a vast biodiversity at present, so do Brazil, Malaysia and Indonesia. The species composition of these countries varies greatly. We must therefore preserve our own diversity to the maximum extent possible. As we have advanced much in biotechnology and genetic engineering, we can make better use of this natural resource too, compared to others.

The importance of preserving biodiversity has come to be realized globally. Many international agreements have been signed in the past few years to protect this resource. Some of them are Washington Convention on International Trade in Endangered Species of wild flora and fauna (CITES), 1973; Bonn Convention on the Conservation of Migratory Species of wild animals (CMS), 1979; Convention on Biological Diversity (CBD), 1992 etc. These treaties gain importance in a situation when many animals and plants are nearing extinction worldwide. For example, the global tiger population is only about 7,000; giant pandas are only 1000 in the wild today, while only 1,700 individuals of horned parakeet are alive today, worldwide. Ginseng root grows only in small areas of some countries and may disappear in a few decades.

India has signed many international conventions on biodiversity including CITES. The Government of India has also declared many areas as protected. Project Elephant, Project Tiger etc. are actively being implemented in India.

Jharkhand is one of the forest states of India. The forest cover of the state is a healthy 31% at present. Tree plantation is being done on a large scale. Saranda forest, with sal as its major tree, is the home of the endangered flying lizard and elephants. Betla National Park boasts of gaur, chital, panther, wild bear and mouse deer. It was the site of the first tiger census in the world. In the Dalma wildlife sanctuary, one can see elephants, barking deer, sloth bear, leopard, tiger, porcupine, pangolins etc. Many rare medicinal plants grow in the forests of Jharkhand.

4.5 MEGA BIODIVERSITY ZONES

Not all countries in the world are rich in biodiversity. Some countries do not have climates suitable for the development and sustenance of diverse forms of life. In some others, extinction rates have been rapid. At present, twelve countries have been listed as mega diversity countries in the world. They are: 1. Mexico 2. Columbia 3. Brazil 4. Peru 5. Ecuador 6. Zaire 7. Madagascar 8. Indonesia 9. Malaysia 10. India 11. China, and, 12. Australia. These countries together contain 60 to 70% of the world's biodiversity.

4.5.1 India as a Mega Diversity Nation

India is one of the mega diversity centres of the world. Three hundred and twenty species of wild crop relatives and several species of domesticated animals have originated here. The genetic diversity within these species is astounding. Examples of domesticated species given here are from National Bureau of Plant Genetic Resources and other sources.

- Rice .... 50,000 varieties
- Mango .... 1,000 varieties
- Sorghum ...... 5,000 varieties
- Pepper .... 5,00 varieties
- Cattle .... 27 breeds
- Goats .... 22 breeds
- Sheep .... 40 breeds
- Poultry .... 18 breeds

This amazing diversity is not only natural, but also the result of careful selection and cross-breeding done by India's farmers and scientists over the years. Cultural customs have helped conserve this diversity against the forces of destruction. Livelihood of millions in India depend on the forests, ponds, rivers, coasts, fields and mangroves. Therefore people should not be removed from an area to protect wildlife.

Table 4.1

Biodiversity Element	*Range Native to India*	*Destroyed / Under threat*
Ecosystems		
Forests	200 types, scrub to rainforests	Approximately 50% wiped out over the last century; efforts on to increase forest land by Afforestation and re-forestation.
Wetlands	8 types, seasonal floodplains to lakes	One-third drained out, 70% polluted
Agro-ecosystems	20 agro-ecological zones	Mass homogenization across plains
Coasts	Several types of beaches, mangroves, coral reef systems	40% of mangroves wiped out; major portion of coral reefs bleached or silted
Wild Species	47,000 plants and 89,000 animal species	
Flowering plants	17,500 species	At least 20% extinct; 10% threatened
Mammals	390 species	2 or 3 extinct; 20% threatened
Birds	1,232 species	2 extinct; 5 to 10% threatened
Insects	57,000 species	No estimates

(Information from Zoological Survey of India, Botanical Survey of India and others) Some ecosystems and wild species, with their threat levels, native to India.

Many villages in the country are engaged in self-initiated community forest protection. The villagers will destroy the biodiversity of an area only if their means of livelihood are snatched away. The main reasons for destruction lie elsewhere.

India is a land of diversity in ecosystems and wild species too. Some species have been destroyed or are under threat.

The geographical history of the world shows that about 70 million years ago, the single giant continent that the earth was, split into 2—the southern and the northern continents. India was part of the southern landmass, called Gondwanaland. Australia, Africa and Antarctica were also part of the southern landmass. Later, India moved to the north to join the northern Eurasian continent. The sea between these two landmasses was called Tethys Sea. When it dried up, plants and animals that had evolved in Eurasia and also the Far East migrated into India before the Himalayas were formed. A final lot came from Africa. These were plants adapted to dry regions. Thus, India got the advantage of being in the centre, among 3 different centres of plant and animal origin. These geological events are the single largest cause of the biodiversity we see in our country today. High percentage of endemism is also seen here. About 18% of our species are endemic, seen nowhere else. For example, 62% of our amphibians (animals that can live on land and water, like frogs) are endemic, while 50% of our lizards are endemic. Many marine worms, insects, centipedes etc. are also endemic to different parts of India.

India has a good world ranking in species diversity.

	India's world ranking
Mammals	8th
Birds	8th
Reptiles	5th
Amphibians	15th
Angiosperms (Flowering, fruiting plants)	15th to 20th

Moreover, our gene banks have collected more than 34,000 cereals and 22,000 pulses and many others grown in India. Research works in genetic engineering and biotechnology are going on in many laboratories in our country using materials from the gene banks. It is expected that better varieties of many cultivated species will be released sooner.

4.6 HOTSPOTS OF BIODIVERSITY

Of more than thousand major eco regions in the world, 200 are very rich. These are called the 'Global 200'. Most of these are under some threat, but some are under immediate threat and need immediate attention. Such regions in the world that have rich biodiversity and high endemism and are under threat of extinction and habitat destruction are called hotspots for preservation. Norman Myers (1988) identified 12 hotspots of rainforests. These hotspots cover only 0.2% of the total land surface, but contain 14% of the world's plant species. Later, in 1991, Myers expanded his study and declared 8 non-forest areas as hotspots—4 in the tropics and 4 outside the tropics. Fifty thousand endemic plants comprising 20% of global plant population live in these 20 hotspots. As per different environmentalists, the total number of hotspots is 18, 24 etc. Such variations are due to international priorities and overlapping.

Of these twenty-four hotspots of the world, twelve are rainforest hotspots as identified by Myer in his first list. They are seen in different countries or regions like North and South America, Philippines, Australia, the Himalayas etc.

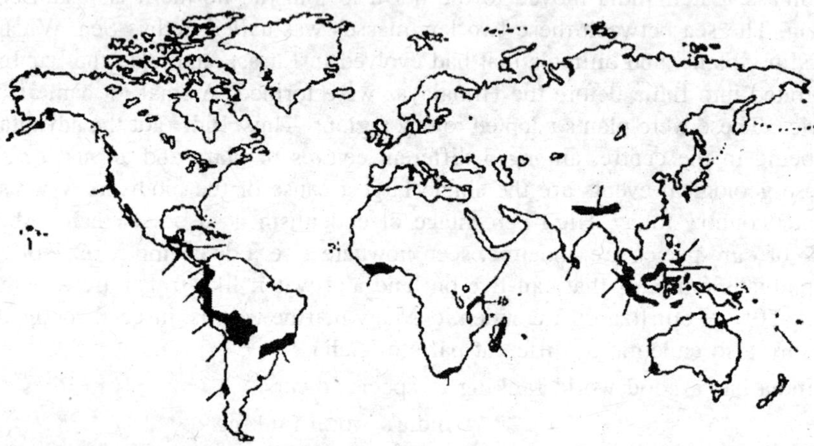

1 to 12 – Rainforest hotspots; 13 to 20 – Eight hotspots in other climates;
21 to 24 – Four more zones recognized as hotspots
The 12 rainforest hotspots are:

1. Hawaian Islands, west of North America
2. Colombian Choco
3. Western Ecuador
4. Western Amazonia
5. Atlantic forests of Brazil
6. Eastern Madagascar
7. Peninsular Malayasia
8. Eastern Himalayas
9. Philippines
10. Northen Broneo
11. Queensland, Australia
12. New caledonia (Oceania)
13. South West Australia
14. Sinharaja Forest, Srilanka
15. Western Ghats, India
16. Eastern Arc Forests, Tanzania
17. Cape Floral Province, South Africa
18. Upper Guinean Forests (Ivory Coast)
19. Central Chile
20. Califonia Floral Province

In addition to these twenty hotspots, four more hotspots have been recognized by many. These are:

21. Central American Highland Forests
22. Central American Lowland Forests
23. Tropical Andies, and,
24. Indonesia

Figure 4.2 Hotspots of the world

As mentioned earlier, opinions differ regarding the number of hotspots. So the regions may also differ slightly. However, the fact is that over the world, biodiversity is under threat in many regions, spread in many countries. In India, we have 2 hotspots—the Western Ghats and the North east, in the Eastern Himalayas.

The Western Ghats lie parallel to the west coast of peninsular India. Its length is about 1,600 km. It covers parts of Maharashtra, Karnataka, Tamil

Nadu and Kerala. Evergreen and semi-evergreen forests are found here. Altitude is up to 1,500 meters. The main centre of biodiversity is the Silent Valley in Kerala.

Eastern Himalayas of the north-eastern India were an active centre of evolution. Many endemic species (species growing specifically in some regions and not anywhere else) grow here. Mountains as well as valleys are rich in flora and fauna. The altitude is up to 3,500 meters. Many primitive angiosperms are found in this region. (See Biogeographic classification of India).

4.7 THREATS TO DIVERSITY

Biodiversity of any region is threatened in different ways. The most serious threat is from the human beings. Destruction done to nature and natural systems by us is of the worst kind.

4.7.1 Habitat Loss

This includes habitat destruction and habitat fragmentation.

(a) *Habitat destruction*: Habitat means the place where an organism lives. In nature, we see that organisms live in a variety of habitats. If these are destroyed due to any reason, the organisms will have to move to places more suitable for them or perish. Thus habitat destruction is a very serious threat to biodiversity. Humans destroy habitats in different ways, the commonest of which is deforestation.

Deforestation: The valuable timber is cut to many times more than the permission given by governments. Cutting down the trees by stealth is common in some forests. Both these take place where protection is not up to the mark. In poor countries around the world, the rural poor are pushed onto marginal, often hilly lands from which they have to cut trees and clear land for cultivation just to survive. In such cases, poverty plays a major role. Forest fires—natural or induced by man for cultivation—often destroy vast areas of forests. Urbanization is the cause in some cases. Villagers encroach into nearby forests as population increases. Roads and sometimes railways are constructed through forests. For dams and hydroelectric projects on rivers flowing through forests, large areas are cleared. Thus, there are many reasons for deforestation. Deforestation results in floods, droughts, reduced rain, soil erosion, pollution etc. in addition to flight or death of many animals and birds, reduction in the yield of river fish etc. The damage caused by Orissa's killer cyclone would have been much less if the mangrove forests, which once protected the coastline, had not been cut for 'coastal development'.

Draining up of marshes: This is another reason for habitat destruction. Marshes are like sponges. They absorb rainwater in the monsoon and hold it for the whole year. In summer, when the myriads of organisms living in the marshes are in need, water is released. Moreover, estuarine and wetland ecosystems are the most productive ecosystems in the world. The highest agricultural production is in rice fields, which are wetlands. If these are drained to make way for human settlements or prawn farming, one can very well imagine the level of damage this will inflict on the natural habitats.

Bleaching and destruction of coral reefs: Polyps are small aquatic, mostly marine animals, usually fixed and colonial. These belong to a group of animals called Polyzoa. Coelenterata is another group of similar animals, simpler in structure. These animals secrete skeletons of calcium carbonate around themselves. The skeletons of numerous animals are joined together to form what we see as corals. The animals live inside the corals. Corals in continuous populations are called coral reefs. They are very beautiful and may be white or of different colours. They grow in unpolluted waters of the sea. Innumerable small fish of various species go to coral reefs for breeding as big fish cannot enter the reefs and so the reefs are safe and secure. Thus, coral reefs are ecosystems by themselves. Around the world, coral reefs are disappearing from seas and oceans due to pollution, siltation from the sea coasts and a rise in temperature of the surface water. In many places, sewage is just released into the sea, agricultural runoff containing pesticides flows into the sea, sedimentation is increased due to erosion caused by deforestation.

Soil degradation: The fall in the quality of soil is called soil degradation. This is caused by many factors. Addition of fertilizers to an undesirably high extent leads to the death of soil microbes, many of which are useful. Biological nutrients do not get replenished as before. This has led to the degradation of millions of hectares of agricultural land in Punjab, Haryana, Eastern Uttar Pradesh etc. The practices of monoculture in U.P., shifting cultivation in Nagaland etc. also degrade once fertile land. Other reasons for land degradation are water-logging caused by floods resulting from deforestation, mining, dam building, increased salinity, acidity etc.

Pollution: Air, water and soil pollution cause untold damage to habitats all over the world. Dust, smoke, sulphur dioxide, fluorides etc. in the air are injurious to plants. Many air-borne microbes cause diseases in human beings, animals and plants. Water is polluted in different ways. Water pollution decreases or even destroys entire aquatic ecosystems. Soil is also polluted in many ways. Vast fields are threatened as the natural enemies of their pests have been killed off by pesticides. All these cause great harm to biodiversity. (For details, see chapter on Pollution.)

(b) *Habitat fragmentation*: Before the advent of urbanization and industrialization and the like, before the formation of thousands of settlements all over the land, natural ecosystems were vast expanses. Now these habitats have been fragmented by roads, rail lines, canals, power lines etc. This has disturbed the ecosystems by affecting the dispersal of fruits and seeds of plants by changing territoriality shown by animals etc.

4.7.2 Poaching of Species

Certain wild animals in the forests are killed stealthily and illegally so that the parts of their bodies can be sold. These are used for various purposes.

Small birds and other animals are captured alive and sold as pets. Different plant species are also carried away from forests for such trade. Such an activity is called poaching. It is a grave threat to animals and plants. As this is a big source of income, poachers are ready to take great risks in realizing their objective. The skin and bones from tigers, ivory from elephants, horns from rhinos and perfume from the musk deer are extensively used abroad. Bears are killed for their gall bladders. Corals and shells are also collected for their export or sold on several beaches in the country. Tortoises, exotic birds and other small animals are packed in tiny containers and smuggled abroad for pet trade. A variety of wild plants with medicinal values are over-harvested from the forests and elsewhere. The commonly collected plants include Rauwolfia (Sarpagandhi), Nux-vomica, Datura etc. Garden plants collected for illegal trade include orchids, ferns and mosses.

4.7.3 Introduction of Alien Species

Species native to any region are adapted to the environmental conditions in that region perfectly. Alien species brought to any geographical region may harm the indigenous ones. They may change the habitat also to such an extent that the native ones gradually disappear. This is true of plants and animals. Some fish, on being introduced into ponds and lakes eat the native fish so much, beyond their capacity of multiplication that the latter disappear completely. Plants like *Eupatorium ayapana* have become so widespread in many regions in India that it has already replaced many other plants. *Parthenium hysterophorus*, an alien plant introduced into India is seen growing almost everywhere. Its pollens are powerful allergens. People who are unaware of this do not destroy the plants and get affected by lung diseases. Disease causing microbes introduced into new localities may cause havoc not only by their own multiplication, but also by modifying the indigenous microbes genetically.

4.7.4 Desertification

The effect of wind erosion is very clear in the deserts of Rajasthan. The land west of the Aravali Mountains is a sandy waste. Over the years, the desert has spread due to reckless tree felling by man and overgrazing by camels, sheep and cattle. Large towns near the desert are threatened by the advancing sand dunes.

4.7.5 Wrong Policies

In many developing countries, the policies formulated by the governments are faulty. Environmental issues are not given the importance due to them. This is mainly due to the lack of knowledge on the part of the policy-makers and also due to their indifferent attitude while planning the economic system to be adapted by the country.

4.7.6 Disease

Sometimes, plants and also animals get infected by different fungi and bacteria. If these diseases do not get healed in time, they may spread rapidly in the community and become an epidemic. In such cases, innumerable plants and animals may die. Some rare species among them also become extinct.

4.7.7 Extinction of Species

This is the most serious threat to biodiversity. Biological extinction had taken place in nature many times, in many places. Natural extinction is a slow process because of environmental changes. If a large number of species become extinct together due to some natural calamity, it is called mass extinction. But anthropogenic extinction, i.e., the extinction due to man's activities is the worst as it is very, very fast and is going on and on. Thousands of species are becoming extinct. As species in the tropics show more variety, greatest losses are in the tropics. Selective and excessive exploitation of species is going on continuously in many countries. In addition to this, proper management practices are not done for their regeneration. When a species becomes extinct, that combination of gene pool is lost permanently. Along with each disappearing species, its dependent species of higher animals, insects etc. also disappear. Therefore conservation of endangered species is a must. The world may lose half of its biodiversity if extinction is not checked now.

As far as our country is concerned, at least 10% of our plants and animals are threatened with extinction. Much of the crop and livestock diversity characteristic of Indian farming is on the decline. Some areas have lost all their seed diversity, and all the 18 breeds of poultry are threatened by the breeds introduced from outside India. Thus extinction is taking place not only in the wild, but also among the crop plants and the domesticated animals.

4.8 MAN-WILDLIFE CONFLICTS

The instances of man-animal conflicts have become more and more in recent years in many parts of the country. Many villages bordering forests are facing the wrath of wild animals which stray into the villages. For example, in Orissa, wild elephants and bear stray into villages at regular intervals and kill and injure human beings. They also carry cattle away or kill them. Crops and houses have been reported to be damaged by animals venturing into villages at night. Retaliatory killing of wild animals by the villagers is also common. Other than elephants and bears, tigers and crocodiles have also been reported intruding into human settlements. Growing anthropogenic pressure exerted on wildlife habitats is the main reason for conflict with wildlife. The other reason is fragmentation of wild animal habitats. The loss of walkways and migratory routes of animals like elephant is the major reason for increasing man-elephant conflict in Orissa.

Total number of human beings killed in Orissa between 1998 and 2004 due to elephant attack is 228, with Keonjhar district alone accounting for 93. Other forest divisions have also recorded instances of elephant menace.

During the same period, 24 human beings were killed by bear in Keonjhar and Angul Districts. Other forest districts also have recorded instances of human killing in bear attacks. In most of the cases, victims had gone for collecting mohua flowers/ kendu fruits / other such items from the areas.

Crocodile attacks are comparatively rare, but 8 persons have been killed in the same period in Orissa. Instances of crocodile attack or both human beings and cattle have taken place along many rivers. Bhitarkanika mangrove is home to 1358 estuarine crocodiles as per census 2004. The crocodiles had moved away from their original home as the buffer along the rivers has been lost due to human activities.

Man-leopard conflicts have also taken place regularly in many villages. Injury to men, cattle kill and human kill have taken place frequently. Tigers have killed cattle too in these villages.

Monkeys have become a great menace in many places. Many instances of monkey-menace have been reported to the Chief Wildlife Warden from various districts like Kendrapada, Puri, Cuttack etc. and cities like Bhubaneswar, Cuttack etc. Sometimes these monkeys go out to the nearby village and town areas and bite people. They are also responsible for the destruction of fruit orchards, vegetable fields, and the thatched and asbestos roofs of houses. Complaints about monkey menace rose very high in 2000 and this caught the attention of concerned district administration and ultimately the state administration. The reason for such rise was the loss of habitat and scarcity of food.

4.8.1 Measures Taken

Various measures have been taken to prevent, control and handle situations arising out of man-wildlife conflict. In order to prevent the conflict, the methods used include (elephant-proof) trenches, rubble-walls, energized fences etc. Before the season when conflicts are anticipated, due publicity is given and education programs are launched through newspapers, electronic media and by distributing pamphlets. It is now planned to launch a scheme to organize watch committees in the line of Village Forest Protection Committees. These committees are expected to keep track of the problem animals or animal groups and inform the villagers and the Forest Department in case of approaching emergencies. A few such squads have been operating at divisional headquarters of vulnerable districts like Dhenkanal, Angul, and Keonjhar. Measures taken to ward off the problem-causing animals from the site include the use of high-intensity focus light, crackers, drums and fire-torches. Whenever some particular animal is linked with a series of incidents, it is generally removed from the site. This has been done for leopards, bear, elephants, crocodiles and monkeys. Apart from the above, compassionate payments have also been made to victims sustaining severe losses of various kinds.

In Jharkhand, there is a peculiar custom among the tribals. It is called "Jani Shikar", which means hunting by women. Hunting job is widely considered men's work because it involves lots of violence and physical strength. But during this special festival, women wears men's cloths, carry traditional weapons and go to jungle in a group for hunting. They keep on hunting hens, goats and whatever comes their way. Villagers do not mind if their animals are also killed on this occasion. The women hunt rabbits and other eatable animals in the jungle and celebrate a bonfire in the jungle. They cook the hunted animals, feast and dance all night. Men are not allowed in this celebration.

A few years back, Government of India banned killing certain species of animals, and hence tribal people now celebrate this festival in a low key. It still exists in the remote tribal locations of Jharkhand, but most of the celebrations are limited within the village areas.

The annual hunting festival of tribals, Bishu Shikar or Sendra, is celebrated to protect themselves from wild animals. Traditionally, tribals celebrate Vishu by chasing and killing animals in the wild. In recent years, killing of animals during Sendra has also reduced considerably all over India. Conservation efforts are thus intricately associated with conflicts between villagers and Forest Officials in many parts of the country. And, fortunately, villagers have begun to co-operate with the officers.

4.9 ENDANGERED AND ENDEMIC SPECIES OF INDIA

The Red Data Book is published by IUCN (International Union for Conservation of Nature and Natural Resources). It lists plants and animals that are at different levels of extinction. In 1998, IUCN published the Red Data Book of threatened animals. Earlier, in 1978, IUCN had already published the Red Data Book of plants. IUCN is also known as the World Conservation Union. Its Headquarters is at Gland, Switzerland. Using IUCN categories, WCMC (World Conservation Monitoring Centre) described threats to about 60,000 plant species and about 3,500 animal species around the world.

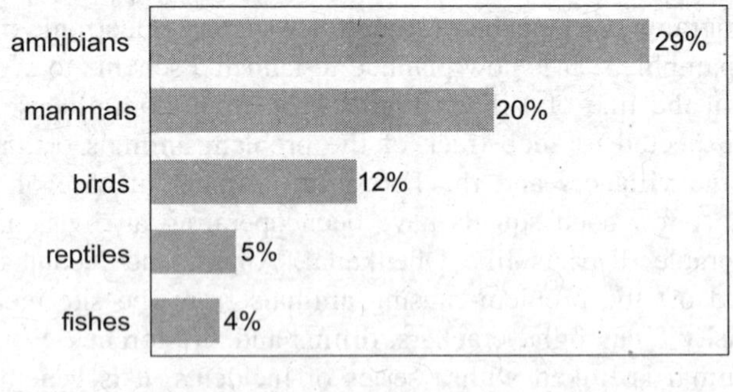

Figure 4.3 Species threatened with extinction as a percent of total number of known species (Source – IUCN, 2008)

There are 5 levels of extinction, including the extinct species. Extinct species are no longer known to exist in the world. Search of localities where they were once found and of other possible localities has been reported to be futile. This is the first category.

The second, third and fourth categories are collectively known as 'Threatened'. Threatened species have thus been classified into 3 groups—rare, vulnerable and endangered. If the rare species are not protected, they will become vulnerable. If the vulnerable species are not protected, they will become endangered. If the endangered species are not protected, they will become extinct.

Those which are suspected to be in any of these 3 classes, but the status of which is no clearly specified, are called insufficiently known. This constitutes the 5th group.

India is a country of great biodiversity. It has not been fully explored yet; its diversity has not yet been assessed fully, nor has any complete list been prepared of all its flora and fauna. Yet, for the past two or three decades, research works have been going on, and as a result about 1500 rare and threatened species of both flowering and non-flowering plant groups have been identified. The rarity of or threat to a majority of them is a result of anthropogenic factors like habitat destruction due to grazing, urbanisation and other developmental activities and over-exploitation. After a critical evaluation of their status and threat levels, the scientists of the department have already compiled Red Data Sheets. These are either already published or under publication in five volumes of the Red Data Books of Indian Plants. Besides, a tentative list of species that need conservation has also been drawn, and efforts are being made to introduce them in the Indian Botanic Garden and the chain of experimental botanic gardens of different regional circles for their *ex situ* conservation, multiplication and scientific study.

In the Red data book prepared by the Govt. of India, 435 species of endangered plants have been listed. Some examples are *Psilotum nudum* (an ancestor of ferns), *Nepenthes khasiana* (Pitcher plant), *Rauwolfia serpentiana* (sarpagandhi - medicinal), *Atropa acuminata* (medicinal), *Dioscorea deltoidea* (source of steroids), *Santalum album* (sandalwood or chandan), *Cypripedium himalaicum* (a beautiful ornamental orchid) etc. In addition to major trees, shrubs and climbers, there are hundreds of small herbs that are greatly threatened due to habitat loss. Several orchids are under severe threat in the states of the North East.

One hundred and seventy two species of animals considered threatened globally are found in India. Among the endangered animals are tiger, elephant, rhino etc. Himalayan wolf, Giant squirrel, free tailed bat, Asiatic lion etc. are highly threatened and are found in India. There are also endangered bird species like the Great Indian Bustard, the Giant tailed macaque (a rare parrot) etc. During the recent past, vultures, which were common even a decade ago, have suddenly reduced greatly in numbers and are now highly threatened. Equally threatened are several species of reptiles and amphibians, and invertebrates like many species that inhabit the coral reefs of the Indian coast.

Endemic species of India: Usually, plants are spread over large areas as they have different methods of dispersal of fruits and seeds. In some cases, however, plants of one species are restricted to small areas. This may be a district, a hill, a valley, an island etc. For example, *Sequoa*, a giant sized gymnosperm, is restricted to California in America. In many cases, these are the last remnants of a group, which is becoming extinct. Such species are called endemic species. They have little variability due to poor muatability and are adapted to only one particular environment. Even if they migrate to other areas, they cannot survive there. Existence of great natural barriers like high mountains, large deserts, wide rivers etc. is another reason. This is why islands in the middle of oceans, like the Hawaian Islands, have many endemic species. In India, species of *Saxifraga, Anemone, Primula* etc. are endemic to the high altitudes of the Himalayas. Many less known mammals are seen restricted to single areas. Some such endemic mammals of India are the Indian wild ass, Hangul or Kashmir stag, the golden langur etc.

There are three main endemic centres in our country—Eastern Himalayas, Western Himalayas and the Western Ghats. Each of the three has more than a thousand endemic plants. There are many microendemic centres also in India, like the Andaman & Nicobar Islands, Agasthyamala and Aanamala (2 groups of hill ranges) in Kerala, Palani Hills in Tamilnadu, Mahabaleshwar-Khandala and Konkan-Raigarh in Maharashtra, Tirupathi Hills in Andhra Pradesh, Shimoga-Kanara in Karnataka, Chota Nagpur Palteau in Orissa and Jharkhand, Aravali in Rajastan, Kathiawar-Kutch in Gujarat etc.

4.10 CONSERVATION OF BIODIVERSITY

4.10.1 Conservation of Germplasm

It is very essential to maintain the equilibrium of the ecosystems in nature. By this, we are conserving valuable germplasm. The reproductive cells of any organism together make its germplasm. It is the germplasm that later gives rise to the whole plant or animal. It is through germplasm that reproduction of species and continuation of generations take place. So conserving germplasm is of utmost importance. Germplasm is stored mostly in the form of seeds in gene banks or seed banks. This way, no particular gene combination is lost forever. If germplasm is not conserved, many species of the present-day will gradually become extinct.

4.10.2 Conservation of Biodiversity

The existence of a large variety of living organisms in any locality is called biodiversity. It is conserved by:

- Protecting natural habitats
- Limited exploitation of species
- Maintaining species in protected areas like national parks, sanctuaries etc.

- Establishment of biosphere reserves for plants and animals
- Strict implementation of laws
- Restricting export of rare species
- Educating people for environmental protection etc.

There are many non-governmental organisations (NGO's) in India like Bombay Natural History Society, Wildlife Preservation Society of India etc. which are very active in conserving biodiversity and also spreading awareness regarding this among the masses.

To conserve biodiversity, there are two methods.

1. *In situ conservation*—This is the conservation of plants and animals in their own habitats. For this purpose, a Protected Areas Network has been made in India. This includes National Parks, Wildlife Sanctuaries, Nature Reserves, Biosphere Reserves etc.
2. *Ex situ conservation*—This is conservation outside the original habitats of the organisms. Conservation done in zoos, aquaria, botanical and zoological gardens etc. are examples of *ex situ* conservation.

4.10.3 Conservation of Fish and other Aquatic Animals

Fish is an important renewable resource in freshwater and sea water. There are 30,000 types of fish in the world. Japan is a major fishing nation.

Fish catch in the world has been increasing steadily. Conservation of fish has become necessary as the catch is exceeding levels of natural regeneration. As a result of overfishing, fish like Halibut, Salmon etc. are gradually becoming endangered. To prevent overfishing, the following should be done:

- Ban catching young fish. For this, sizes of the holes in the net can be fixed
- Punish offenders who pollute freshwater and sea water
- Encourage pisciculture (rearing fish)

Conservation of biodiversity can be done by conservation of species and ecosystems. It is not advisable to protect a single species, though this is done in special cases when it is protected for aesthetic or medicinal or other purposes. The whole ecosystem of which it is a part must be conserved. Natural areas need protection as wild stocks cannot be maintained in zoos and parks. The attention they receive there results in a rapid loss of variability. In addition to the protection and conservation of threatened species, habitats that are particularly interesting can be given the special status of National Park or Nature Reserve to prevent further deterioration. Areas of outstanding scientific importance can be conserved in this way for future generations.

4.10.4 *In situ* Conservation

Complete protection of one species may result in its overpopulation. This shows that management of a reserve is not easy, especially as man has greatly disturbed the natural balance by eliminating predators. A nature reserve is an area where a particular habitat is completely protected together with its flora and fauna. Human activity is reduced and scientific workers' access is also restricted. Industrial and commercial activities are prohibited.

Such conservation of biological diversity in its natural surroundings is called *in situ* conservation. These are of different kinds.

(a) *Protected areas*: These may be land or water. These are protected legally by the Government or community efforts. Government action may be at the national or local levels. Sometimes, land is bought and conservation is carried out by private organizations. Frequently, the governments of developing countries tie up with international organizations, multinational banks or governments of developed countries for funds and training of conservation experts.

Table 4.2 System of Classification of Protected Areas Developed by IUCN (1985)

S. No.	*Type of Protected Area*	*Characteristics of the Protected Area*
1	Scientific Reserves and Strict Nature Reserves	Strictly protected areas. Maintained for education, environmental monitoring and scientific study. Undisturbed.
2	National Parks	Large areas. Picturesque. Maintained for scientific, educational and scientific use. Not exploited for commercial uses.
3	National Monuments and Landmarks	Area smaller. Preserves unique areas of national heritage
4	Wildlife Sanctuaries and Nature Reserves	Limited human population and harvesting allowed
5	Protected Landscapes	People residing here can use the environment, but no destruction allowed. Tourism and recreation permitted.
6	Resource Reserves	Controlled resource use allowed, in accordance with the law of the country
7	National Biotic Areas and Anthropological Reserves	Traditional societies permitted to live here. Resources can be used. Hunting of common animals permitted.
8	Multiple-use Management Areas	Sustained use of all natural resources like timber, water, wildlife etc. allowed. Activities like fishing, tourism, grazing permitted.

The first five categories in this classification by International Union for Conservation of nature and natural resources (IUCN) are fully protected areas. The main purpose of their maintenance is conservation of biodiversity. However, in the last three categories also, biodiversity is observed.

India has about 580 protected areas. These cover only 4.7% of the land area, while the recommended figure is 10%. Yellow Stone National Park of US is very famous. Important National Parks in India are Kaziranga National Park in Assam, Sundarbans in West Bengal, Corbett National Park in Nainital etc. Wildlife sanctuaries protect only animals. Animals are allowed to live in natural surroundings. India has 492 sanctuaries. Important ones are Keoladeo

Ghana Bird Sanctuary in Bharatpur, Rajasthan; Periyar Sanctuary, Kerala; Chilika Lake Bird Sanctuary, Orissa; Manas Wildlife Sanctuary, Assam etc.

(b) *Biosphere reserves*: These are special protected areas where people form a component. In the centre of the reserve is the undisturbed Core Zone. Outer to this is the buffer zone where educational activities and research are allowed. In the outermost Transition Zone, management of the reserve interacts with the local people.

(c) *Sacred forests, lakes, groves etc.*: These are maintained by local communities. Those are areas where many rare and endangered plants and animals are protected. Aquatic flora and fauna are protected in sacred water bodies. Such centres are seen in many states in India like Kerala, Karnataka, Maharashtra, Meghalaya, etc.

(d) *Legal protection of species and habitats*: This can be done at local, national and international levels. In many countries, private conservation organizations buy land and conserve species. At the international level, CITES (Convention on International Trade in Endangered Species) was established in 1973. Countries that signed the Convention agreed to restrict trading of animals, plants and their parts. There are international agreements for protection of habitats etc. One example is the Ramsar Convention on Wetlands of International Importance, 1971, signed by 123 countries.

Chilika Lake in Orissa is a Ramsar site. This lagoon is situated along the east coast of India. It is a mixture of marine and freshwater ecosystems — an area of great biodiversity where many endangered species listed in the Red Data Book live. To restore the lagoon to its original status, the Government of Orissa created Chilika Development Authority (CDA). Then the Authority decided to involve various institutions and the local communities in the restoration work. The result has been amazing. Chilika Lake (lagoon) stands completely restored now. For wetland conservation and sustainable use in any part of the world, CDA got the Ramsar Wetlands Award in 2002, which is given only once in 3 years. This is also an example of *in situ* conservation.

4.10.5 *Ex situ* Conservation

As *in situ* conservation is not possible in many cases, *ex situ* conservation is also done. Conservation of genetic resources outside the natural habitats of organisms by maintaining sample populations is called *ex situ* conservation. Genetic drift and inbreeding, environmental variations, habitat loss, competition, diseases, over-exploitation etc. lead to decline and extinction of species in the wild. So species have to be maintained in artificial conditions under human supervision. This *ex situ* conservation is therefore also called off site conservation. Zoos, aquaria, game farms etc. in the case of animals and botanical gardens, seed banks etc. in the case of plants are established for this purpose. Tissue culture and genetic engineering are very helpful in *ex situ* conservation.

Ginkgo biloba (maiden hair tree) is a type of gymnosperm (plants with naked seeds, i.e. seeds not covered by fruits are called gymnosperms). It was once known as 'living fossil' as it was seen only in some countries and was about to become extinct. However, serious efforts were made to save the species all over the world. Now Ginkgo is a common plant grown in many gardens. This is a very good example of conserving genetic resources by *ex situ* conservation.

(a) *Gene banks*: Plants with desirable characters have to be stored in live condition for use by plant breeders and biotechnologists. Institutions where such materials are preserved are called gene banks. In gene banks, living materials are stored in the form of seeds or genetic materials or both. It can be used by scientists in any part of the world.

Seed is the precursor of plant. Inside seeds, embryos remain in an inactive state for variable lengths of time and later germinate into plants. The process of germination needs many external and internal factors. The main external factors are water, air and suitable temperature. The internal factors necessary for seed germination are the required enzymes and hormones, and release of energy through respiration. If these factors are denied to seeds artificially, they do not germinate; at the same time, they remain alive too. This is done by storing seeds in a dry condition (about only 5% moisture) and at low temperatures (–10 to –20°C).

(b) *Orthodox seeds and recalcitrant seeds*: Not all seeds can tolerate drying and freezing. Seeds of some species are not damaged by drying and freezing. Such seeds are called 'orthodox seeds'. Cereals and legumes come under this category. Seeds that get killed by drying and freezing are called recalcitrant seeds. Examples are coconut, litchi, jackfruit and many other fruits and timber plants. They can be stored only for short periods after giving them fungicides, required moisture and oxygen. Such plants can be conserved using tissue culture.

(c) *Tissue culture and storage of germplasm:* Growing isolated cells or tissues or organs or embryos in an artificial medium under aseptic conditions outside the parent organism is called tissue culture. In cases where seeds are recalcitrant or are not formed at all, tissue culture is very helpful. Many plant organs are used for tissue culture, but shoot tips are the best. Shoot tips are easy to regenerate and are free of viruses. The main steps in tissue culture are as follows:

1. Mother plant is selected
2. Growing tissue (e.g., shoot tip) is dissected out
3. Surface is sterilized
4. Tissue is introduced into culture medium (nutrients dissolved in distilled water)
5. Tissue grows and forms a lump, called callus, which is cut into pieces
6. Pieces are grown in many tubes containing culture medium (subculturing)

7. Process is repeated many times
8. Shoots are then rooted by applying hormones
9. Transferred to pots
10. Transferred to fields

Thus, tens of thousands of plants are produced in a very short time from one shoot tip or embryo. In this way, plants facing the danger of extinction can be multiplied and propagated. This helps in the conservation of germplasm and biodiversity.

(d) *Cryopreservation*: Storing germplasm at very low temperature of –196°C (temperature of nitrogen) is called cryopreservation. No biological activities take place at this temperature. Therefore genetic changes are not possible. Animal cells, sperms and ova, tissues of embryos, whole embryos etc. can be preserved in this manner.

(e) *Zoos as ex situ conservation centres*: Zoos form very good examples of *ex situ* conservation. In India, it was in the 1960's that scientists realized the gravity of the problem of rapid biodiversity loss in the country. This led to zoos being converted into conservation centres in addition to their original role of education and entertainment. Thus, zoos came to have 3 additional functions:

1. Captive breeding
2. Conservation
3. Research

Indian zoos have successfully bred some species like tiger, wild dog, leopard, crocodile and many such other animals. Managers of zoos have come to realize the role of zoos in *ex situ* conservation.

However it is no use protecting some areas while the rest of the country is allowed to deteriorate. This can happen especially outside the protected areas. In such areas, conservationists are engaged in restoring the degraded environment. Scientific measures taken by experts in conservation leading to restoration of degraded or damaged areas surrounding protected areas is called Ecological Restoration. The branch of study dealing with conservation of species outside protected areas is called restoration ecology. In fact, the protection of nature must be extended to cover the whole country. It is possible to protect nature only with public awareness. If the diversity of nature is conserved, man can become fully integrated with nature.

QUESTIONS

A. Very short answer type:

(a) Multiple choice – tick the right answer

1. Maintaining biodiversity in natural habitats is called:

(a) *Ex situ* (b) *In situ*

(c) *In vitro* (d) *En masse*

2. Cryopreservation is preserving germplasm at:
   (a) High temperatures (b) Intense light
   (c) Very low temperature (d) High pressure
3. Endemic plants are those which:
   (a) Grow in cold countries (b) Grow in tropical countries
   (c) Live exclusively in some areas
   (d) Are distributed in all types of localities
4. Sundarbans is the name of a place in:
   (a) Assam (b) West Bengal
   (c) Karnataka (d) Orissa
5. Mangrove forests are seen in:
   (a) Mumbai (b) Nicobar Islands
   (c) West Bengal (d) All of the above
6. Forest cover of Jharkhand at present is:
   (a) 18% (b) 31%
   (c) 45% (d) 12%
7. Rearing fish is called
   (a) Pisciculture (b) Sericulture
   (c) Fishiculture (d) Horticulture
8. Western Ghats are located in:
   (a) North-East India (b) Maharashtra
   (c) Peninsular India (d) Gujarat
9. Red Data Book was prepared by:
   (a) IUCN (b) UNEP
   (c) IBP (d) UNESCO
10. Earth Summit was held in:
   (a) 2000 (b) 1996
   (c) 1995 (d) 1992

(b) Define the following:
1. Biodiversity
2. Species
3. Hotspots
4. Ecological restoration
5. Cryopreservation
6. Soil degradation
7. *Ex situ* conservation
8. Endemism
9. Orthodox seeds
10. Recalcitrant seeds

(c) Write very brief notes on:
1. Earth Summit
2. Trans-Himalayan zone
3. Desert zone of India

4. Semi deserts in India
5. Gangetic zone
6. Deccan plateau
7. Coasts of India
8. North-east zone of India
9. Islands of India
10. Consumptive use value of biodiversity
11. Productive value of biodiversity
12. Social value of biodiversity
13. Ethical and moral value of biodiversity
14. Aesthetic value of biodiversity
15. Option value of biodiversity
16. Megabiodiversity zones of the world
17. Draining of marshes
18. Habitat fragmentation
19. Red Data Book
20. Biosphere Reserves
21. Cryopreservation
22. Conservation of germplasm

(d) Differentiate between the following pairs:

1. Natural extinction and anthropogenic extinction
2. Orthodox seeds and recalcitrant seeds
3. Habitat destruction and habitat fragmentation

B. Short answer type:

Write short notes on:

1. Species diversity
2. Community or ecosystem diversity
3. Deforestation
4. Himalayan ranges
5. Megadiversity nations
6. Destruction of coral reefs
7. Extinction of species
8. Gene banks
9. Poaching of wildlife
10. Hotspots

C. Long answer type:

(a) Write an essay on the biogeographical regions of India.

(b) Give a detailed account on the values of biodiversity.

(c) Write long notes on:

1. Genetic diversity 2. *In situ* conservation

(d) What are the different levels of biodiversity? Elaborate.

(e) What are the threats to biodiversity? Discuss.

(f) Give an account of the endangered and endemic species of India.

(g) Differentiate between *in situ* and *ex situ* conservation of biodiversity with the help of suitable examples.

Unit 5

Environmental Pollution

5.1 INTRODUCTION

Life on the earth depends on the environment. The environment gives energy from Sunlight, suitable temperature, water, nutrients etc. for the living components of the ecosystem. There is a balance between the living organisms and the environment on the earth. If any component—living or non-living—is disturbed, it causes a harmful effect on the entire system.

5.2 DEFINITION, CAUSES AND EFFECTS OF ENVIRONMENTAL POLLUTION

Any physical, chemical or biological change in the environment that leads to its deterioration is called environmental pollution.

The agent that causes environmental pollution is called a pollutant. Kitchen refuse, garbage, smoke, pesticides, noise, emissions from industries and automobiles, effluents from industries etc. are all pollutants. Pollutants are mostly residues of things we make, use and throw away. They are by-products of man's actions. They rise with a rise in human population. Pollutants disturb the ecological balance and create serious problems on earth.

Pollution is a man-made problem. Developed countries are exploiting available natural resources to produce luxurious goods for using and exporting them to the developing countries. During this process, many things are dumped in the environment. In the developing countries, industrialization is catching up, and because of this, pollution is increasing. Moreover, due to the lack of proper management measures, waste materials accumulate in the environment. Garbage dumps become breeding grounds for mosquitoes and flies, which cause many diseases.

In short, pollution is an undesirable change in the air, water or land of our environment that harmfully affects human, animal and plant life in different ways.

5.2.1 Classification of Pollution

Pollution can be classified in many ways.

Based on the part of the environment being polluted, pollution is of 3 types:

- Air pollution
- Water pollution
- Soil pollution

Based on the nature of the pollutant, pollution can be of several types like:

- Lead pollution
- Carbon dioxide pollution
- Smoke pollution
- Noise pollution
- Pesticide pollution
- Radioactive pollution
- Mercury pollution
- Thermal pollution
- Plastic pollution

Pollution can also be classified into:

(a) *Natural pollution*: There are some pollutants in nature itself. These are oil and natural gas; mercury, lead, sulphur etc. in river waters; lava from volcanoes etc.

(b) *Man-made pollution*: Wastes from factories, exhausts from automobiles, pesticides from farms, smoke from chimneys etc. cause man made pollution.

Based on the area affected, pollution is of different types:

- Local
- National
- Continental
- Planetary

Air and water pollution have no boundaries. Sulphur dioxide from chimneys of factories spread far and wide. Rivers, carrying pollutants, may flow through different countries before emptying into the sea.

Pollutants can be classified into two types:

(a) *Biodegradable pollutants*: Pollutants that can be easily decomposed by natural processes are called biodegradable pollutants. Examples are various types of kitchen wastes like skin and seeds of vegetables and fruits etc. Sometimes, however, they too accumulate and create problems. This happens when more is dumped in the environment than can be degraded by nature.

(b) *Non-biodegradable pollutants*: All materials that do not degrade or degrade very slowly in nature are called non-biodegradable or non-degradable pollutants. Aluminum cans, mercuric salts, long chain phenolic compounds, plastics, DDT etc. are non-biodegradable. Similarly, the detergents that we use for more lather cannot be degraded by natural processes. The fumes produced during the burning of plastics remain in the atmosphere, as nature cannot degrade them. The wastes produced during the manufacture of many luxury goods are blown by wind or carried by water to far-off places. These non-degradable pollutants are not recycled naturally and get accumulated.

Environmental pollution results in different types of diseases in humans, animals and plants. It makes the environment dirty and damages useful articles. Smoke and dust in the air reduce the visibility, leading to traffic accidents. Pollutants cause genetic changes in microbes also, which become resistant to various drugs. Many plants and animals fall prey to pollution and become extinct, thereby reducing biodiversity and disturbing the balance of nature. Pollutants damage all types of buildings including national heritages etc. It leads to decrease in tourism.

Pollution is a very costly affair. Its cost can be measured in different ways:

1. Loss of resources due to wasteful exploitation
2. Cost of pollution control—money, manpower etc. to dispose of the pollutants
3. Cost of medical care for different diseases
4. Loss in working hours due to illness
5. Cost of repairing buildings etc. damaged by pollutants etc.

How to minimize pollution: Various strategies can be adopted to reduce the level of pollution in the world at present.

1. *Educating people at all levels*: To a great extent, pollution occurs because man makes too much use of natural resources and throws away too many unused things into the environment. This upsets the natural balance. To maintain the balance, intake and output should be equalized; equilibrium should be maintained.

People should be made to realize that they can do quite a lot to prevent pollution. Some methods to educate the people about pollution:

- Reading materials and audiovisuals like newspapers, radio and television have a major role in creating awareness among the general public.
- Conservation clubs can be established in rural and urban areas.
- Anti-pollution weeks can be observed.
- Demonstrations can be organized at the sources of pollution.
- Youth can impart information at school and college levels.

2. *Finding ways to reduce the already existing pollution hazards*: Concentrated efforts must be made to minimize the existing pollution levels in both rural and urban areas. For this, governments, NGO's and the people must work in unison. Here is a list of what can be done. The list is in no way complete. It is just an indicator.

- Sewage treatment plants have been set up in some cities like Jamshedpur. Many more must be set up in all towns and cities.
- Industries should not be allowed to set up near residential areas.
- Around the sources of pollution, a chain of monitoring stations and equipment should be set up.
- There must be a massive drive to pressurize industries to adopt pollution preventive measures at all levels.
- New projects should not be cleared if they have not included provisions for pollution prevention.
- Offenders must be punished without any discrimination.
- Proper traffic control must be done.
- Automobile engines that pollute less must be designed.
- Farmers must be encouraged to use organic manures and biopesticides.
- Solid waste management must be done in rural as well as urban areas according to the nature of the waste generated.
- Organizers of festivals etc. must be told firmly not to cross permissible levels of sound.
- Tree planting must be encouraged in all suitable locations etc.

5.3 AIR POLLUTION

Air is a mixture of many gases, and is composed of 3 components—major, minor and trace. Major components are nitrogen, occupying 78.09% and oxygen, occupying 20.94% of air by volume. The minor component, carbon dioxide occupies 0.03% of air by volume. Argon is also a minor gas. The percentage of water vapour varies greatly from 0.1 to 5%. In addition to these major and minor gases, there are many gases present in traces in the air, in different proportions such as hydrogen, helium, methane, ozone, nitrogen dioxide, sulphur dioxide, carbon monoxide, ammonia, neon etc.

Air is never found clean in nature as it contains various natural and man made pollutants. Gases like carbon monoxide, sulphur dioxide, hydrogen sulphide etc. are continuously released into the atmosphere due to natural activities like volcanic activity, decay of dead plants, forest fires etc.

However, the gases released into the air by the activities of man are present in much higher concentrations. They upset the natural equilibrium and cause great harm to plants, animals and human beings. These are the result of chemical and biological processes used by man.

There are 5 primary pollutants in the air. Together, they account for 90% of air pollution on the earth. These are:

- *Carbon monoxide* (CO): This is a colourless, odourless poisonous gas. It is released by the incomplete combustion of coal, wood, natural gas etc. and also from automobiles.
- *Nitrogen oxides* (NO_x): These are found in the exhaust of vehicles.
- *Sulphur oxides* (SO_x): Sulphur oxides are produced by the burning of fuels that contain sulphur. Bursting of firecrackers during the festivals releases large amounts of SO_2 into the atmosphere.
- *Hydrocarbons* (HC): These are compounds made up of carbon and hydrogen. They are produced during incomplete combustion of various fuels.
- *Particulates* (tiny particles of solids in the air): Particles suspended in the air are known as aerosols. These include particles of smoke, dust, asbestos, ash, soot etc.

Industrialization is the major cause of air pollution. Industries are ever expanding in all countries. In the developing countries, especially, many industries have been set up in or near cities. Larger quantities of polluting gases are released into the air by industries. They are also carried to far-off places by air currents. For example, a large power station produces 500 tonnes of sulphur compounds and dust each day. Ten per cent of these pollutants may be carried over a distance of 5 km. Atmospheric pollutants produced in Texas have been detected in Ohio (both in the US), 1,600 km away.

Automobile exhausts are another major source of air pollution. There has been an enormous increase in the number of automobiles in the past few years. These are essential for:

- personal conveyance;
- passenger traffic; and
- goods transport.

The exhausts from the internal combustion engines of the motor vehicles foul the air. Motor vehicles contribute about 60% of air pollution. Many harmful gases are emitted by vehicles. Most important of these are carbon dioxide, carbon monoxide, several unburnt hydrocarbons and various compounds of sulphur, chlorine and nitrogen. Atmospheric levels of tetraethyl lead, a compound added to petrol before selling, are increasing rapidly. Each vehicle emits about 1 kg of lead per year.

The oxygen content of the atmosphere is about 20% by volume. A single aeroplane uses 35 tonnes of precious oxygen in crossing the Atlantic Ocean. The plant cover of the earth is being depleted fast too. The major consumers of carbon dioxide in the air are the green plants. When they are destroyed, there is an increase of carbon dioxide in the air. When the carbon dioxide emitted by the vehicles and factories is added to this, the total amount becomes highly disproportionate to the levels of oxygen, now depleted by aeroplanes in the air and automobiles on the land.

The carbon dioxide content of the air is increasing slowly. Over the past century, it has increased by about 15% (15% of 0.03%). Some amount of CO_2 in the atmosphere is absorbed by the oceans. But as industrialization and the number of automobiles have increased, the rate of CO_2 release has become more than the rate of absorption by the oceans (see biogeochemical cycles in Unit 7).

Normally, carbon dioxide is 0.03% by volume in the atmosphere. In such normal condition, the temperature at the surface of the earth is maintained properly. The sun rays that strike the earth and the heat that is radiated back into the atmosphere are balanced. However, when there is an increase in CO_2 concentration, it prevents heat from being re-radiated out. The thick layer of CO_2 functions like the gas panels of a greenhouse. It allows sunlight to pass through, but prevents the heat from re-radiating out. This is called the greenhouse effect.

The result of this greenhouse effect is that the air becomes warmer near the earth's surface. The earth also becomes warmer by re-radiation from the air. The whole process is called global warming. Global warming will increase the temperature of the biosphere as a whole. This can melt the polar ice caps, leading to flooding of the coastal countries (See Unit 3).

All gases that contribute to the greenhouse effect are known as greenhouse gases or GhG's. These are:

- CO_2 (Carbon dioxide)
- N_2O (Nitrous oxide)
- CH_4 (Methane)
- HFC's (Hydrofluorocarbons)
- PFC's (Perflourocarbons)
- SF_6 (Sulphur hexafluoride)
- SO_2 (Sulphur dioxide)
- CFC's (Chlorofluorocarbons)

These are discharged into the atmosphere mainly by industries. Some greenhouse gases like carbon dioxide, nitrogen oxides, sulphur dioxide etc. are also emitted by automobiles. Both nitrogen oxides and sulphur dioxide are harmful to human health.

Nitrogen oxides cause:

- Irritation of the lungs
- Asthma
- Chronic bronchitis

It also makes the affected persons more susceptible to influenza, common cold and all other types of infections of the respiratory system.

Sulphur oxides, especially sulphur dioxide, have the following harmful effects:

- They irritate the respiratory tissues
- They damage lungs and cause problems like bronchitis

SO_2, NO_x and other acid forming chemicals are emitted by automobiles and industries. In the air, these gases combine with water vapour to form acids. These fall back to the earth along with rain. This is called acid rain. Acid rain increases the acidity of soils, damages crops and forests and causes harm to plants and animals living in water. This problem is seen more in the industrialized West (See Unit 3).

Diesel and petrol produce similar pollutants on burning, but petrol produces more gases. The gases given out during idling and acceleration are more harmful than those produced during moderate speed. (A type of cheap non-polluting biofuel, a substitute for petrol and petrol, can be extracted from a plant called Jatropha, belonging to family Euphorbiaceae. Many acres of land in India are under Jatropha cultivation these days.)

One of the main pollutants emitted by automobiles is carbon monoxide (CO). It has many negative effects on human health:

- Headache
- Irritation of mucus membranes
- It combines with haemoglobin of blood, reducing its oxygen carrying capacity.

  This results in:

  (a) Difficulty in breathing

  (b) Bronchial asthma

  (c) Fainting

  (d) Coma etc.
- Thinking power is reduced
- Reflexes are slowed down
- Drowsiness
- Dizziness
- Higher concentrations may be fatal

Effects of other chemicals are as follows:

1. Incomplete combustion in the internal combustion engines of automobiles releases 3, 4, benzpyrene, which causes lung cancer.
2. Oxides of nitrogen cause irritation of eyes and nose.

5.3.1 Control of Air Pollution by Automobiles

Air pollution by automobiles can be controlled by various methods:

1. By traffic control
2. By designing automobile engines with reduced amounts of unburnt hydrocarbons
3. By fitting vehicles with exhaust emission controls
4. By using additives like barium compounds in fuels
5. By control of idling

6. By developing non-polluting engines for automobiles. (Recently, hydrogen fuelled engines have been introduced on an experimental basis in Germany and Japan. They will emit only water vapour into the air.)
7. Using compressed natural gas (CNG) in vehicles. This has reduced pollution to a great extent in New Delhi.

5.3.2 Control of Air Pollution by Industries

Following are some of the methods to control air pollution by industries.

1. Factories should be far away from residential areas.
2. Old machineries should be modified and new and more efficient ones should be set up.
3. Pollution control equipment like scrubbers, dry and wet collectors, filters etc. must be installed at the design stage itself.
4. Chimneys should be high so that the fumes are emitted as far away from the ground as possible.
5. Raw materials that cause more pollution should be replaced by less polluting ones.
6. Round the clock measurement of pollution must be done.
7. When tolerance limit is reached, the factory must be shut down.

Smoke Pollution: Smoke is a major pollutant of air in rural as well as urban areas. This is called smoke pollution. About 10 to 15% of air pollution is caused by smoke. Smoke is produced from various sources. Smoke from chimneys of industries, automobiles and homes causes air pollution. It is mostly a man-made chemical pollution of air. Smoke contains lead (Pb), sulphur dioxide (SO_2) and carbon monoxide (CO). All the three are poisonous to animals and man. SO_2 gets oxidized in the air to sulphuric acid (H_2SO_4). This is seen in the air in the form of small droplets. SO_2 is formed mostly during the combustion of coal and fuel oils. Incomplete oxidation of domestic fuels gives out CO.

Smog formation: One very bad effect of smoke is smog formation. Dust and soot (powdery carbon) particles from factories provide the nuclei (central cores) for smog formation. Particles of coal dust, SO_2 and CO form the nuclei around which other chemicals collect to form smog. Smog is fog made darker and heavier by smoke and chemical fumes (smoke + fog = smog).

Photochemical smogs have also become common now. When hydrocarbons and nitrogen oxides react in the atmosphere in the presence of ultraviolet (UV) rays in the sunlight, ozone (O_3) and peroxy acetyl nitrate (PAN) are formed. These together produce a photochemical smog. Both ozone and PAN are bad for health. Their harmful effects are:

- They cause watering in the eyes.
- Lead to respiratory diseases.
- They are toxic to plants also.
- PAN blocks the Hill reactions of photosynthesis (making of food by green plants).

- Both smoke and smog are very bad for health.
- They irritate the eyes and nose.
- They constrict throats.
- Lung diseases like bronchitis, asthma etc. are widespread in areas where there is smoke pollution.
- Smoke, smog, dust, soot etc. scatter and absorb light, which reduces visibility and is a major cause of traffic accidents. In December 1952, more than 4,000 people died in London in a 5-day long smog.
- Smog also damages agricultural crops.
- Damage livestock.
- Corrodes:
  (a) Metals
  (b) Building materials
  (c) Painted surfaces
  (d) Textile
  (e) Paper
  (f) Leather etc.

The dust and soot particles in the smoke:

- make our surroundings dirty.
- spoil our material goods.
- rob us of our pleasure etc.
- in many localities, rooms and all the articles in them get covered by a layer of black dust every morning. This is the soot deposit. Those who walk barefoot get their soles blackened. Dusting household articles becomes a problem; and
- soot also settles on uncovered water surfaces, thereby polluting the water.

Effect of air pollution on plants: Some air pollutants break down the cuticle (waxy covering layer) of leaves. This has many harmful effects on the plants.

- It makes the leaves susceptible to various types of infections, drought, frost etc.
- More water is lost from the plants.
- Photosynthesis is reduced and leaves turn yellow and fall.
- Growth becomes stunted.

Air quality monitoring: Pollution by various pollutants in the air must be monitored properly, only then can the right pollution control measures be taken. In India, all air pollutants are not being monitored at present. Only common ones like carbon monoxide, sulphur dioxide, nitrogen oxides etc. are monitored effectively; benzene, small particulates etc. are not monitored.

National Environment Engineering Research Institute (NEERI) located at Nagpur has prepared a database of ambient air quality in different cities of India. But this does not give a complete picture. The State Pollution Control

Boards (SPCB's) of various states also collect data and send it to the Central Pollution Control Board (CPCB). Even then much more work—more extensive and widerspread—is to be done in our country in air quality monitoring.

5.4 WATER POLLUTION

The quality of water is very important for mankind. It is directly connected to human health. Pollution of drinking water is the single major cause of water-borne diseases like diarrhoea, dysentery, cholera, jaundice and many others. These diseases kill many children in the developing countries. Adults also fall prey to these diseases.

Water that has lost its pure nature is called polluted water. Polluted water can be distinguished from pure water by observing one or more of the following signs:

- Mild or stronger colouration (yellowish, brownish etc.)
- Bad taste
- Offensive smell
- Too many water plants in the water body
- Reduction in the number of fish and other aquatic animals
- Oil and grease floating on water surface

Polluted water is harmful not only for human beings, but also for animals, plants and industries. Pollution affects freshwater bodies like ponds, lakes, streams, rivers etc. and marine water bodies like seas, oceans, estuaries etc. In polluted water, BOD (Biochemical Oxygen Demand) should be above 30 mg/L as per Indian standard for drinking water.

The major sources of water pollution are:

- Domestic wastes, mainly sewage
- Industrial wastes
- Agricultural wastes

5.4.1 Freshwater Pollution

Groundwater, streams, rivers, lakes, ponds etc. are polluted in different ways.

(a) *Pollution by sewage*: The pollutants of sewages may be solid wastes or liquid wastes. Household sewage and laundry detergents add sulphates (SO_4), nitrates (NO_3) and chlorides (Cl) to water. This dirty water flowing down the sewages and drains also contaminates groundwater before it reaches rivers. Sewage water contains large numbers of coliform bacteria.

In many localities, the supply of drinking water is obtained from wells—either dug wells or tube wells. Such water is unfit for drinking unless purified or boiled. Many water borne diseases like cholera, dysentery, typhoid, jaundice, diarrhoea, amoebiasis etc. are caused by drinking such polluted water.

Filtration is not enough as only the suspended impurities can be removed by it. Purification is a better process if done effectively. Boiling kills the germs, but some useful minerals are lost. Moreover, water develops a certain taste by boiling, whereas pure water is always tasteless.

In sewage some chemicals promote the growth of small non-flowering aquatic plants—blue green algae. When they grow very fast, the whole water surface is covered by them. This is called water bloom. The algae consume a large quantity of oxygen in the water for their respiration. Some blue green algae also release certain toxic substances into water.

The algae release oxygen during photosynthesis. This oxygen is used up by bacteria to oxidize organic matter in the sewage. This also reduces the oxygen content in the water. Gradually, aerobic bacteria (bacteria that need oxygen for respiration) are totally excluded from the sewage. Only blue green algae and anaerobic bacteria (bacteria which do not need oxygen for respiration; they use other chemicals for energy) can live in such dirty sewages. These emit a foul smell due to rotting organic matter.

In villages, towns and cities where sewage treatment plants have not been set up, this stinking sewage is released into rivers, streams or other water bodies. The water in these water bodies thus gets polluted. If poisonous substances are also present in the sewage, which does happen frequently, many aquatic plants and animals die.

To fight sewage pollution, septic tanks and specially designed sewage treatment plants must be set up to remove the excess organic matter. If septic tanks are not properly maintained, water contaminated with faeces leak out and gets mixed with ground water. Drinking such water causes many water-borne diseases.

1. *Sewage Treatment Plants*: In a typical sewage treatment plant, purification of water is done in 2 stages. First of all, the sewage water coming from various sources collects in a stilling chamber. Here, the velocity of the flow is reduced. A flow meter checks the velocity. From here, the sewage moves on to the screening chamber.

Stilling Chamber → Screening Chamber → Aeration tank → Secondary Clarifier → Clear water discharge into River
↓
Sludge Drying Bed
↓
Manure to Parks and Gardens

The screening chambers are fitted with mechanical bar screens. Here large sized suspended matter such as stones, plastic pieces, sticks, pieces of cloth etc. are filtered out. This whole process is called primary treatment. After filtration, the sewage is taken to large aeration tanks, where huge fans rotate continuously. This churns and aerates the sewage, making aerobic bacteria grow fast and oxidize organic matter. From the aeration tanks, water moves to a secondary clarifier or settling tank, where the remaining impurities settle down. The clear water from the top part can now be released into nearby rivers after chlorination, if necessary. This series of processes together is called secondary treatment.

The sediment waste matter in the secondary clarifier is taken to sludge drying beds, where it is dried and made into manure. Thus a sewage treatment plant serves a dual purpose.

This is the basic principle of sewage treatment plants (STP's). In some plants, there is provision for tertiary treatment also.

(b) *Pollution by Industries*: While setting up mills or factories, owners see to it that there is a water body like a river or stream nearby. Into almost all rivers in the LDC's, industrial wastes are dumped continuously. For example, this is done at short intervals in the River Yamuna, which is one of the primary sources of Delhi's water supply.

When industrial wastes are emptied into a river, its self-purification mechanism is lost. Common industries polluting river water are petrochemical complexes, fertilizer factories, oil refineries, plants for making drugs, rubber, fibres, plastics etc. A synthetic drug factory's wastes may contain up to 70 different chemicals and many concentrated acids. These wastes pollute the river water, and also filter through the soil to pollute the groundwater.

River water pollution by different industries is very injurious to the aquatic animals and plants as many chemicals are poisonous to them. This indirectly affects human beings through the food chain. For example, the effluents from paper mills contain mercury (Hg). When dumped into rivers, Hg enters the body of the fish that man eats. Mercury causes nervous disorders in humans.

Coke works and plastic units produce many phenolic compounds as wastes. Metal finishing plants release heavy metals and even cyanides. Caustic soda and chlorine factories produce wastes rich in Hg. Many aquatic organisms die due to these toxic industrial pollutants. This upsets the natural balance of the rivers.

Several mollusks (snails) in streams take in copper (Cu), cadmium (Cd) etc. and accumulate them in their bodies. This happens because of bioaccumulation. This affects their metabolism (life activities in the cells). Rivers take these toxic substances to the seas or oceans. Cadmium (CD) and chromium (Cr) are toxic to many marine animals.

Wastes produced during the manufacture of paper, sulphur, cement, sugar etc. when released into streams or rivers get oxidized into various compounds, utilizing a large quantity of oxygen in the water. Due to oxygen deficiency, many fish die. This problem is very severe in Kali Nadi in Meerut. Several wastes are released by paper mills into River Gomti in Lucknow. Industrialisation also leads to thermal pollution of water.

(c) *Pollution by nitrates from fertilizers*: Chemical fertilizers are used widely to increase farm yield. These seep into groundwater below the farms. It has been estimated that since the 1950's, the use of nitrogen fertilizers has increased 20 times all over the world. It was much later that a study in China showed that plants absorb, on an average, only 40% of this increased nitrogenous fertilizers applied. In the US, between 33% and 50% of the applied nitrates are lost. All the unused nitrates dissolve in rain and irrigation water and filter through the soil to finally reach the underground water.

In Punjab and Haryana, where nitrates are added in large quantities to crop plants, nutrient level of well water has been found to be 5 to 15 times higher than the safe limit. Extra nitrate in water leads to many problems, like:

- Certain type of anemia in babies that causes suffocation and death
- Miscarriages (natural abortion) in women
- Cancer of digestive tracts
- Too much nitrogen is not good for plants too. The plants become susceptible to diseases. Their roots are damaged, and their stems and branches become thin.

(d) *Pollution by pesticides*: Chemicals that kill pests are called pesticides. These chemicals are sprayed in croplands. In India, many farmers use pesticides indiscriminately. Moreover, much higher quantities than needed are sprayed in the fields. Both these practices are harmful to the environment.

Pesticides were first introduced in the 1940's. But it was only after many years that it came to be known that they were highly injurious to non-target organisms including human beings. Groundwater pollution by pesticides has become a serious problem now.

Pesticides are used not only in farmlands to kill insect pests attacking plants, but they are also used to kill harmful insects like mosquitoes. So the effect of pesticides is seen in rural and urban areas. They accumulate on the soil surface or surface waters where they are sprayed to control mosquitoes. From here, they seep down and pollute underground water. Sometimes, they remain in the underground water for long after the usage of the pesticide is stopped. For example, even 30 years after banning, DDT (used earlier to kill mosquitoes) is still found in the underground water in some parts of US.

DDT is an organochlorine pesticide and has been banned all over the world as it is a non-biodegradable compound. It pollutes air, water and soil. In Bihar and West Bengal, CPCB (Central Pollution Control Board) has found that DDT is present in groundwater many thousands of times the permitted level. This has happened after stopping the spraying for a few decades.

Even low concentrations of organochlorine compounds inhibit aquatic photosynthesis in phytoplanktons (microscopic algae on the surface of water). As phytoplanktons are the food of zooplanktons (microscopic animals on the water surface), they decrease in population. The population of small fish that depends on the planktons falls as a result. This affects the population of big fish. In this way, the whole food chain is affected, and finally we suffer due to shortage of fish.

One consumer consumes many producers. So, the amount of pesticides in one consumer is always more than that in the producer. This goes on at all steps of the food chain. This is called biomagnification. It is very harmful to the organisms at the top of the food chain. And in many food chains, human beings are at the top.

In the recent decades, many new pesticides have been found, that are highly toxic. Drinking water contaminated with organochlorines affect the hormonal balance of human beings. It may also cause cancer.

Natural recovery process in rivers: Usually, if the pollutants are not much in quantity, the rivers have a natural recovery process. This is by a combination of dilution and decay by bacteria. Thus, rivers that are not polluted much recover fast from degradable, oxygen demanding wastes. This is what is happening at present in the rivers of US and many MDC's.

Control of freshwater pollution: After the water pollution control laws were passed in the 1970's, the number and quality of sewage and industrial waste water treatment plants has increased greatly in more developed countries. Still, some incidents do occur, like accidental spillage, or rarely, deliberate release or malfunctioning of treatment plants.

In the 1950's, the famous Thames River in Great Britain was highly polluted. But because of consistent efforts by the government and the industries, the river has recovered now. Commercial fishing is going on and many water birds have reappeared. There is an urgent necessity for the same strategy being worked out in India.

- We have the Ganga Action Plan, the Yamuna Action Plan etc. But these plans should be implemented with more force and dedication.
- Action plans should be formulated and implemented for all rivers in India.
- Wastewater treatment plants must be set up in all towns and cities. (details given earlier)
- New industries should not be given permission to start production unless they have set up waste water treatment plant.
- The general public must be educated on the need to keep water clean by not polluting water bodies etc.

5.4.2 Marine Pollution

Many waste materials are deposited in the seas and oceans. These are mainly municipal wastes, oil, minerals, radioactive substances etc.

- Municipal waste and sewage from small towns are often discharged directly into the sea.
- Lubricating oils are discharged into the sea.
- Accidental spilling of oil from the oil tankers is a matter of grave concern. Oceans and beaches are becoming more and more polluted by oil residues. In 1968 alone, 5 million tonnes of oil was deposited in the sea. Once, in an island near France, over 100 tonnes of oil waste was seen on a single day.
- Fertilizers and pesticides from agricultural lands are washed away finally into the sea.

- Ships carrying toxic substances may meet with accidents in the sea.
- When ships are washed and oil tankers emptied, the sea water is polluted.

Marine pollution is harmful in many ways:

- Many birds are affected by marine oil pollution. The future of birds like divers and sea gulls is threatened by such oil waste discharges in the sea.
- Oysters, mussels and fish will ultimately become unfit for human consumption, as they consume the polluted water.
- Fish stock may also decline through the destruction of eggs and fry (baby fish).
- Food chains are disturbed due to reduction in planktons.
- Fishing nets are damaged by oil.

***Control of marine pollution*:**

There are some methods to reduce marine pollution:

- Setting up of sewage treatment plants at the required sites
- Spraying chemical dispersants on the oil on the surface
- Use of slick lickers—materials that absorb the oil slick, i.e., the layer of oil on the water surface etc.

5.5 SOIL POLLUTION

Soil is polluted due to wastes produced by industries, automobiles and agricultural and domestic practices.

5.5.1 Soil Pollution by Industries

Combustion of sulphur containing fuels in factories emits sulphur dioxide. This is converted into sulphates, which are deposited on the soil. Nitrates from industries are also deposited on soil. In highly industrialized Poland, people living in areas where the soil was highly polluted by mercury had to be evacuated.

Sulphur dioxide and nitrogen oxides emitted from industries dissolve in water vapour and are converted into sulphuric and nitric acid respectively. These fall on the soil along with rainwater. This is called acid rain. Acid rain makes the soils acidic. This is a serious problem in the West. In some parts of India also, there have been reports of acid rain. Many crops connot grow on acidic soils. Such soils become degraded and need extensive reclamation to make them cultivable.

5.5.2 Soil Pollution by Automobiles

Lead added to petrol (to prevent knocking) is emitted in vehicle exhaust. It is toxic and settles along soil on both sides of the roads with heavy automobile traffic. It causes paralysis and many other disorders. Tissues of people living near roads have a higher lead content than those living in other places. In Netherlands, Great Britain etc., it has been reported that cancer is twice as much in towns as in villages.

5.5.3 Soil Pollution due to Wrong Agricultural Practices

Pesticides, herbicides etc. are retained in the soil for many years. They seep into the soil and become part of the environmental cycle.

5.5.4 Soil Pollution by Pesticides

Till a few years ago, DDT (Dichloro Diphenyl Trichloroethane) was used as a pesticide on crops and also a powerful chemical that kills mosquitoes. DDT is a non-biodegradable compound and gets fixed in the soil. It also collects on the surface of plants. It is dispersed in the whole area through water and air currents. DDT and other synthetic organochlorine compounds like polychlorinated biphenyl etc. have been in use since 1940.

The initial success of organochlorine compounds encouraged the belief that pesticides can solve any pest related problem. The possible ecological consequences were ignored. The wide use of pesticides is very harmful to man, animals and plants. Some of their adverse effects are as follows:

1. Many harmless and often beneficial species are harmed. This reduces the diversity of species in the habitat. Honey bees and other insects that pollinate different plants have been killed by pesticides.
2. Resistant strains develop that become increasingly difficult to control.
3. Residues accumulate in the soil, which may persist for many years.
4. Fish are very susceptible to DDT.
5. Bird population, especially birds of prey, has been reduced by pesticide poisoning.
6. In birds, ovulation (formation of ovum or egg) is delayed and gonad (reproductive organs) development is inhibited.
7. Mosquito control programmes using the wrong type of pesticide may be one of the reasons for the reduction of fauna (animals) in the treated areas.
8. Pesticides in the soil are consumed by earthworms. Earthworms, called 'friends of farmers', increase the soil fertility by their worm castings.
9. They also make the soil porous. Porous soils contain more air and are better for seed germination and root growth. Mass killing of earthworms reduces soil fertility.
10. The dead earthworms are eaten by birds; these birds also get poisoned and die.
11. Pesticides destroy many aquatic organisms by killing their larvae.
12. Use of pesticides kills parasites of some other pests.

13. Pesticides from the soil seep into underground water, polluting it and making it unfit for drinking for humans and animals alike.

To demonstrate the harmful effects of pesticides, some incidents are described below:

- In the highlands of Scotland, the proportion of breeding pairs of Golden Eagles was 72% of the total pollution in the 1940's. The percentage fell to 29% by the 1970's.
- The Bald Eagle has become threatened in the US. Fatal doses were found in 25 of the 26 species.
- Once, the Clear Lake in California was sprayed with TDE (a compound similar to DDT) at 0.014 ppm (parts per million). Soon, zooplanktons had 5 ppm, small fish feeding on them had 22 to 221 ppm and large fish feeding on small fish had 40 to 2,400 ppm of TDE in their bodies.
- In Spain, pesticides were used to control some pests of oak trees. This programme was successful, but parasites of other pests were also killed. In the absence of their parasites, these pests increased greatly in numbers, and turned out to be a big problem.
- In 1961, a mean value of 925 mg/person of organochlorine residues were recorded in the adipose (fat) tissue. In France, this was 370. Such concentrations do not kill or harm immediately, but result in chronic (slow, continuous) poisoning, just like poisoning by fluorides and some carcinogenic (cancer causing) hydrocarbons.

5.5.5 Soil Pollution by Herbicides

Like pesticides, herbicides have also become major pollutants of soil. They are frequently used in clearing the sideways of roads and rails, to clear new tracks for new roads and rails, for weed control in agriculture and for forest management. After the discovery of natural phytohormones (plant), many organic compounds have been synthesized artificially resembling the natural hormones. These are widely used in agriculture and horticulture (cultivation of fruits and vegetables). Weed control is one of them.

Herbicides have selective action. For example, the synthetic herbicide 2,4,D kills weeds with broad leaves, but does not kill plants with narrow leaves. So it is used in fields of rice, wheat, corn etc. where the crop plants have narrow leaves. Other herbicides having the opposite effect are also available. They kill plants with narrow leaves only and do not kill those with broad leaves. Such herbicides are used to kill grasses (which are narrow-leaved) in fields of vegetables like potato, brinjal, cabbage, cauliflower, tomato etc. In both these cases, death is due to increased respiration.

Herbicides cause great harm to plants, animals and humans.

- Herbicides like Monuron, Simazin etc. interfere with photosynthesis.
- 2,4,D and 2,4,5,T block the transport of food in plants.

- Herbicides kill plants selectively, thus modifying natural communities.
- When plant communities are changed, herbivores (animals eating plants) also change, and then carnivores (animals eating other animals) also change in a community.
- Herbicides modify the functions of important systems of all organisms.

In short, pesticides and herbicides can be considered to be harmful drugs introduced into nature and natural systems. Both these are sprayed on plants for specific benefits, but they pollute air, water and soil.

(See concerned Units for conservation, land degradation and soil erosion)

Control of soil pollution: Some methods of controlling soil pollution are:

- Prevention of erosion
- Soil conservation
- Prevention of land degradation
- Right agricultural practices, minimizing the use of artificial fertilizers and chemical pesticides

5.6 NOISE POLLUTION

Unwanted or undesirable sound is noise. It is harmful to human ears. Noise is increasingly becoming a major pollutant. As noise pollutes the air, noise pollution can in fact be included in air pollution. Just like smoke, dust, toxic fumes etc. pollute air, noise also pollutes air. This is why noise is included as an air pollutant in Schedules 3 and 7 of EPA (Environmental Protection Act). On 14th February, 2000, Noise pollution (Regulation and Control) rules were enforced, supplementing EPR. These rules stipulate ambient air quality standards with respect to noise. Industrial, commercial and residential areas have been covered under these rules. Some areas have also been fixed as silent zones. As per the rules, there are restrictions on the use of loud-speakers and public address systems. It is the duty of district magistrates and police commissioners to enforce the noise pollution rules. They must see that noise levels do not exceed the stipulated norms.

The sound measured at the source is meaningless. There must be a distance at which the hearing process is taking place. This is because the sound becomes lesser and lesser as the source is farther and farther away. Therefore, all measurements must also indicate the distance. Thus there is no single number of measurements—it is always decibels along with distance.

Bel is a unit of sound invented by the engineers of the Bell Telephone Laboratory.

It is the decrease in the level of sound over 1 mile (1.6 km) length of standard telephone cable. It was initially called transmission unit (TU), but later, to honour Alexander Graham Bell, who had invented the telephone, this unit was renamed as Bel. As the Bel was too large for everyday use, decibel (dB), which is equal to one-tenth of a Bel (0.1Bel) became popular.

Table 5.1 Sound levels in comparison with sources and distances

dB	*Source with Distance*
194	Theoretical limit for a sound wave at 1 atmosphere environmental pressure more intense waves function as 'shock waves'
180	Krakatoa (an island near Java) volcanic explosion at 1 mile in air
160	MI Garand (a semi-automatic US rifle) fired at 1 metre
150	Jet engine at 30 metres
140	Low caliber rifle at 1 metre
130	Civil defence siren at 30 metre – threshold of pain
120	Train horn at 1 metre – pierces ear drum
110	Football stadium at kick off at 50 yard line
100	Jack hammer at 2 meters; inside Disco
90	Loud factory, heavy truck at 1 metre
80	Vacuum cleaner at 1 metre
70	Busy traffic at 5 metres
60	Office or restaurant inside
50	Quite restaurant inside
40	Residential area at night
30	Theatre, no talking
20	Whispering
10	Human breathing at 3 metres
0	Threshold of human hearing with healthy ears; sound of a mosquito flying 3 metres away

Measurements of noise that damage the ear are taken near the ear itself. Measurements of ambient noises do not need a distance, as they spread uniformly in a small area. In general, healthy human ears cannot perceive sounds of less than 3 dB.

Noise pollution is partly a result of technological progress. Aggressive noise is produced by road and rail transport, in factories, business centres, industries, space rockets, aeroplanes, sirens, different machines etc. Televisions, radios, telephones, alarm clocks, calling bells and even kitchen appliances sometimes produce sound too loud to our liking. If produced continuously, these irritate us.

The commonest direct effect of noise pollution is temporary deafness. This is medically known as temporary threshold shift (TTS). People affected with this problem cannot hear the sound of low volume. It is seen in people living very close to 'pooja pandals'. After a month or so, the symptom disappears. Noise induced permanent threshold shift (NIPTS) is much more serious. People affected with this problem lose their hearing capacity forever. In addition to these, noise pollution has many negative effects on human health. These are as follows.

1. Loud noise strains the senses and the nervous system.
2. The most evident effect is on the central and sympathetic nervous system.
3. Refreshing, undisturbed sleep becomes impossible.
4. Causes neurosis and headaches.
5. Blood pressure rises.
6. Heart beat becomes faster.
7. Leads to palpitation.
8. Constricts stomach muscles.
9. Flushes skin.
10. Results in depression.
11. Produces ulcers.
12. Leads to general fatigue.
13. Affects emotional aspects.
14. Reduces intelligence.
15. Sleeplessness leads to tiredness.
16. People become short tempered.
17. Memory is affected.
18. Concentration power is lowered etc.

In general, intermittent (discontinuous) noise is more harmful than continuous noise.

Control of Noise Pollution and its effects: There are many methods of controlling noise pollution.

1. *Reduce noise at source*: This is the best method. To reduce the noise of machineries, they can be covered by sound absorbing material placed in sound proof enclosures, kept high on absorbent mounts etc. Machineries must be maintained properly. Noise from vehicles can be controlled by maintaining them properly, by proper traffic control etc.
2. *Block the path of noise*: Noise from construction equipment can be blocked by temporary barriers. To block the noise from vehicles, along the sides of highways, vertical barriers can be built.
3. *Increase the length of the path of noise*: Machineries, construction equipment etc. can be kept as away as possible from residence. Alternative routes are usually given to trucks. Airports are always constructed away from residential areas.
4. *Protect the recipients*: Workers working in high sound situations should wear ear plugs and ear muffs. Good ear muffs can reduce sound up to 40 dB.

5. *Plant trees*: Trees can prevent noise pollution. Therefore, plant trees around houses.
6. *Use absorptive finishing materials*: Use finishing materials that are highly absorptive on ceilings, walls and floors inside houses, offices etc.

5.7 THERMAL POLLUTION

It should be understood that in the case of air pollution, water pollution and soil pollution, we are discussing the medium that is being polluted, whereas noise pollution and thermal pollution indicate the type of pollutant. Noise pollutes air; similarly, heat pollutes water. Thermal pollution is the degradation of water quality by any process that changes ambient water temperature.

A common cause of thermal pollution is the use of water as a coolant by power plants and industrial manufacturers. Many industries use water to cool the plants. Here, water is being used as a coolant. Water becomes hot during this process. This hot water is usually released into rivers. The rise in temperature of water leads to thermal pollution. Power plants that use turbines to generate electricity take water from water bodies. The steam that leaves the turbines is condensed into water. This water, which is at least 15° hotter than the normal water, is released back into the water body. In cities, when storm water, i.e., the water that flows off along roads and footpaths and parking lots is discharged into surface waters, the temperature of the surface water increases.

The solubility of oxygen in water (DO, i.e., dissolved oxygen) is reduced when water becomes warmer. The decrease in the levels of DO can harm aquatic animals such as fish and amphibians. Thermal pollution may also increase the metabolic rate of aquatic animals, because increased enzyme activity in these organisms leads to their consuming more food in a shorter time. This results in lesser quantity of food in the medium. Moreover, food chains will be different in the old and the new environments. In such a situation, the more adapted organisms move into this part of the water body. They will have an advantage over organisms that are not used to the higher temperature. This results in a change in the species composition of the water body. Biodiversity can be reduced as a result of enzyme metabolism.

These cellular level effects can adversely affect the mortality and reproduction. Primary producers are affected by warm water because higher water temperature increases plant growth rates, resulting in a shorter lifespan and species overpopulation. This can lead to an algal bloom, which reduces oxygen levels still further.

In the United States, thermal pollution from industrial sources is generated mostly by power plants, petroleum refineries, pulp and paper mills, chemical plants, steel mills and smelters.

When a power plant opens or shuts down for repairs or other causes, an abrupt rise in water temperature results. This kills fish and other organisms adapted to a particular temperature range. This is known as 'thermal shock'.

Moreover, a large quantity of heat is also generated in the nuclear power reactors. Thermal pollution is thus one of the harmful effects of radioactive pollution. At higher temperatures, aquatic organisms become more susceptible to toxic materials. Also, less desirable blue green algae replace the desirable green algae.

In limited cases, warm water has no bad effect. It may even lead to improved function of the receiving water body. This phenomenon is seen especially in seasonal waters and is known as thermal enrichment.

Release of unnaturally cold water from reservoirs can dramatically change the fish and large sized invertebrate fauna of rivers, and reduce river productivity. This has been reported in Australia.

Control of thermal pollution:

- Build cooling ponds
- Construct bodies of water designed for cooling by evaporation, convection and radiation
- Construct cooling towers, which transfer waste heat to the atmosphere through evaporation or heat transfer or both
- Practise co-generation, a process where waste heat is recycled for domestic and/or industrial heating purposes
- To prevent thermal pollution by storm water runoff, install facilities that absorb it and direct it into ground water

5.8 NUCLEAR HAZARDS

Nuclear hazards are a result of radioactive pollution, which affects air, water and soil. During their decay, some elements emit high energy radiations called ionizing radiations. This is because electrons are removed from stable atoms in these elements. There are 3 types of ionizing radiations of ecological importance:

1. Alpha (α) particles
2. Beta (β) particles
3. Gamma (γ) rays

α particles travel only a few centimeters, and when stopped produce a lot of local ionization. These are formed from the nucleus of the helium atom and are, positively charged.

β particles are high speed electrons and are negatively charged. They penetrate further, travelling a few feet in the air, giving energy over a long path. In living tissues, they travel up to 8 cm.

γ rays are non-particulate electromagnetic radiations like the X- rays. As they have no electrical charge, they penetrate deep. They have shorter wavelengths and ionize biological materials through which they pass.

The effect of radiation depends on its energy and intensity. The intensity is determined by the number of particles emitted in unit time. Radiation intensity is measured in Curies (Ci). Smaller units used in biology are millicurie (mCi) and microcuries (μCi).

The radiation dose received by an organism is expressed as the amount of energy absorbed by a given mass of an organism. Units used are roentgen (R), rad (radiation absorbed dose) and rem (roentgen equivalent, man).

The maximum dose to which man can be exposed without injury is 0.3 rad/ week for repeated exposures and 25 rad for a single dose.

Radiosensitivity varies from one organism to another. The lethal dose is 10^6 rads for bacteria, 10^5 rads for insects and 10^3 rads for mammals. It varies with age. Children under 4 are more sensitive. They accumulate Strontium^{90} more rapidly than adults.

Isotopes of elements that emit radiations are called radioactive isotopes or radionuclides. These escape frequently into the atmosphere.

Wastes from nuclear power reactors add such isotopes to water reservoirs, rivers etc. Nuclear tests emit large quantities of isotopes into the air.

The most important pollutants are Strontium-90, Iodine-131 and Caesium-137. Non-radioactive forms of all these elements occur in the human body. Iodine (I_2) is found in the thyroid, while strontium accumulates in the skeleton as its chemical properties are similar to those of calcium (Ca). Accumulation of radionuclides in human tissues leads to development of cancer. Mutations (changes in the structure of genes) in the reproductive cells are hereditary. This results in the birth of abnormal babies with malformed organs.

This danger is made more serious by the concentration of radionuclides along food chains. The quantity of radioactive substances within the bodies of different organisms thus increases along the food chain by biomagnification. When radioactive phosphorus (P) was added to a pond, insects concentrated it by 500 times, and the ducks that consumed the insects concentrated it further 75,000 times. The concentration in duck eggs was up to 200,000 times the original concentration.

Radioactive pollution also results from industrial wastes. Disposal of wastes in the sea in water tight barrels is also a dangerous practice, as corrosion by sea water over the years can liberate the wastes before radioactivity has fallen to a safe level. Burying it in the soil is not a solution too. Once, isotope of Ruthenium-106, placed in a waste pit, moved through the soil into oak trees 80 metres away. The trees concentrated the radionuclides in leaves and branches. Levels were higher in the leaf litter (fallen and decomposing leaves).

Thus, waste disposal is a serious problem. It is because of this problem that full utilization of nuclear (atomic) energy for peaceful purposes has not been possible yet. Moreover, a large quantity of heat is also generated in the nuclear power reactors. Thermal pollution is thus a result of radioactive pollution. Thermal pollution increases water temperatures. At higher temperatures, aquatic organisms become more susceptible to toxic materials. Also, less desirable blue green algae replace the desirable green algae.

Biological effects of ionizing radiations on humans are mainly of 3 types:

1. 'Acute radiation sickness' results from brief exposures to high doses.
2. Chronic exposure to low doses has many effects like cancer.
3. Production of mutations in sperms and eggs affects future generations.

The bombing of Hiroshima and Nagasaki and the Chernobyl disaster (given as case studies later in this unit) show that nuclear accident may take place anywhere in the world. The harmful effects are seen in very vast areas around the site. This is the main lesson from nuclear hazards. This is also the major obstacle in attempts to tap nuclear energy as an alternative energy source

5.9 SOLID WASTE MANAGEMENT

Waste materials may be solids, liquids or gases. These waste materials pollute our environment. Therefore all types of wastes must be disposed of harmlessly, or better still, utilized for various purposes. This is called waste management. If not managed properly, solid wastes accumulate in the environment and cause several problems. Solid wastes are of different types. They may be domestic, industrial, agricultural or plain natural. The most common examples are:

- Garbage/rubbish such as wastes from kitchens, slaughter houses, broken glass, plastic, crockery etc.
- Large wastes such as bricks, concrete, sand, furniture etc. from demolition/construction works
- Sludge/settled components of sewage
- Industrial wastes like chemicals, paint etc.
- Mining wastes from coal mines etc.
- Agricultural wastes like crop residues, pesticides, farm animal wastes etc.
- Ash
- Biomedical wastes
- Dead animals and animal excreta
- Fallen leaves, branches etc.

Generation of solid wastes is an important issue and has to be given due importance because:

- the volume of solid wastes generated is very large;
- they are major sources of pollutants; and
- they are unhygienic and a health hazard.

Solid waste management must be practised in all rural and urban areas, because:

- There are many known eco-friendly methods of management

- Enough land is available for the purpose
- The cost of disposal is not high compared to the direct and indirect costs that will have to be borne if management is not done
- In industries, management of wastes by recycling, recovery, reuse etc. generates revenue, thereby increasing profit

Solid wastes are produced in rural as well as urban areas. Some important solid wastes and their place of generation are listed below:

- *Municipal (domestic) solid wastes*: These are produced in all places, rural and urban.
- *Industrial solid wastes*: These are produced in various factories located mostly in urban areas.
- *Biomedical wastes*: These are generated in hospitals and clinics, seen mostly in urban areas.
- *Agro-wastes*: These are produced in rural areas where agriculture is practised widely. The methods of management of these solid wastes are highly varied.

5.9.1 Urban and Industrial Wastes

Towns and cities are included in urban areas. Cities are highly populated than towns. People live in towns and cities crowded together. In many, due to the shortage of space, buildings are high rises instead of being single storey houses. Garbage generated by all the residents in any urban locality amounts to hundreds or even thousands of tonnes per day. To keep the city clean and to maintain the health of the residents, management of these solid wastes is most essential. If there is no proper garbage management systems in urban areas, the garbage heaps become breeding grounds for mosquitoes, flies, rats and other agents that cause diseases.

Waste materials found in urban areas in India are:

- Empty bottles
- Broken crockery
- Plastic containers
- Polythene and other packing materials
- Automobile spares
- Damaged machines
- Cycle parts
- Wastes from building construction and demolition
- Dead animal skeletons
- Rotten fruits and vegetables and their skins etc.

Solid wastes were also causing problems in effluent countries of the West. Along highways were seen:

- discarded paper cups and glasses;
- empty cigarette packets;

- paper cartons;
- magazines;
- maps;
- coffee cans;
- shirts;
- tyres;
- bags;
- shoes;
- comic books etc.

But of late, regular voluntary companies have been set up in the US and European countries for collecting these waste materials and taking them to recycling plants.

Used vehicles were also creating a serious problem in the western world. In New York City, as early as 1966, 40,000 abandoned cars were counted by the police!

In India, several million tonnes of solid wastes are dumped along roadsides, rail tracks and other places in towns and cities. In Class 1 cities, an average of more than 2 million tonnes of solid wastes is thrown in the environment every year; in Class 2 cities, it is about 0.3 million tonnes.

***Major steps in municipal solid waste management in urban areas*:**

1. *Collection*: Collection is done mostly by trucks. Much time and labour is spent in loading, transporting and unloading, which increases the cost of collection. To avoid this, some methods have been suggested. These are being implemented in the developed countries. We are also using some of them.

   To reduce the volume of the wastes, garbage grinders are used. In US, Japan, Sweden etc., the garbage from homes are sucked through underground pneumatic pipelines. Compactors have been set up in many urban commercial establishments. In larger communities, transfer stations are set up. In one city, there are many transfer stations. Trucks from different parts of the city/town bring the refuse to the transfer stations. From here, bulldozers cram the refuse into large vans, which take it to the final disposal site, which are mostly landfills.

   $$\text{Town} \xrightarrow{\text{Trucks}} \text{Transfer Station} \xrightarrow{\text{Large Vans}} \text{Landfills}$$

2. *Disposal*: The compressed garbage is used to fill huge gaping pits formed during mining away from the city, or in sanitary landfills. Sanitary landfills are seen near many cities in the world.

   A modern sanitary landfill is not just a lowland, but a depression in an impermeable soil layer with an impermeable membrane. This is important because in the absence of such layers, pollutants from the lower layers of the wastes leach into the ground below and pollute the groundwater. There must also be a proper ventilating system, so that the gases formed escape safely into the atmosphere. However, in some advanced countries, methane, which is the major gas formed in landfills is collected, purified and used for various purposes. In such countries in the west, vast areas are demarcated for this purpose.

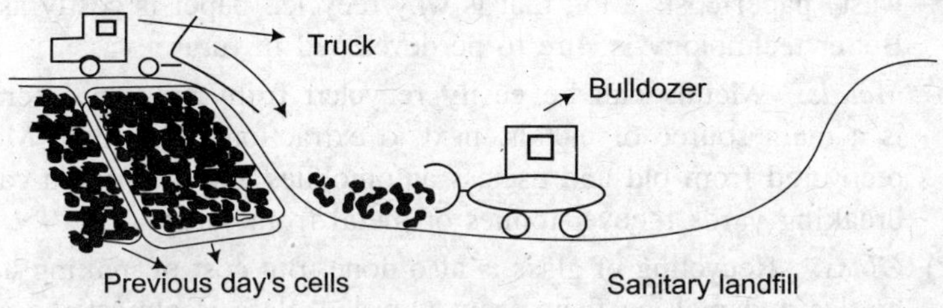

Figures 5.1

To reduce the volume of wastes, before depositing in the landfills, different methods are used:

(a) *Incineration*: Incineration means burning at a specific temperature in specific conditions. An incinerator is a suitably designed furnace. The collecting trucks unload the wastes in a pit. From the pit, cranes lift the waste and put it in the combustion chamber. Most of the refuse burns. The gases formed are collected, separated and used for different purposes. The unburnt wastes are compressed and used as manure. Incineration can reduce the volume of solid wastes by 90%. Disadvantages are that some gases formed may be harmful, and the ash might contain toxic heavy metals. So proper arrangements should be made to dispose of these end products.

(b) *Shredding*: Shredding is also called pulverizing. Here, solid wastes are torn to small pieces. Then they are spread in fields.

(c) *Pyrolysis*: Combustion (burning) in the absence of oxygen is called pyrolysis. The end products of this process are combustible gas, tar, charcoal etc. These are used as raw materials in different industries. Pyrolysis reduces the volume and also produces stable end products.

3. *Resource Recovery*: Resource recovery is commonly called recycling. Paper, metals, glass and organics are usually recycled. The first step is collection of garbage by hawkers. The hawkers sell garbage to the industries. The industries clean it, sort it out, recycle it and resell it to the customers.

   Recycling is beneficial in many ways. It helps in:

   - conservation of resources;
   - reduction in volume of waste to be disposed; and
   - enabling selling of remanufactured items in the market.

Resource recovery of certain solid wastes

(a) *Papers*: Recycling of paper can partly save the forests. Recycled paper must be made popular. Used paper should be torn and not crumpled. Recycled paper made out of crumpled paper is not of good quality as fibres tear during crumpling. Collection, sorting and transportation of waste paper costs a lot; that is why recycled paper is costly as of now. Better technology is sure to be developed in future.

(b) *Metals*: Metals can be easily recycled from industrial scrap. This is a main source of metals, next to extraction by mining. Metals are recovered from old and useless automobiles and aluminum cans. Ship breaking yards recover tonnes of metal from scrap.

(c) *Glass*: Recycling of glass is also done. But cost of making new glass bottles and making them from recycled glass is almost the same. So there is no profit.

(d) *Organics*: The commonest recovery process from organic wastes is composting. This is an aerobic process, i.e., it takes place in the presence of oxygen. In this process, first of all, metals, glass, plastics etc. are sorted out. The remaining organic wastes are ground in a grinder. Then they are sent to the 'wind rows' where they get enough ventilation, sufficient moisture and are kept at about 140°F. Aerobic bacteria grow in such condition by themselves and convert the wastes into compost manure.

Sludge from Sewage Treatment Plants (STP's): Sewage is the waste carried by water. It is generated from homes and food processing factories. Domestic sewage contains human excreta, cloth, soap, detergents etc. Sewage should never be directed into rivers or any other water body. It should be treated in sewage treatment plants. The clear water on the top can be released into water bodies. The sludge that settles at the bottom is dried and made into manure.

Biogas Plants: Biogas plants are very useful in managing solid wastes by converting them into energy and manure. During anaerobic decomposition (decomposition in the absence of oxygen) of organic wastes by bacteria, methane (CH_4) is produced. Some carbon dioxide (CO_2), hydrogen (H_2) and nitrogen (N_2) are also released along with methane. All this together is called biogas.

Biogas contains 60 to 70% methane, 25 to 30% carbon dioxide and the rest is hydrogen and nitrogen. The composition and the amount of biogas depend on the composition of the wastes and the temperature.

In villages, usually, dung is the main organic matter decomposed in biogas plants, but wastes from kitchens can also be fed into biogas plants. Recently, ARTI (Appropriate Rural Technology Institute), Pune, designed a modified biogas plant in which sugarcane waste produced during the making of sugar cane juice, and such other waste materials are used in biogas plants as raw materials. The biogas produced is carried by pipelines to households. In Pune, this modified version of the original biogas plant has become very popular.

Biogas is an ideal alternative energy source for rural and urban people alike. Biogas has many advantages. Some of them are listed below:

- Prevents environmental pollution
- Improves family health by avoiding smoke, dust, soot and coal in houses
- Prevents deforestation, as firewood is not needed in homes any more
- Improves economic condition

Some cities have found ways to manage municipal solid wastes successfully. (See case study No. 3.)

Solid Waste Management in Industry: In industries, solid waste management is based on the following approaches:

- Source reduction (waste minimization)
- Segregation (separation of different types of wastes)
- Recycle
- Recovery
- Reuse
- Environment friendly disposal

Source Reduction: Waste reduction at source has become a 'greening business strategy'. (The term 'Green' has come to mean 'environment-friendly'.) Since the past few decades, companies have realized that pollution and waste are signs of inefficiency. Many companies give incentives to employees who find the methods to prevent pollution and minimize waste as early in the production process as possible. This saves money.

***Waste minimization in industry is based on the following principles*:**

- Find out and make a list of sources, quantities and types of waste material being generated in the industry
- Collect information on the various operations, the raw materials needed for these operations, the products and the wastes generated in these processes
- Find out the inefficiencies in the various processes
- Find out the areas of poor management
- Fix certain targets for minimization of wastes
- Do research and development to find out the best methods of waste management
- Make the workers in the company aware of the benefits of waste minimization

Segregation: Many industries have realized that segregation of waste at source saves time, energy and money. Accordingly, wastes are classified, colour codes fixed for bins/collecting areas, and specific departments/agencies given the full responsibility for the removal of these wastes within a stipulated time period.

For example, the wastes may be:

- Scrap
- Plastics
- Wood
- Used oil
- Rubber material
- Biodegradable wastes like cut grass
- Wastes from canteens
- Electrical wires etc.

The colour codes may be:

- Any one colour
- A combination of colours
- Any one colour with stripes of any other colour etc.

The departments/agencies responsible may be:

- Any engineering department
- Canteen services etc.

Recycle, Recovery and Reuse: After segregation, wastes are sent to different sites for appropriate recycling, recovery and reuse. These are known as the 3 R's. Industries that practice the 3 R's are more efficient, more cost effective and less polluting. For example:

- Used plastics, paper etc. can be recycled
- Metals can be recovered from scrap
- Oil soaked jute and all types of oily wastes can be mixed with coal to be burnt for power generation
- Used acid from old batteries can be used to neutralize alkaline waste water
- Biodegradable wastes can be made into compost manures for the gardens etc.

It has been proved beyond doubt that as solid waste utilization goes up in any company, the purchase of raw materials goes down. The companies that do not waste anything are called zero waste companies.

Environment-friendly disposal: Different industries use different environment-friendly methods of waste disposal/management. In Japan, landfill space is very little. Therefore, Japanese companies have thought of many other waste management plans. Plastics are recycled into carpets, cardboard cartons are reprocessed into paper etc.

Companies can also use environment-friendly waste disposal mechanisms to improve their public image and to reduce the popularity of their competitors who do not use such methods.

Hazardous substances as wastes: These are wastes generated by certain industries like chemical manufacturing companies, paper mills, petroleum refineries etc. They cause death or very serious health problems to human beings or severely damage the environment permanently. Hazardous wastes can be included in one or more of the following types:

- *Toxic*: Toxic wastes may kill immediately or cause slow poisoning or severe illnesses. In Japan, mercury waste from a plastic factory dumped into the nearby Minamata Bay entered into the body of fish and killed many people who ate these fish. Arsenic is another toxic heavy metal, which is polluting groundwater in many parts of the world. In India, Midnapore in West Bengal is a highly affected district. Lead is also a toxic heavy metal. It is deposited on roadside soils, as automobile exhausts contain lead.
- *Highly reactive*: Some wastes react strongly with air or water when exposed to them. They may produce toxic gases on heating. They may explode while handling. Nitroglycerine, gunpowder etc. are examples.
- *Igniting rapidly*: Some waste materials burn by themselves at low temperatures. They may catch fire in storage, during transport or disposal. Examples are gasoline, alcohol etc.
- *Corrosive*: Acids and bases corrode materials on coming into contact with them. These are examples of corrosive wastes.
- *Infectious*: Many components of biomedical wastes are infectious. Examples are: used dressing materials, bandages soaked in blood, injection syringes and needles, human body tissues discarded after operations, wastes from pathological laboratories etc.
- *Radioactive*: Wastes generated in nuclear power plants are radioactive. These remain in soil, water and air for many years. For example, uranium ore mined out from the earth is of low grade, containing only 0.2% of uranium by weight. It is processed (enriched) so that the uranium – 235 content is increased to 0.7 to 3%. The enriched uranium is used for nuclear fission; part of it remains unused. This is a radioactive nuclear waste.

There are many toxic wastes in the articles of daily use, like various solvents and cleaners. Hazardous wastes pollute all parts of the environment, but the most affected is groundwater. This is because wastes from the soil also finally reach groundwater.

Control of hazardous waste pollution: Hazardous chemicals like poly vinyl chloride (PVC) used in the manufacture of plastics cannot be eliminated. Similarly, many such wastes are bound to be formed during the manufacture of many utility articles. Therefore, this type of pollution can only be minimized and not fully stopped. Methods of minimization are:

- Reduce the quantity of wastes
- Encourage recycling
- Use substitutes
- Educate the public etc.

E-wastes (Electronic wastes): Electronic wastes are increasing very fast in today's progressive world. Old and useless computers, televisions, mobile phones etc. are the important e-wastes. In the US, more than 3 million tonnes of e-wastes are deposited in landfills every year. Some data are given here to create awareness regarding the extent of this problem:

- Between 1997 and 2007, there will be 500 million obsolete computers in the US.
- Total e-waste generated in India at present is 1,46,000 tonnes per year. This is expected to go up to 16,00,000 tonnes by 2012.
- Eight thousand tonnes of computer waste is generated in Bangalore alone every year.

The main constituents of e-wastes and their health effects are given below:

- Lead—Damages nervous and circulatory systems and affects kidneys
- Cadmium—Accumulates in kidney and liver, causes nerve disorders
- Mercury—Damages brain and skin
- Barium—Causes muscle weakness; damages heart and liver
- Beryllium—Causes skin diseases and cancer
- Plastics—Burning releases poisonous dioxin, which damages reproductive system and hormones

When computer wastes are deposited in landfills, the toxins leach into the groundwater. If they are burnt, toxic pollutants are emitted into the atmosphere. Therefore, some other methods must be thought of:

- Upgrade old computers as far as possible
- Buy only those brands that can be upgraded
- Give old ones for recycling
- Metals, plastics and other materials can be recovered from e-wastes. Send the e-wastes for landfilling only after the recovery of reusable substances.

5.9.2 Composting and Vermicomposting

Composting and vermicomposting are two very good methods of managing domestic biodegradable solid wastes. These can be employed in urban areas also, as a community initiative, because large quantities of municipal solid wastes are generated in towns and cities. Both compost and vermicompost are organic manures. They can be absorbed by plants easily. Moreover, the heat generated during their formation kills many harmful insects, worms and eggs. Seeds of many weeds are also destroyed in this way.

Composting: The most popular method of composting in India is by the Marathi model. This is known as the NADEP method. It was named after the designer, Narayana Rao Pandari Pande, who was addressed by the villagers as Nadep kaka. Steps to make a compost pit of this model are given below (dimensions can be changed as per requirements, but it is advisable to keep the depth as suggested).

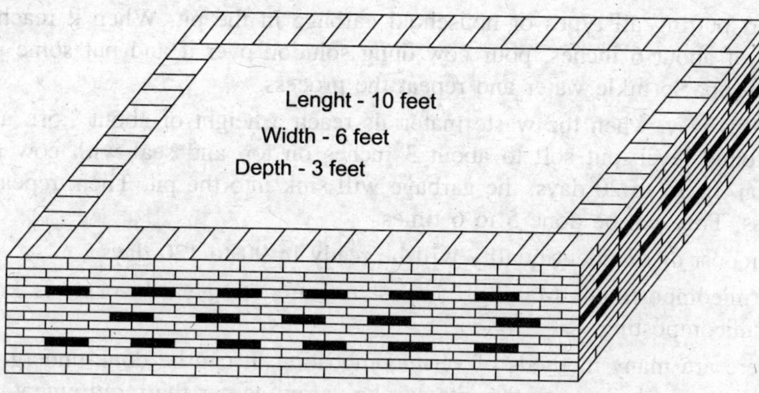

Figure 5.2 Compost pit—Maharashtrian model

1. Select a high level ground, free of water logging.
2. Clear the ground of weeds and stones.
3. Dig a pit of 10 × 6 × 3ft.
4. Spread a layer of bricks at the bottom.
5. Bind the sidewalls with bricks.

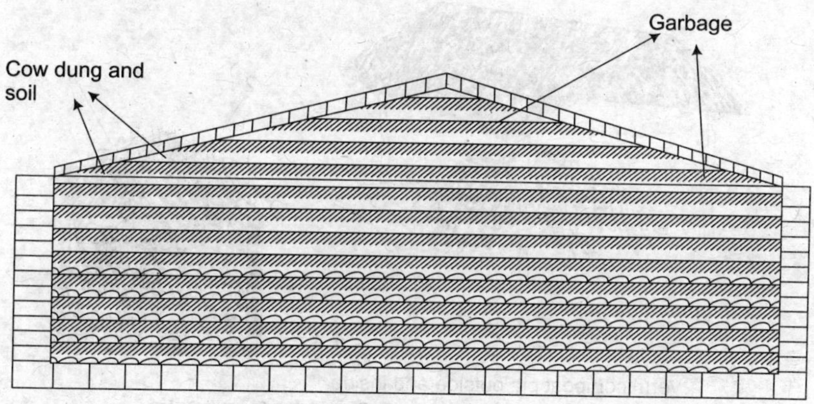

Figure 5.3 Compost pit—vertical section

6. For the side walls, set the bricks without gaps in the lowermost 2 rows and uppermost 2 rows. In all the intermediate rows, leave gaps between bricks.
7. Make a low wall of 2 rows of bricks above ground also, and cement it smoothly.
8. Apply a paste of cow dung in water all over the inside walls and floor.

9. Let the paste dry. This will take 1 or 2 days.
10. The compost pit is now ready to use.

Start putting all types of household garbage in the pit. When it reaches a height of about 6 inches, pour cow dung solution over it and put some good soil on top. Sprinkle water and repeat the process.

In this way, when the waste materials reach a height of about 1.5ft. above the ground level, put soil to about 3 inches on top and seal with cow dung solution. In 15 to 20 days, the garbage will sink into the pit. Then, repeat the process. This can be done 5 to 6 times.

Compost of excellent quality will be ready in 90 to 120 days.

Vermicomposting: Making compost manure using earthworms is known as vermicomposting.

There are many methods of vermicomposting. It can be done underground or above ground. Sizes of the pits can be varied as per the requirement.

A. *The Underground method*: Here, the vermicompost pit is underground. The steps in making such a pit are given below:

1. Dig a pit of 2m × 1m × 1m.
2. Put good soil to a height of about 15 to 20 cm in the pit.
3. Water well

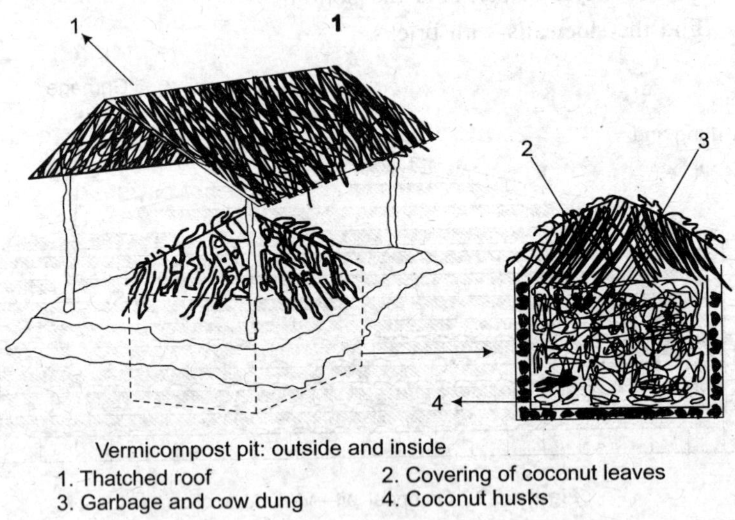

Figure 5.4 Vermicompost pit: outside and inside

4. Put 100 earthworms in the pit along with some soil from their habitat.
5. Put some cow dung on the earthworms and then spread some dry leaves or hay to a thickness of about 6 cm.
6. Leave it undisturbed for 30 days. Within this time period, the earthworms will multiply. This worm rich bed is called vermi-bed.

7. On the vermi bed, garbage and cow dung in a ratio of 8:1 can be put twice a week.
8. After 2 weeks, mix well, sprinkle water, put cow dung and cover with hay or dry leaves.
9. Continue the process.
10. In 60 to 75 days, the vermicompost will be ready. Take it out without disturbing the worms.
11. Dry the vermicompost in the sun and pack in polythene bags.

From the next time onwards, time taken will be only 45 days as the vermi-bed is already made. Similarly, if more number of earthworms is used in the beginning, at the rate of 1000 worms (about 1kg) in a 2 m × 1 m × 30 cm pit, vermicompost will be ready in 45 days.

B. *Overground method*: Vermicompost pits can be overground also.

1. Clean the floor of a firm ground, plough it, remove the stones etc. and press the floor firm.
2. Make low walls all around with bricks and cement.
3. Spread coconut fibres, sprinkle water and put garbage.
4. After 2 days, put 30 earthworms for every sq. ft. area.
5. Then put wastes and cow dung mixture in a ratio of 8:1.
6. When the wastes reach about 1.5 ft. aboveground, cover with wet sacs.
7. Moisten the sacs daily.
8. In 2 months, the vermicompost will be ready. Take it out and heap it in the sun in a few places. The worms move down.
9. Take out the manure from top
10. Redeposit the worms in the pit.

Whatever be the method, some points should be kept in mind:

1. Make a tent over the pit to protect it from sun and rain.
2. Use good species like *Eudrelus euginae, Lampito mauriti, Dravida wills* etc.
3. Do not pour too much water as it will reduce the reproduction.

Making vermicompost in houses: To make vermicompost in small amounts in individual houses, use barrels with both ends open. Tie a plastic sheet with holes in it (for ventilation) at the bottom. Excess water will flow out through these holes but the worms cannot move out.

Benefits of compost and vermicompost:

1. Yield of vegetables, fruits etc. increases.
2. Quality and taste improve.
3. Soil texture improves.
4. Water holding capacity of soil increases.

5. Soil erosion is prevented.
6. Pollution is minimised.
7. Purchase of costly artificial manures can be minimised.
8. Both composting and vermicomposting can be developed into cottage industries for the economic upliftment of the rural poor.

In today's world of rapid urbanisation and unplanned population growth, accumulation of solid wastes has become a grave problem in urban areas. This pollutes air, water and soil. It also causes social, aesthetic, ecological and economic problems in society. It causes serious health problems, especially in the slums of big cities. Open garbage dumps are converting more and more land into wastelands. Per capita solid waste generation in India is about 500 grams/ day. The composition of this garbage is as follows:

Moisture content—50 to 55%

Fruit pulp—35 to 45%

Other organic and inorganic waste—8 to 15%

Earthworms consume organic substances rapidly. They have an organ called 'grinding gizzard', which fragments the consumed particles into much finer particles. It is to increase the rate of decomposition that cowdung is added to the solid wastes in vermicompost pits. The most suitable pH to be maintained in the pits differs depending on the nature of the wastes and the species of the earthworm used. Some species like *Lampito mauritii* increase the pH of the wastes during conversion to vermicompost. The percentage of increase differs with different types of wastes.

Water passed through vermicompost is called vermi wash. This solution also has high nutritive value. To prepare vermi wash easily at home, the following steps can be taken.

1. Take a flower pot about 12″ in height.
2. Place pieces of stone or bricks about 4″ in height.
3. Place a piece of muslin cloth or filter paper on the stones.
4. Above this, deposit a mixture of degradable wastes and cow dung in an 8:1 ratio along with earthworms.
5. Keeping only one hole at the bottom open, close the others with cement.
6. Fix a rubber tubing in the open hole. When not in need, this tube can be folded back.
7. Every day, pour three table spoons of a solution of cow dung in the pot, on top of the garbage. Also, pour 100 ml of water in the pot daily.
8. Collect the vermi wash through the rubber tubing.
9. Pour the vermi wash directly at the base of plants.
10. Vermi wash can also be sprayed on leaves, after diluting it four times.
11. Vermi wash will make leaves greener and plants healthier.

12. By pouring vermi wash in vermicompost pits, formation of vermicompost can be speeded up.

In short, composting and vermicomposting are two excellent methods of solid waste management that can be followed in rural as well as urban areas. Technology for large scale use of earthworms for converting all types of organic wastes into vermicompost and the final application of the same in croplands is being developed. This technology is known as vermitechnology.

5.10 ROLE OF INDIVIDUALS IN PREVENTION OF POLLUTION

Pollution is a global problem, and it has reached dangerously high levels. Therefore, for controlling pollution, we just cannot and should not depend fully on governments. Since a few decades, in addition to governmental departments and agencies, non-governmental organisations (NGO's) are also doing a lot of work in this direction. Yet the role of individuals in preventing pollution can never be underestimated.

Individuals should take more interest in pollution control activities due to many reasons:

1. First of all, it is ultimately the individuals who have to endure the harmful effects of pollution.
2. By actively participating in pollution control activities, individuals set an ideal precedent.
3. The benefits of clean and unpolluted environment will be enjoyed by individuals first.
4. If one or more individuals start any work on controlling pollution, more and more will emulate them and ultimately this will give rise to a movement.
5. 'Think Globally, Act Locally' has become a famous slogan in all environment related issues. To act locally, individuals and collections of individuals are best suited.
6. Controlling pollution is a huge task. It cannot succeed unless everybody puts in his / her effort.
7. Even if the Government and the NGO's are working wholeheartedly towards pollution control, a lot of their efforts will go waste if the individuals in the society are not aware about the seriousness of the work.
8. No individual should think that he/she has only rights and no duties. The society is giving the individuals a lot; it is the duty of the individuals to reciprocate. They must help in keeping the neighbourhood clean and pollution free.

A society, a state, a country, the whole world is a collection of individuals, in the final analysis. Therefore, if each and every individual does his/her duty to perfection, the whole world can theoretically become pollution free. It is the duty of all of us to spread awareness among the people in this regard.

Following is a list of what individuals can do to control different types of pollution and to conserve natural resources, as depletion of these resources is one of the most harmful effects of pollution.

1. Plant as many trees as possible in as many places as possible.
2. Do not use plastic glasses, cups, plates etc.
3. Burning of petrol pollutes the air. It releases many harmful gases into the atmosphere. Therefore, use petrol sparingly.
4. Get your vehicles checked for pollution control regularly. If there is any defect, get it repaired immediately.
5. Service your vehicles periodically. This will help to keep them in a good condition.
6. Do not keep cars switched on while getting out for even a short time. This state, called idling, releases more harmful fumes into the air.
7. During acceleration also, more harmful fumes are emitted. Therefore, do not increase and reduce the speed of your vehicle frequently while driving. Try to go at the same speed if possible, accelerating only when necessary.
8. Do not honk the horn unnecessarily. This is a very bad habit with Indians in general. There is no meaning honking the horn while stuck in a traffic jam; anyway, everybody wants to move! Just imagine, if every vehicle stuck in traffic jam blows its horn, how much noise pollution it will cause! So, be patient, the jam will clear anyway. It is not going to clear any sooner if you honk your horn!
9. Know the sound levels permitted in different places so that you can act accordingly. Hospitals are silent zones; so are libraries. Do not disturb the peace of these places by shouting or talking loudly.
10. If you know the rules regarding noise pollution, may be you can politely tell organisers of festivals etc. about them.
11. Use pesticides in your homes only when necessary. Try biopesticides instead of chemical pesticides as much as possible. Keep neem leaves between the folds of your woollen garments this winter, in place of naphthalene balls. Even after winter, while storing your woollens away for next year, just use neem leaves. No insects will damage your woollens; they will remain as fresh as ever and will also be free of dust. In your garden also, you can use neem seed oil, and neem oil cake to protect your plants against pests. Popularise these concepts among your friends.
12. Do not throw garbage outside your house.
13. Practise segregation at source, i.e., keep different dust bins for biodegradable and non-biodegradable waste materials. In the former, put kitchen wastes like skins of fruits and vegetables, tea leaves, waste

paper etc. In the latter, put plastics, aluminium foils, broken glass, crockery etc.

14. If you live in single houses with some gardening space, have compost pits dug for preparing compost manure from kitchen wastes, used paper, cloth, torn jute bags etc. This way, you will be managing the biodegradable solid wastes also.
15. Vermicomposting can also be done, in suitably sized barrels, in houses.
16. If you are living in flat complexes or in houses with no garden space, convince the landlords/neighbours to have community compost pits in a common area.
17. By managing the degradable solid wastes, you are doing a great service to the whole locality—you are safeguarding yourself, family members and neighbours against sanitation related diseases. You are also keeping the neighbourhood clean.
18. Throw only the non-biodegradable wastes in the municipality dumpers.
19. Do not use plastic bags as far as possible. Take bags of cloth or jute for shopping.
20. Do not throw toffee wrappers etc. on the roads or footpaths or college grounds if dustbins are not seen nearby. Put them in your bags/pockets and bring them home to be put in dustbins.
21. Do not throw garbage into the sewer. It will block the flow of the sewage and create problems.
22. Do not throw garbage into any water body. It will harm the aquatic plants and animals.
23. While buying new computers, choose equipment that can meet your long term needs or can be easily upgraded.
24. During new purchase, trade in your old computer to be recycled.
25. If you are buying a better model, donate your old computer to a school or any other charity organisation, if possible.
26. Try to contact any recycling company nearby to handle your recyclable equipment.
27. For items like glass, aluminium etc. also, try to contact dealers, if possible.
28. Do not buy new batteries every time; use batteries that can be recharged.
29. Spread environmental awareness by organising various activities in your locality.
30. Write articles on pollution control and publish in local newspapers.

31. Understand that though we are superior to all other animals in intelligence, still, we are part of nature. By polluting nature, we are harming ourselves along with the other parts of nature.

5.11 CASE STUDIES

5.11.1 Bhopal Gas Tragedy on Dec. 2–3, 1984

Methyl isocyanate leak from Union Carbide India Ltd. (UCIL) pesticide plant in Bhopal killed 15,274 people according to government figures. Amnesty International says 8,000 people died in the first three days and another 15,000 in the aftermath, because of the long term effects. Thousands of deformed children were born. About one lakh people continue to suffer from chronic diseases and thousands still drink polluted water. This was the world's worst industrial disaster. The case lingered on and on for years. The initial charge was "culpable homicide" which can carry a life term. But, on an appeal by the accused, it was changed under Supreme Court orders in 1996 to "causing death by negligence", which is usually invoked for traffic accidents. As a result of this gross injustice, on 7th June 2010, after 26 long years, seven men were awarded sentence of just two years in jail and a fine. Mr. Warren Anderson, the American, who was the chairman of UCIL at the time of the accident, had fled the country after paying a bail of Rs. 25,000 and an undertaking that he would return to India. He never returned. He is still in hiding and has gone unpunished. The Bhopal gas tragedy compensation award of $470 million finalised works out to a lowly Rs. 12,000/- a victim.

5.11.2 The Weldon Spring Ground Water Pollution

In 1940, during World War II, the US Army had established the world's largest TNT producing facility on 70 km^2 of land around a small town called Weldon Spring and its neighbouring areas. Here, toluene was treated with nitric acid to produce more than a million of the explosive Tri-nitro-toluene (TNT) each day at the peak of production. Part of the manufacturing process involved purifying the TNT, thus generating wastes in huge quantities. Some of the red coloured muck was treated at wastewater plants, but much of it ran off from the facilities and finally got mixed with underground water. In 1980, the US Government ordered to clean up several sites in the country. The engineers assigned this task had expected to find traces of the discarded nitro-aromatic waste at Weldon Spring, but they were astonished to find the chemicals in peoples' wells, in towns several miles from the site.

5.11.3 The Case of the Minamata Disease in Japan

A crippling deformity due to consumption of contaminated fish from Minamata Bay of Japan showed up among the Japanese in 1952. Soon, it was found that the Bay was polluted with mercury from a plastic factory. The disease hence came to be known as Minamata disease.

5.11.4 Manure and Electricity from Solid Waste

Cases of pollution can be cited in large numbers; at the same time, cases of treating the wastes, cleaning up the environment etc. are also common. For example, within the city limits of Vijayawada in Andhra Pradesh, 500 tonnes of solid wastes are generated every day. Of this, the Vijayawada Municipal Corporation (VMC) is disposing of 350 tonnes of waste every day with the help of 2 innovative projects. The Corporation has involved the private sector in the process to put into practice the concept of 'Wealth from Waste'.Them first project is run involving Excel Industries Ltd. The company generates organic manure from waste. The other is run by Sriram Energy System Pvt. Ltd. It generates power from waste. VMC has given the 2 companies land on lease and the necessary subsidies. It supplies waste to them for a royalty.

5.11.5 Ban on Plastic in Dhaka, Bangladesh

In our neighbouring country, Bangladesh, a survey showed that of the 10 million plastic bags used daily, 9 million were dumped every day in the city of Dhaka alone. Only 10-15% were put in dust bins; the rest finally landed in drains and rivers, ponds and lakes. Eight per cent of the drains in Dhaka got clogged because of plastic bags. The bed of river Buriganga flowing by Dhaka got littered with plastic bags to a height of 5-6 feet. All the water bodies started getting choked with plastic bags.

The Govt. understood the seriousness of the problem and issued a ban on the use of plastics within the city of Dhaka, with effect from 1st Jan. 2002. Later, the ban was made effective throughout the country. The Govt., various NGO's and environmental activists undertook a massive campaign to create awareness. Those breaking the law were punished.

Media played an important role too. They circulated survey reports from the West that workers of plastic factories are more susceptible to cancer, skin diseases and other health problems, and that fish and meat wrapped in polythene become poisonous. As people gradually understood the seriousness of the problem, they started paying heed to the Govt. ban. Due to all these determined moves of the Govt., Bangladesh is now largely free from poly bags, if not fully. Shopping bags made of paper, cloth and jute are catching up fast.

5.12 DISASTER MANAGEMENT

Environmental hazards or disasters are extreme events, natural or man-made, which exceed tolerable magnitudes. The results include loss of lives and property. They become big news at local, national and even international level. Disasters occur suddenly and indiscriminately.

In the world, many environmentally extreme events take place regularly. But we consider them as disasters only if they affect human beings in some way. For example, a very severe cyclone may develop in the middle of an ocean and die off before it reaches the shore and cause damage to human beings. This is not considered a disaster. On the other hand, if a cyclone originating in an ocean reaches a coastal area where people live and causes loss of lives of human beings and animals and destroys plants and affects buildings and other property, it is called a disaster. Similarly, a volcanic eruption in an island where nobody lives is an environmentally extreme event, but not an environmental disaster. Only if a volcano erupts in a well populated area, it is called a disaster.

Thus, all environmental disasters are extreme events, but all environmentally extreme events are not disasters.

The intensity of any disaster is proportional to the seriousness of the damage done to:

- lives of human beings, animals and plants, and
- property

Environmental disasters are broadly of two types—natural and man-made.

Examples of natural disasters are:

- Volcanoes
- Earthquakes
- Landslides
- Cyclones
- Droughts
- Floods etc.

Examples of man-made disasters are:

- Earthquakes
- Landslides
- Nuclear explosion etc.

Floods and droughts are also sometimes man-made. For example, when dam gates are opened, (many times, this happens without any prior warning to the affected people), flash floods occur in the downstream areas. Floods are also caused by human activities like sand mining from the river beds, siltation of the river banks due to deforestation etc. Droughts can be man-made too as a result of deforestation or overgrazing. But generally speaking, floods and droughts are natural disasters.

5.12.1 Floods

Flood can be defined as submergence or inundation of vast areas of land under water for many days in continuation.

Floods are generally associated with rivers. In India, the rivers that get flooded most frequently are the Ganges and its tributaries like Gomti, Gandak, Ghagra, Kosi etc. and Brahmaputra, Mahanadi, Krishna, Narmada, Godavari etc. Rivers Missouri and Missisippi in the USA, Yangtze in China, Irawadi in Myanmar, Indus in Pakistan etc. are also regularly affected by floods.

Floods are the result of accumulation of large amounts of water in rivers, which overflow the river banks. These are in fact natural phenomena. But when they cause damage to life and property, they become environmental disasters.

To understand the most suitable disaster management methods to adopt, it is essential to know the causes of any disaster.

(a) *Causes of floods*: There are many causes of floods. These causes act in various combinations and thus become very complex. It is because of the complex nature of their causes that floods are difficult to manage and control. (See Floods in unit 6 for details)

***Natural causes*:**

- Heavy rainfall
- Winding course of rivers
- Occurrence of vast floodplains
- Sudden change in slopes
- Landslides

***Anthropogenic causes*:**

- Deforestation
- Urbanization
- Changing the course of rivers
- Construction of dams
- Some agricultural practices
- Man-induced landslides

In India, most of the flood-prone areas are in the northern parts. The Gangetic plains of UP, Bihar and West Bengal get flooded very frequently. Floods are frequent in Andhra Pradesh also. These states account for more than 60% of the damages caused by floods in our country.

The frequency and severity of floods have increased in India over the years. The losses caused by floods to lives and property and the expenses incurred in trying to control floods have become enormous. Therefore, it is high time for us to take steps in preventing, controlling and managing floods.

(b) *Control of floods*:

There are various methods to control floods. By implementing these, the occurrence of floods can be minimised. So these can be considered as flood management measures.

1. *Slowing down runoff to rivers*: After a heavy rainfall, most of the water flows off along the slope. This is called runoff. By reducing the speed of runoff, more water can be directed underground. For this purpose, reforestation and afforestation should be done on a large scale on the hills from where the rivers originate. By this method, soil erosion is reduced. As a result, siltation of rivers is slowed down.
2. *Reducing the number of twists and turns of rivers*: By doing this, water flows straight and fast to the ocean, thereby reducing floods. But this is an expensive process. It has been successfully done in the US.

3. *Constructing diversions in the floodplains*: In this method, the flood water is diverted to low lying areas or artificially constructed channels.
4. *Constructing flood control reservoirs*: These are large reservoirs that store water. They controls floods and also help in irrigation. In India, we have many such reservoirs and dams on rivers like Damodar, Tapti etc.
5. *Constructing flood walls*: Artificial walls of earth or stones or concrete are built on the sides of the river, where necessary, according to the volume of water and its force. This helps to prevent water from flooding the plains.
6. *Forecasting and warning*: Central Flood Control Board was set up in India in 1954. Later, State Flood Control Boards were also set up. These boards have implemented many flood control measures in our country. In 1959, flood forecasting and warning systems were started in Delhi. Soon, many such systems were set up in different parts of the country. These centres continuously monitor the river basins of the Ganges, Brahmaputra and their tributaries, Damodar, Subarnarekha, Tapti, Narmada, Godavari etc. The centres collect data on:
   - Rainfall
   - Rate of discharge
   - Flood level etc.

These data are passed on to the local authorities who warn the local people about the chances of floods well in advance. Thus, people get opportunity to move to safer places with their belongings before the flood arrives. Loss to life is prevented fully and loss of property is greatly reduced by this method.

5.12.2 Earthquakes

An earthquake is a motion of the ground surface caused by wave motion starting from a focal point. It may be a mild tremor or a severe motion, leading to collapse of buildings and formation of deep fissures in the ground. The quake is the result of a form of energy transmitted through the surface layer of the earth.

The intensity of the energy released during earthquakes is measured on the Richter scale. This scale was devised by Charles F. Richter in 1935. The scale is made between 0 and 9. A very small tremor, which can be detected only by a seismograph, is taken to be of 0 (zero) magnitude. Depending on the severity of the quake, they are described as feeble, moderate, strong, very strong, destructive, disastrous etc. Disastrous earthquakes measure 7 and above on the Richter scale. Above 8, the quakes are classified as catastrophic.

The place of origin of the quake is called the focus. It is also called hypocentre as it is always deep, even a few hundreds of kilometres deep, below the surface of the earth. The place on the earth's surface, right above the hypocentre, is the epicentre. The seismic waves of energy move in widening circles away from the hypocentre. The instrument used for recording these waves is called the seismometer.

Most earthquakes are of small magnitudes. Those of magnitudes 5 or above cause serious damages, while catastrophic quakes, measuring 8 or more on the Richter scale, lead to complete destruction.

The degree of destruction caused by earthquakes depends on many factors:

- Magnitude
- Distance from the epicentre
- Type of ground
- Duration
- Materials used in building construction etc.

(a) *Causes of earthquakes*: Any disturbance in the equilibrium in any part of the earth's crust results in an earthquake. Many factors may cause such disequilibrium:

- Volcani`c eruptions
- Fractures and faults in the earth's crust
- Pressure of man-made dams, reservoirs and lakes
- Movements of plates of earth etc.

As per the most popular theory, the earth's crust is made up of 6 major plates and 20 minor plates. These are constantly in motion. These movements are the result of convective currents within the earth.

For example, the Asiatic plate is gradually moving southwards. At the same time, the Indian plate is slowly moving northward. The resulting collision is the cause of the slow rise of the Himalayas. This rise is taking place at the rate of 50 millimetres per year. When there are sudden variations in this slow movement, earthquakes occur in North India.

(b) *Disastrous effects of earthquakes*: An earthquake that strikes an area where many people live has many disastrous effects. During the 1934 earthquake in Bihar, the epicentre was located near Darbhanga. A total of 10,700 people belonging to Bihar, Nepal and Tibet lost their lives. Landslides blocked roads for 250 km. Ground was cracked at many places. The magnitude of this earthquake was 8.4. The earthquake at Latur, Maharashtra in 1993 measured 6.3 on the Richter scale. Losses were great as most of the houses were made of mud. The disastrous effects of earthquakes are as follows.

1. *Landslides due to slopes being unstable*: When there is an earthquake, masses of land fall down from unstable slopes. This takes place more in the rainy seasons as soil is soft during the monsoon. Such landslides damage houses and block roads and rails. It may take days to clear the debris.
2. *Damage to man-made structures*: Buildings, roads, railway lines, bridges, industries etc. are damaged to different extents. If the grounds

on which buildings are built are not firm, buildings collapse faster, e.g., buildings on drained marsh. Mud huts collapse more while buildings of cement and concrete last longer.

3. *Loss of human life and property*: Thousands die in earthquakes. The worst disaster was in China in 1976, measuring 7.8 to 8.1 on the Richter scale. It caused the death of 750,000 people. Whenever there is loss of human life, there is a corresponding loss of property also.
4. *Fire inside houses, mines etc.*: When quakes shake buildings, gas cylinders upturn and burst, leading to fires. Fires are also caused by quakes in factories, mines etc.
5. *Damages to towns and cities*: When large buildings shake and collapse, many people are buried under the debris. Uprooted trees and broken branches block traffic. This results in the delay in rescue teams reaching the site late. Underground water pipes are bent and spoilt. Sewers are destroyed and blocked at many places. Clean water and dirty water mix and cause epidemics. Electricity supply and telephone lines are cut off. Communication systems are totally disrupted.
6. *Damage to ground surface*: Strong quakes damage the surface of the ground, creating fissures and deep clefts. Masses of ground are displaced.
7. *Flash floods*: Severe earthquakes result in breach of dams, which lead to floods.
8. *Tsunamis*: (In Japanese, *Tsu* means harbour and *nami* means wave). Seismic waves caused by earthquakes under the sea travel through water. This gives rise to very high sea waves, which lash at the coasts. The untold misery and destruction caused by the tsunamis that lashed the eastern coast of India in 2004 are well known.

(c) *The importance of acting on warnings*: The quake that occurred on 26th January, our Republic day, in 2001 in Bhuj, Gujarat was a killer earthquake, measuring 8.1 on the Richter scale. Ninety per cent of the of houses in Bhuj were destroyed completely. Almost 1,00,000 people died. A moderate quake had occurred just a few kilometres away from Bhuj a month ago in December, but no one had paid any heed to this. Damage would have been much less if timely evacuations had been done. This was a failure on the part of the Disaster Management System.

5.12.3 Cyclones

Tropical cyclones are deadly atmospheric storms blowing in a spiral manner. They are closed low pressure systems with an average diameter of 650 km. Cyclones are very destructive. They are known as hurricanes in the Atlantic Ocean and typhoons in the Pacific Ocean. Cyclones are different from local storms which also cause great harm. Tornadoes of USA, thunderstorms with heavy rains, dust-storms of deserts etc. are all different types of storms, but not cyclones.

(a) *Characteristics of cyclones*:

- A cyclone is a wind blowing spirally inward towards a centre of low barometric pressure.
- Wind speeds of tropical cyclones range from 180 to 400 km per hour.
- A very heavy, continuous rainfall accompanies cyclones.
- Atmospheric pressure dips very low, leading to a rise in sea levels. Vast areas near the sea are flooded.
- Great damage is done to life and property.

Tropical cyclones trouble the eastern coasts of India and the southern coasts of Bangladesh regularly. The most disastrous cyclone that occurred in Bangladesh and West Bengal put together in 1970, killed three lakh people (ten lakh as per some sources, including missing persons). Most of the deaths were by drowning.

(b) *The importance of setting up monitoring, predicting and forewarning systems*:

The cyclone that struck the Andhra Coast in 1977 caused sea waves 6 metres high and killed 55,0000 persons. The speed of wind in this cyclone was 175 km/hr. Homes of 2,000,000 people were destroyed. 1,200,000 hectares of croplands became useless due to thick salt deposition after the sea waves receded. It took 3 years to desalinate (remove salt from) this land.

A more powerful cyclone struck coastal Andhra again in 1990. But by then, advance cyclone monitoring system had been set up. From the time it originated in the Bay of Bengal, the movement of the cyclone was closely monitored every minute and predictions and warnings given from time to time. Therefore, though 300,000 people became homeless and 90,000 cattle died, number of human deaths was just 598.

(c) *The importance of reaching the warning to all concerned*:

In 1999, Orissa had to bear the colossal damages caused by a super cyclone. The cyclone ravaged coastal Orissa for 3 days. It started as deep depression in the Andaman Sea, evolved into a cyclonic storm, became very severe and finally became the super cyclonic storm, moving at a velocity of 260 km/hr.

Indian Meteorological Department (IMD) issued an alarm, but the Orissa Government could not inform the general public in time due to lack of radio network. Finally, the super cyclone took people unawares. The cyclone at 300 km/hr speed remained within 10 coastal districts for 8 hours. Tidal waves 9 metres high travelled up to 20 km towards land.

Damages brought about by the super cyclone of Orissa:

- Though official estimates are 10,000 deaths, unofficial estimates are about 100,000.
- Many hundred thousand cattle lost their lives.
- Croplands spread in 1.75 million hectares were destroyed.

- Innumerable homes were flattened.
- Water and electricity supplies were totally disrupted.
- Communication systems stopped working.
- Road and rail transport stopped completely for many weeks.
- Paradeep Port became non-functional for weeks.
- Dead bodies accumulated in several places could not be burnt due to lack of dry wood.
- The decomposing dead bodies of people and animals raised such a stink that relief operations had to be done with difficulty.
- There was no clean water to drink.
- Epidemics broke out.
- Most of the trees were uprooted.
- Total loss of property amounted to rupees 10,000 crore.
- Suffering from intense mental agony, many people went into depression.

India has more than one hundred disaster warning systems. There is need to set up more. More important than setting up systems is to maintain them in working condition at all times and to pass the information well in advance to the local people.

5.12.4 Landslides

Tumbling down of huge landmasses is called a landslide. They may be natural or man-made. Landslides may be a natural phenomenon caused by heavy rain and the resulting softening of the soil, earthquakes etc. They can be anthropogenic (man-made) also. For example, mining on hills makes the ground shaky, leading to landslides. Deforestation on slopes results in severe soil erosion. In the absence of trees and other plants which fix soil in its place by their roots, huge masses of land tumble down.

If landmasses from hill slopes fall into rivers, the flow of water is blocked. This may result in flooding. If the landmasses fall from slopes on to passing vehicles on roads, loss of life and damage to the vehicles result. Roads also get blocked, and till the time as the blocks are removed, all transport comes to a standstill.

(a) *Causes of landslides*: There are many causes of landslides:

1. Natural causes: Sometimes, earthquakes occur on hilly or mountainous localities. If these hills or mountains are made of comparatively soft and weak rocks, the slopes are unstable and they fall down in a mass, causing landslides. In 1970, 25,000 people were killed in a massive landslide caused by an earthquake in Peru. More landslides occur in the rainy season.
2. Mining: Deep underground mining also leads to landslides.
3. Blasting of rocks for construction of roads, dams and reservoirs.

4. Dam building can also lead to landslides. In major dams, enormous quantities of water are stored. This water causes the building up of great hydrostatic pressure underground. This may result in landslide and flow of mud along the reservoir walls.

(b) *Disaster management*: Natural disasters will happen; there is no way to prevent them. But it is possible to reduce the level of risk by developing efficient early warning systems. By doing this, the local people will be prepared for the approaching disaster. Plans can be finalised and implemented accordingly. The losses resulting from the disaster can be reduced and relief measures carried out properly. It will also be a good idea to bring those affected by disasters under insurance cover. There must be provision to insure not only human life, but also the lives of animals and also houses, household goods, crops etc.

It has been estimated that about 60% of the area of the Indian subcontinent is vulnerable to seismic activity of different degrees. We have a long coastline of 5,700 km and cyclones occur in large coastal areas. The maximum damage is done by floods in different rivers, which occur with a regular periodicity. There are mainly four aspects of management of environmental disasters:

1. To provide immediate relief measures to the people affected by the disaster: The victims of disasters need immediate help. This must reach to them as soon as possible. There may be more deaths and more trauma affected people if help does not reach the victims in time.

If it is an earthquake, there will be enormous quantities of debris all over the place. Many people may be trapped under the debris. To rescue them is an urgent need. For this, various equipment will be needed like gas cutters etc. Skilled people must be present at the sites. Those who are taken out of the debris will be injured and will also be in a state of fear. They need physical as well as psychological treatment. They have to be carried to the nearest hospitals for such medical help.

There will be many others who are injured. They have to be carried to the hospitals too. Many of them will survive if help is provided at the right time. It is due to delays on such occasions that the number of deaths shoots up. People who know first aid are of great help on such occasions.

After a disaster, the scenes will be horrible everywhere. There is a need to cremate the dead too. There is the risk of people who are almost dead being mistaken for dead. So doctors will be required to certify death. The dead must be cremated without much delay. The more the delay, the more are the chances of spread of epidemics. Many animals will also be dead. Disposing of their dead bodies must also be taken on a war footing.

All the survivors will need drinking water, food, clothing and shelter. It may become really difficult to provide potable water to the survivors. Special tankers carry food packets from helicopters in the affected areas. This is the only practicable method in many cases as roads and rails will be blocked by uprooted trees or debris from fallen buildings, or totally damaged by cracks, deep fissures etc.

Regarding clothing, it usually takes time. The news of the disaster spreads far and wide in no time in these days of fast communication methods. This happens immediately after the disaster. But as communication links from the site of the disaster are usually completely damaged, news broadcast directly from the site becomes next to impossible in many cases. In such cases, satellite communication may become the most important method of communication.

The role of media in disasters is very important. It is the duty of responsible media to give correct coverage. Under-coverage is as bad as over-coverage or incorrect coverage. If the coverage is not proper, sufficient help will not reach the site. But more often, problems are created by over-coverage and incorrect coverage. The worst effect of over-coverage is that too many agencies, organisations, other states and even other countries send relief materials to the affected site. This creates a huge problem of storing them safely and distributing them properly. More often than not, large quantities of relief materials are wasted.

There have been cases of unscrupulous elements taking advantage of the situation and taking away the relief materials meant for the victims. The presence of police and army helps to a great extent to tackle this menace. Unthinking acts of charity cause problems too. Many times, uncooked rice and dal reach the sites in tonnes, sent by neighbouring states and foreign countries. But there is no way the victims can cook them in the absence of dry fuel, stoves, vessels etc. Hence the system of air dropping prepared food packets is the best, till relief camps with cooking arrangements are set up. Once the camps are set up, raw materials can be provided at the site of disaster. This can continue till the affected people start normal lives, which may take weeks or months.

Just after the super cyclone of Orissa in 1999, clothes had piled up in all the affected districts. For the women everybody had sent saris. Nobody had sent blouses or petticoats. This had become a real practical problem. Such unthinking acts of charity should be avoided.

In some cases, as soon as there is a news of a disaster anywhere, doctors and nurses arrive from different parts of the country in large numbers. They also bring with them enormous quantities of medicines. Many of these medicines may not be the ones required. They also cause storing and handling difficulties. The agencies and countries send the relief act out of their assumptions and not on the real needs. This adds problems to the already existing ones. Once it had so happened that after a disastrous earthquake in Mexico City, Mexico in 1985, an aeroplane full of doctors and nurses landed in the city. Loads of medicines also arrived. It all resulted in great confusion as all the doctors and nurses needed place to live; moreover, they did not know the local language and so interpreters had to be arranged for them. Thus well meaning measures had resulted in adding to the woes of the local populace.

Social responses: It is always best for others to wait for appeals of help from the concerned governments. However, the government must do its duty efficiently, i.e., gauge the effects of the disaster correctly, and determine the type and quantum of help needed and pass on the information to the organisations/states/countries willing to help in cash or kind or both.

In this context again, the media has a most sensible and responsible role to play. This is because the social response to disasters depends on the media reports. If the reporter's assumption of the situation is wrong, people may get the wrong message. The reporters will never deliberately misguide the public, yet their assumptions may not give an accurate picture of the magnitude of the disaster and the types and quantum of help needed. In India, a lot of coverage is given to visuals, showing the scenes of occurrence and the affected people's miseries. This does arouse the emotional part of the human mind, leading to an increase in the flow of relief materials. At the same time, more important is an authentic list of what is needed and where.

Fearing the breakout of epidemics, tonnes of medicines may arrive, when what are needed urgently may be special tools for rescuing trapped persons, communication equipment, heavy machines to remove debris, water pumps etc. along with people skilled in such jobs. Epidemics will not break out at all if blocked roads are cleared, people are taken to hospitals and dispensaries according to the seriousness of their injuries, extra water is pumped out etc. and hygienic conditions are restored fast.

Help pours in from all quarters. There are many voluntary organisations, including various NGO's doing commendable work immediately after any disaster. The help provided by the Red Cross can never be overemphasised.

2. To predict the disaster beforehand and forewarn the people in the area of the time and intensity of the disaster: All said and done, preventing disasters is always more desirable than managing them after their occurrence. It is not always possible to predict all types of natural disasters, but there are excellent methods of predicting some like cyclones, floods etc. The simplest method is to study past records carefully. From such studies, the periodicity and magnitude of disasters can be predicted to fairly good accuracy.

Sometimes, minor events may precede major events. For example, in the case of earthquakes, minor tremors may be felt a few days before a major tremor. Soil may tumble down in small amounts and slowly before a major landslide. Heavy incessant rainfall is a forewarning to impending floods. A fall in the atmospheric pressure over a particular ocean indicates the building up of a cyclone. What is important in all these cases is analysing the minor phenomena correctly and passing on the information promptly without addition or deletion to the concerned authorities and the local people.

For predicting disasters, monitoring stations have been set up in different countries of the world. Round the clock weather monitoring is most essential in this regard. This is done very efficiently and effectively in the US. Though we have many weather forecasting centres in India, none are as well equipped or as effective as those in the US. The forecasting is so accurate in the developed countries that every day, even kindergarten students see the internet on their own to decide whether they should carry their raincoats to school on that day or not! The author has seen this happening in the US, and knows officials in Indian ships who find global weather forecast by the US more reliable than the forecasts by others. We Indians are in no way less intelligent than our counterparts anywhere in the world. What we need is more hard work, dedication and devotion to the cause.

***Applications of Remote Sensing and GIS*:**

Remote sensing is the science of identification and classification of the surface features of the earth using electromagnetic radiation. Every object reflects or scatters some electromagnetic radiations. The pattern of this radiation is different for different objects. These are understood and classified by the sensors of satellites left by different countries in the space. Indian Remote Sensing Satellite (IRS-1A and 1B) were launched successfully in 1988 and 1991 respectively. More such satellites were launched by India later.

With the help of images from the remote sensing satellites and aerial photos, vegetation, water, geology etc. can be mapped. The scales of these maps can be from entire continents to details of a few square metres. Satellites can detect the events that occur before many disasters like floods, cyclones etc. By remote sensing, the event can be monitored while it is occurring in full force. Thus, because of the positioning of the satellites in the space we are able to understand the disaster in detail and plan management methods.

After the disaster also, remote sensing is of great help. It gives a clear picture of the damages and the area affected. From this, the quantum of relief needed can be assessed. After that, the new situation can be mapped and the data stored. Based on this data, reconstruction can be done. Finally, the whole procedure can help in preventing the repetition of such disasters in future.

Geographic Information System (*GIS*): This is a very good tool in predicting and monitoring environmental disasters. GIS is a system for creating, storing, analyzing and managing geographical information. It can be used to analyze data from the space (spatial data) and non-spatial data also. The processed data can be used at different stages of disaster management. With the help of GIS, plans can be made about the emergency measures to be taken during natural disasters. Maps can be made of areas that are prone to earthquakes.

In any country, information on environmental disasters must be regularly included in planning any development project and making plans for various types of investment. Projects on development and investment should include a cost/benefit analysis, i.e., the cost of installing management measures should be compared with the losses that would occur if these measures were not introduced.

Remote sensing and GIS can be used at many levels:

A. *National level*: At the national level, the government as well as the public can get useful information on many aspects of disaster management with the help of GIS. This will guide governments in formulating policies and creating awareness among the public. This is done by two methods:

(a) By making a map of all disaster-prone zones of the country

(b) By providing a list of all previous disasters

This information can be used for many purposes:

- Developing disaster-free regions
- Avoid most of the development in zones of high disaster
- Take measures to minimise losses in disaster-prone zones that have been already developed
- Investigate reasons for occurrence of disasters in different parts of the country etc.

B. *Regional level*: GIS can be made use of at the regional level by urban and rural planners before giving final shape to various developmental activities. For example, engineers can decide the location of dams, the height of buildings to be constructed etc. Information on the firmness of the ground, nature of rivers, weather patterns of the area and the like will be of great help in this.

C. *Medium level*: At inter-district or inter-municipal level, extending over a few hundred square kilometres, studies on how far the planned project is suitable to the geography of the place and the chance of various environmental disasters are most essential. These should be completed before starting the project. GIS analysis of intensity of possible disasters in the area is of great use in such situations.

D. *Local level*: A single district or municipality is the target area at the local level. Here, GIS is used for deciding the suitability of various developmental works. However, more than this, it is used to prepare local risk maps for existing settlements and cities and to plan how to be prepared for disasters and how to carry out relief measures. The data will include results of field tests of materials and field measurements. Mapping will be accurate to one hectare or less. This is of great help at a local level.

Thus remote sensing and GIS have become very valuable tools in all stages of disaster management. Before the disaster, there are 2 phases:

1. Disaster prevention
2. Disaster preparedness

After the disaster, work has to be done in three phases:

1. Relief
2. Rehabilitation
3. Reconstruction

In both the phases, i.e. before and after the disaster, and also during the occurrence of the disaster, as explained earlier, remote sensing and GIS are very useful:

3. To educate the people on how to manage disasters: Disasters occur on a very large scale. Naturally, most people will not know how to tackle them. Many of them will be so shocked by the destruction all around that they will be at a complete loss as to what is to be done to relieve their own as well as others' sufferings. In such a situation, there is urgent need of help from trained persons. Specialists in disaster management should be rushed to the site of the event.

However, the real process of educating the people starts much before the disaster strikes. People must always be prepared for any disaster, especially those living in risk zones. In addition to the topic taught in schools and colleges, this can be achieved by creating awareness among the general public. For this, the media must be employed, like newspapers, television, radio, documentaries etc. People must be educated on the need to avoid risk zones for developmental activities etc.

In general, it is the duty of the government to warn the people about the approaching disasters. Therefore, scientists and research workers must inform the concerned government departments all their findings promptly. The government can then take steps in the following directions:

(a) Making decision makers and the public aware of the gravity of the disasters
(b) Training the decision makers in handling the situation
(c) Informing the people about the disaster well in advance
(d) Providing the concerned people with the required maps
(e) Explaining to the people how to reduce damages, like constructing quake proof houses etc.

4. To do research on improving the management methods available at present and find out better methods of management: Many organizations in the world, like the International Council of Scientific Unions (ICSU), are doing research programmes in disaster management. This organization studies man-induced environmental changes, the origin of natural disasters etc. It has founded many subcommittees and established research programmes. Some of these are:

- The Scientific Committee on Problems of the Environment (SCOPE)
- The Global Change Programme (GCP)
- The International Geosphere-Biosphere Programme (IGBP) etc.

Social scientists have also started research programmes like the various Human Dimension of Global Change (HDGC). The UN has also launched such research programmes with the aim of improving early warning systems, relief and rehabilitation in different countries.

Research is done in many fields in disaster management. Some of these studies are:

- The factors that cause the disasters
- The mechanism of action
- Potential of the disasters
- Mapping of land and water
- Population density
- Population distribution
- Systems of transport

- Methods of communication
- Facilities available in clean water and food
- Break out of epidemics
- Best methods of administration
- Role of police and army
- Rehabilitation methods
- Medical care including treatment of trauma etc.

In conclusion, disaster management is a multidisciplinary area. It involves forecast, warning, evacuation, rescue, medical care, relief, reconstruction, rehabilitation etc. It also involves persons from various fields like administrators, scientists, decision makers, volunteers, doctors, nurses, engineers, financiers etc. in addition to NGO's, communities, charitable organizations and the like. When so many agents are working together, there is every possibility of confusion and repetition of work. Therefore, coordination among all the agencies is a must for any successful disaster management/mitigation programme. Disaster mitigation is promoted all over the world these days. Disaster mitigation means lessening the negative effects of the natural disasters.

We have governmental organizations like the Indian Space Research Organization (ISRO), the National Building Research Organization (NBRO) etc., which are doing commendable research and development. We also have the Indian Meteorological Department (IMD), the National Disaster Management Centre (NDMC) etc. as governmental institutional machineries engaged in useful work.

Yet, in general, in India, we have to learn a lot from our past experiences. It will not help if different agencies compete with each other or conduct parallel assessments after the disaster. The efforts have to be cohesive for best results.

[Note: Disaster management discussed in this book is common for all disasters. Therefore, while preparing long answers individually, after describing each disaster, the same disaster management practices are to be written, leaving out the parts specific to other disasters, if present. Time available to write the answer should be kept in mind.]

QUESTIONS

A. Very short answer type:

(a) Multiple choice type: Tick the right answer

1. Radiation is a health hazard because it leads to:

   (a) Typhoid (b) Cancer

   (c) Colour blindness (d) Pneumonia

2. Which of these gases causes air pollution?

   (a) Nitrogen (b) Hydrogen

   (c) Water vapour (d) Carbon monoxide

3. Kitchen wastes contain:
   (a) Non-biodegradable pollutants
   (b) Biodegradable pollutants
   (c) Radioactive pollutants
   (d) None of the above
4. Water pollution is caused by:
   (a) Sewage (b) Industrial effluents
   (c) Discharges from farms (d) All of the above
5. DDT pollutes:
   (a) Air (b) Water
   (c) Soil (d) All of the above
6. Which of the following is not a biodegradable pollutant?
   (a) Plastic (b) Skins of vegetables and fruits
   (c) Dry leaves (d) Paper
7. Minamata disease is caused by:
   (a) Cadmium (b) Strontium
   (c) Mercury (d) Uranium
8. BOD stands for:
   (a) Biochemical Oxygen Demand
   (b) Biological Oxygen Demand
   (c) Biogeochemical Oxygen Demand
   (d) None of the above
9. Greenhouse effect is caused by excess of
   (a) CO_2 (b) H_2
   (c) He (d) O_2
10. Photochemical smog is caused by:
    (a) PAN + O_2 (b) PAN + O_3
    (c) $O_3 + O_2$ (d) PAN + CO_2
11. Which of the following is not a greenhouse gas?
    (a) CO_2 (b) CH_4
    (c) CFC (d) H_2
12. In plants, PAN blocks:
    (a) Reactions of respiration
    (b) Reactions of hormone formation
    (c) Absorption of water from soil
    (d) Hill reactions of photosynthesis
13. Ionizing radiations of ecological importance are:
    (a) α rays (b) β rays
    (c) γ rays (d) All of the above

14. The proportion of methane in biogas is:
    (a) 20 to 30 % (b) 10 to 30%
    (c) 60 to 70% (d) 80 to 90%
15. Which is not a natural disaster?
    (a) Cyclone (b) Nuclear explosion
    (c) Earthquake (d) Volcano
16. Landslides are caused by:
    (a) Earthquakes (b) Dam building
    (c) Mining (d) All of the above

(b) Define the following:

1. Pollution
2. Pollutant
3. Landslide
4. Remote sensing
5. GIS
6. Radioactive isotopes
7. Incineration
8. Vermicomposting
9. Vermitechnology
10. Vermi wash
11. Earthquake
12. Seismometer
13. Flood
14. Environmental disasters
15. Tropical cyclones

(c) Name the following:

1. Two major components of air
2. Five examples of pollutants
3. Five examples of non-biodegradable pollutants
4. The three main types of pollution based on the part of environment being polluted
5. Five types of pollution based on the type of pollutant
6. Three natural pollutants
7. Three man-made pollutants
8. The five primary pollutants in the air
9. Three harmful gases emitted from vehicles
10. Five greenhouse gases
11. Three water-borne diseases
12. Five industries polluting rivers
13. Three types of ionizing radiations of ecological importance
14. Five solid wastes
15. Three hazardous wastes
16. Two disasters which are natural and also man-made
17. Three most flood-prone rivers in India
18. Three man-made causes of floods
19. Two natural causes of floods

20. Two factors which determine the degree of destruction of earthquakes
21. Two causes of earthquakes
22. Five benefits of vermicompost

(d) Write very short notes on the following, in 3 or 4 sentences each:

1. Biodegradable and non-biodegradable pollutants
2. Three ways in which cost of pollution can be measured
3. Methods of educating the public regarding the ill effects of pollution
4. Primary pollutants in air
5. Greenhouse effect and global warming
6. Effects of carbon monoxide on human health
7. Methods of control of air pollution by industries
8. Photochemical smog
9. Methods of control of air pollution by automobiles
10. Water bloom
11. Bioaccumulation
12. Biomagnification
13. Thermal pollution
14. Natural process of recovery in rivers
15. Soil pollution by industries and automobiles
16. Importance of acting on earthquake warnings
17. Concentration of radionuclides along food chain
18. Advantages of recycling
19. Biogas
20. Source reduction and segregation of waste
21. Causes of landslides
22. Recycle, recovery and reuse
23. Minamata disease
24. Hazardous wastes
25. Richter scale
26. Hypocentre and epicentre

B. Short answer type:

Write short notes on:

1. Harmful effects of environmental pollution
2. Methods that can be adapted to reduce existing pollution
3. Sewage pollution
4. Smog and photochemical smog
5. Sewage treatment plants

6. Nitrate pollution
7. Methods to control water pollution
8. Educating people on how to manage disaster
9. Herbicide pollution
10. Harmful effects of noise pollution
11. Control of noise pollution
12. Bombing of Hiroshima and Nagasaki
13. Chernobyl disaster
14. Resource recovery of solid wastes
15. Biogas and its advantages
16. E-wastes
17. Flood control measures
18. Disastrous effects of earthquakes
19. Characteristics of cyclones
20. Applications of remote sensing
21. Fields of research in disaster management

C. Long answer type:

(a) Give an account of industrial and automobile pollution.

(b) Describe in detail air pollution by smoke and smog.

(c) Give a detailed account of water pollution, its harmful effects and control measures.

(d) Which are the methods by which soil pollution takes place? How can soil pollution be minimised?

(e) Describe in detail noise pollution, its effects and control measures.

(f) Write an essay on nuclear hazards.

(g) What do you understand by solid waste management? Elaborate.

(h) Write long notes on any two:

1. Marine pollution
2. Soil pollution due to wrong agricultural practices
3. Harmful effects of pesticides
4. Thermal pollution and its control

(i) Give an account of the major steps in urban and industrial solid waste management.

(j) What is vermicomposting? Explain in detail.

(k) Give a detailed account of role of individuals in prevention of pollution.

(l) Write long notes on any two:

1. Solid waste management in industry
2. Hazardous wastes
3. Earthquakes and their effects

(m) Write an essay on floods, their control and management.

(n) Give an account of tropical cyclones.

(o) Describe in detail the four aspects of management of environmental disasters.

(p) Write long notes on any two:

1. Providing immediate relief measures after disasters
2. Prediction and forewarning of disasters
3. GIS and its usefulness at various levels

Unit 6

Social Issues and the Environment

6.1 INTRODUCTION

It is said, "Man is a social animal". This is a true statement. We have no normal existence separated from society. Therefore, all types of social issues are important for, our well-being. Many of these issues are closely connected to the environment in which we live. Thus, any study of Environmental Science will be complete only if the relation between the environment and social issues is well understood. Some of the most important of these are included in this unit.

6.2 FROM UNSUSTAINABLE TO SUSTAINABLE DEVELOPMENT

The belief that economic development is the only desirable development is wrong. This is clear from the fact that if the economically developed countries do not pay heed to sustainable development, the environment will become impossible to live in. Only economic improvement is not enough, people must have a clean environment to live in. The more developed countries (MDC's) realized this fact by the early 70's. They understood that many of the problems they were facing were caused by indifference to environmental issues. Many of these problems are created by unsustainable developmental processes like:

- Cutting of trees in the forest without replacing them
- Monoculture (planting the same crop year after year in the same area)
- Excess use of artificial fertilizers
- Excess use of toxic pesticides, fungicides, herbicides etc.
- Overgrazing
- Continuous use of hybrid seeds at the cost of traditional seeds
- Over-exploitation of ground water without recharging it
- Over-irrigating croplands

- Over-exploitation of fish and other aquatic animals
- Draining marshes for constructing buildings
- Construction of major dams
- Poaching wild animals from forests
- Illegal trading of wild animals and plants
- Mining
- Shifting cultivation
- Jhum cultivation
- Sand mining from river beds
- Polluting air, water and soil by different means etc.

From this it is clear that any development during which natural resources get depleted faster than their renewal is unsustainable development. This leads to environmental degradation, global climate changes, increase in the occurrence of various diseases etc. To become more industrialized and as a result more economically superior, the MDC's like America, Canada and Western European countries used up a large quantity of their natural resources and exploited the less developed countries (LDC's) in Africa South and South East Asia. The result was that the former became richer while the latter became poorer.

The same happened in different areas of the LDC's. While some areas showed great economic advancement, some others remained poor and undeveloped. This type of discrepancy and inequality in growth led to great discontentment and ill will.

To prevent all these from happening, it is essential to follow sustainable development and discard unsustainable development. Sustainable development can be defined as the development that fulfills the needs of the present generation without harming in any way the needs of the future generations. Such a development will also remove all types of discrimination between:

- Continents
- Countries
- Religions
- Castes
- Classes
- Races
- Genders
- Age groups etc.

All will be equal, as it should be. However, this is practically impossible. Yet we can and should try to reduce the existing gaps between the extremes in the various categories. This is definitely possible by following the principles of sustainable development.

We have to pay a price for development. Thus though industrialization, urbanization etc. are beneficial to mankind, they are not without pitfalls. Social and psychological impacts set aside, developmental activities upset the environment as a whole. The entire biosphere is affected by these. Disturbances in any one part of the biosphere are conveyed to the other parts and cause grave consequences, most importantly, depletion of natural resources. This shows that development should not be at the cost of natural resources. We must develop, but we must see to it that we do not disturb the nature, of which we are essentially a part. This in short is what is meant by sustainable development. It is all about conserving our natural resources along with developmental processes.

India has only 2.4% of the world's land area, but has 16% of the world's population. It is therefore obvious that there is a heavy pressure on the natural resources including land in India. This is so in spite of the fact that an average Indian consumes a much lower amount of resources than an average person in the developed world.

It has been realized over the past few years that technology alone cannot solve our problems. We must combine the best of old practices with the benefits that modern science and technology offer. It is meaningless to imitate the western world. In fact, they themselves have started realizing the follies they have committed. They have become more conscious about the environment now. What we must learn from their experience is that we should not try to be a developed country while ignoring our environment. It is true that right to develop is a fundamental human right. Yet, we must never follow unsustainable patterns of production and consumption. True development is environmentally sound sustainable development. It is not the development followed by the over-consuming wasteful west. We must give utmost importance to the ecosystem balance, in other words, nature itself, while withdrawing resources from the nature, processing them and using the products. For this, a holistic approach to development is a must.

It was in this spirit of sustainable development that India participated actively in the Earth Summit held at Rio in Brazil in 1992. All participating countries realized that protecting and restoring the biosphere to its original state is the responsibility of all countries. The third world felt that since most of the environmental damage has been done by the developed countries, they should bear the cost of the damage control too. It was calculated that $625 billion would be needed to correct the damage done till then. The developed world, led by U.S.A., talked of common responsibility, but it is a fact that the developing world cannot afford to pay an equal share. The words 'differential responsibilities' were included in the document. This was a great achievement for the developing nations.

The four crucial documents adopted by the Earth Summit were:

1. Rio Declaration
2. Conventions for climate change and biodiversity
3. Conventions for forestry
4. Agenda-21, i.e., the Action Plan

Conservation of nature has been ingrained in the minds of Indians for thousands of years now. For example, the earliest reference to forest conservation is seen in the classical Indian epic, *Mahabharatha*. We have worshipped nature as Mother Earth for years. We have given divine status to the sun, the moon and the planets, mountains and rivers and to various plants, animals and birds. The aura of divinity that we have conferred on them prevents us from causing them harm. As a result of this, the ecosystems here were always in balance in the earlier years. Destruction, in the form of deforestation, began first during the British rule. But it is most unfortunate that now our own greedy contractors and poachers are looting our forest wealth. It is high time all this is stopped.

Silviculture is cultivation of trees and forest regeneration. The basic feature of all forest conservation measures is the sustained (continued) yield concept. Using forest produce for making different articles of human use and at the same time ensuring forest regeneration is an example for sustained yield concept. The articles of use may be small like toothpicks, matchboxes or pencils or huge like carts, telephone poles, boats or ships.

Afforestation, reforestation and forest conservation must be done on a very large scale in India. Social forestry is a very good method. It has two aims:

1. The use of public and common land to produce firewood, fodder and small timber for the local poor and thereby achieve soil and water conservation.
2. To relieve the pressure on forest conservation.

The programme involves villages and school children. In India, the area under social forestry has increased substantially in the past few years. Van Mahotsav and other such local programmes and the work of NGO's have helped in this.

Agro forestry is a modification of social forestry. Here, trees are planted in suitable locations on agricultural land. This practice along with integrated animal farming has been an ancient practice in India. Now the activity has been given the name of agro forestry and is being popularized.

Plan of Action to Check Deforestation:

1. Land degraded due to monoculture must be converted into forest.
2. Special species of trees should be planted on saline, alkaline and water-logged soils.
3. Trees must be planted on roadsides, near railwaylines, canal banks, farm boundaries, around villages, school yards, public buildings, market squares etc.
4. Fuel efficient stoves must be developed and alternative energy sources like solar heat and biogas must be popularized.

The main natural resources are minerals, soil, water, forests etc. During developmental processes in a country, degradation and depletion of these will of course take place due to consumption, but these must be immediately replaced. As far as possible, more should be given back to nature than is consumed, because we have already caused immense harm to our environment due to our activities over the past centuries. Thus it has become an added responsibility of each citizen on the globe not only not to disturb nature any further, but also compensate for past mistakes. This is a real sustainable development and real planning for the future.

6.3 URBAN PROBLEMS RELATED TO ENERGY

Some of the most important minerals such as petroleum and coal are used as fuel for power. These and natural gases are fossil fuels. These form a major chunk of the fuel energy used all over the world. These resources evolved millions of years ago. Since the beginning of the machine age, industrial societies have become increasingly dependent on fossil fuels. Most of the industries are located in or near cities and towns. Most of the automobiles consuming petrol or diesel also ply in urban areas. Thus, problems related to energy are faced acutely in the urban areas. This is the cause of concern. There are two main reasons for the concern regarding coal and petrol:

1. Fossil fuel resource (coal and petroleum) is limited in amount. These were produced by solar energy in the distant past, and when they are gone, there will be no more. Fossils are still being formed, but at a very, very negligible rate compared to the rate at which we are consuming them. We first began consuming fossil fuels at an appreciable rate only about 150 years ago. How long will they last? On a global scale, we will still have some coal for a few centuries more, but natural gas and oil will be in short supply in a few decades.

   In the United States, the situation is worse than the global average because they are depleting their resources at a faster rate than other fossil fuel rich countries. In fact, the entire stock of fossil fuel available has been held in storage under the earth's surface for more than a hundred million years, and now it is being completely exploited in only a few centuries.
2. Unexpected environmental consequences result from the extensive use of fossil fuels for everything from cooking in our kitchens to powering our automobiles. When we burn coal, natural gas or oil to obtain energy, gaseous compounds are formed and dumped into the atmosphere. This is causing problems. For many years, it was felt that the emitted gases were not significant as the earth's atmosphere is very vast. But now with increasing world population and industrialization, this is no longer true. The atmospheric pollution is causing health problems and even death. Carbon dioxide emissions from industries and automobiles in the urban areas are threatening to produce climatic changes over the entire globe.

Thus there is a great threat to long term sustainability around the world if proper corrective measures are not taken immediately. In both developing and developed countries, extraction, distribution and use of energy is faulty in urban situations. There is an urgent need to shift to non-polluting renewable forms of energy, which are low in carbon content.

A fact to be remembered is that many of the activities for environmental correction themselves consume energy—we need energy to clean water, to manage solid wastes, to control pollution etc. Thus, saving energy at other levels will be helpful in continuing these activities. A movement to propagate the need to use clean energy is thus very essential.

There are three factors that will provide solutions to urban energy problems. These are:

Equity: Equitable distribution of resources is a burning social issue. Whatever the resource may be, it is not distributed equally. This is true of individuals, societies, localities and nations. In the case of energy also, distribution is not equitable. Research must be done to find out ways of distributing all forms of renewable as well as non-renewable energy on an equitable basis, and the findings of these research works must be implemented.

Efficiency: The present manner of the use of energy is defective. There is much misuse of energy, especially in urban areas. Waste of large quantities of electricity in cities across the world goes on uninhibited. In India and many other developing countries, transmission losses are also great. Even theft goes unpunished. This must be stopped. Efficient usage of energy must be promoted worldwide.

Sustainability: The effect of the present levels of energy consumption in the urban areas and its impact on long time sustainability in the particular area must be calculated. This is essential for planning sustainable development locally as well as globally.

In short, the following plan of action must be implemented to solve urban energy problems:

- Reduce the use of fossil fuels
- Find out suitable forms of non-conventional energy and start using them
- Create awareness among the citizens that local energy issues are linked to climate change, global warming etc.
- Bring about changes in lifestyles so that energy is saved as much as possible (details in 'Role of individuals in prevention of pollution' in Unit 5)
- Try to involve whole communities in all programmes right from the planning stage till the implementation stage and follow up actions.

Cities, especially in Asia, are growing rapidly. Consumption has increased manifold. This is leading to deterioration of the environment. Lifestyles have become lavish and wasteful. This is extremely harmful to the environment. One of the biggest problems in cities is that people have started to depend more and more on automobiles. In the USA, per capita use of energy for transportation is very high. On the other hand, there are many ideal cities like Singapore, which have taken all necessary steps to prevent automobile pollution and at the same time, grow economically. Energy used for transportation here is much less compared to cities of the USA. This is done by maintaining high standards of public transportation in urban centres.

About one-third of the total energy consumed in the world is in the urban areas, on transportation. American cities spend five times, while European cities spend only one and a half times energy than Asian cities. This is because of the much higher use of private automobiles in America.

The major energy, environmental and transportation issues in South Asian cities include increasing energy use, CO_2 emissions and localized air pollutant emissions along with rapid motorization, growing traffic congestion, and increasingly overburdened public transportation systems.

Two- and three-wheelers, the majority of which run on two-stroke engines, dominate south Asian cities. In recent years, some efforts have been made in cities such as Dhaka, Kathmandu and a few others to phase out two-stroke two- and three-wheelers. In the quest for cleaner air for citizens, the judicial system has intervened in ordering the use of CNG in public transportation in New Delhi. In Dhaka, there is a plan at present to convert all 3-wheelers into the CNG type. In Kathmandu, battery operated three-wheelers have replaced smoke belching diesel three-wheelers for commercial operations. These changes represent some transition towards cleaner vehicles and fuel.

The improvements in public transportation systems are progressing at a slower pace in general. At present when private vehicles on the road are increasing greatly in number, public transportation is still inefficient. It does not cover many areas. Road conditions are not good either. Though vehicle emissions standards and regulations are being followed for new vehicles, the same is not true for old vehicles, which are many in number too. Old vehicles perform poorly in terms of energy use and emissions, yet they are not being phased out from the streets. In many major cities, the local air pollution situation has worsened greatly. The automobile exhausts are seen clearly as black smoke in many places due to poor traffic management.

Cities are thus facing the double task of economic development and environmental protection. In fact, policies to reduce energy consumption in cities and towns will bring multiple benefits to the people. It will help solve air pollution and traffic congestion problems, and will also reduce global warming in the process.

Thus, urban environmental problems from energy use are strongly tied to urban management. When environmental problems occur, most policy-makers try to solve immediate issues without paying enough attention to the real issues. This is true all over the world, but in South Asia it is more evident as policy implementation is either weak or does not exist. At times, there is no clear policy at all. Therefore, there is a wide variation in environmental problems between cities with similar income level.

Energy plays an important role in bringing about sustainability in development, since extraction, production and consumption of energy resources are the primary cause of local as well as global environmental problems such as air pollution and the emission of greenhouse gases. Since economic growth of cities is important for the overall growth of the country, policy-makers should pay more attention to urban problems related to energy, especially in the transport sector.

6.3.1 Conservation of Energy

As the fossil fuels that are commonly burnt for energy are limited, we have to depend on alternative energy sources like:

- Solar energy
- Wind energy
- Hydroelectric power
- Tidal energy
- Geothermal energy
- Nuclear energy
- Biomass from municipal waste and agro-waste
- Hydrogen power etc.

We will have to shift to these energy resources sooner than later. The added advantage is that they all are non-polluting.

The solution is that we should turn to alternative energy sources as soon as possible and as much as possible. The main forms of alternative sources of energy are solar and wind energy, biogas from municipal solid waste and agro-waste etc. These are far less polluting than conventional sources. They also help in managing the solid wastes in a profitable way. Preventing pollution is the main aspect of sustainable development.

We must improve our agricultural practices so that precious land is not degraded. For this, crop rotation, use of organic manure and biopesticides, drip irrigation, use of traditional seeds etc. must be encouraged.

Mining sand from river beds should be stopped as it leads to rivers becoming shallow leading to frequent floods. To prevent floods, it is best to have water conserving structures in various places so that not much water flows into the rivers in the first place.

Urbanization is taking place at a very fast pace in many parts of India. This is part of development, no doubt. But, unplanned development is harmful. For example, rice fields conserve precious water and raise the groundwater levels. Permission should not be given to level them for making buildings.

Rainwater harvesting was done in ancient times in many countries, including India by different methods. It is high time we follow these traditional methods and also implement some new methods along with. By practising rooftop rain-water harvesting, we are solving our own water problem. At the same time, we are contributing towards solving a national problem of gigantic proportions. We are also taking part in the global effort towards attaining environment friendly sustainable development in a local way. Localities make a nation and nations make the world. If development goes hand in hand with maintaining the balance of natural resources in all countries, the whole world will show sustainable development.

Educating the common populace about environment is most essential for popularizing the concept of sustainable development, especially in the developing (less developed) countries like India.

6.4 WATER CONSERVATION

'Conservation' in Latin means 'Guard together' (*con* = together; *servare* = guard). Conservation does not mean hoarding or even rationed use of resources. Conservation is a constant effort to increase resources, while at the same time utilizing them for our needs. Conservation of all natural resources is important for us and also for the future generations.

Conservation of Water Bodies: Both freshwater and marine water bodies are important to us and the aquatic plants and animals in various ways. Freshwater bodies are misused more and more by the growing population. Their conservation means managing them properly while at the same time maintaining their high productivity levels. For this, knowledge of the following aspects is necessary:

- Drainage pattern (pattern of water flow from the highest point of the area to the water body)
- Runoff (the water flowing into the water body along slopes)
- Mineral receipts (the types and quantities of minerals carried by runoff water)
- Aquatic life (plants, animals and microorganisms living in the water body)
- Carrying capacity of pollutants (the maximum amount of pollutants the water body can carry without causing harmful effects) etc.

Dams are the largest structures for water conservation. However, it is better to construct small dams in suitable places than to construct 1 or 2 big dams. All over the world, this fact has been recognized now, though India is yet to wake up to the truth that major dams lead to flooding of large areas, deforestation of vast tracts, loss of precious biodiversity, problems of rehabilitation of the displaced rural poor etc.

Interlinking of rivers has been suggested by some in India, but this will be an environmental disaster. It is extremely costly, and will cost lakhs of crores of rupees. Different rivers flow at different levels and it will be difficult to join them. Conflicts between states and also between India with Nepal and Bangladesh will increase. It will also lead to salination, water pollution, sedimentation and waterlogging. Moreover, it will be dangerous to mix waters of different qualities and temperatures as many aquatic organisms will perish.

Another major problem will be displacement of local people. However, interlinking at local levels can be done without grave consequences.

Thus conservation methods adopted must be meaningful. Some such measures are:

- Rainwater harvesting
- Watershed management
- Recycling and reuse of water
- Stopping wastage of water in rural and urban areas
- Adopting drip irrigation in croplands etc.

Water was worshipped by all civilizations in the past as a divine gift of nature. As early as the 3rd millennium B.C., farmers in Baluchistan used stored rainwater for irrigation. Dams built of stone have been found in Baluchistan and Kutch. References to canals and embankments are found in the Vedas, Puranas, *Mahabharata*, Budhist and Jain works and archaecological remains. Inscriptions in many south Indian temples prove the existence of water harvesting systems. Wells were probably a Harappan invention. Every third house in the Indus Valley Civilization had a well.

As society progressed, water was brought to homes by pipes. And then people gradually forgot the importance of water. Today's water shortage is due to different reasons.

Rising population
Urbanization
Industrialization
Deforestation
Over-exploitation
Lack of proper water management policies
Drying of wetlands
Neglecting excellent traditional water conservation methods
Overgrazing and desertification
Siltation of water bodies due to increased soil erosion
Wastage and lack of reuse
Pollution etc.

From this list, it is clear that scarcity of water is mainly a man-made problem. Therefore, it is for us to take corrective measures. If we do not do so now, it may be too late. Thus, there is an urgent need to practise as many water conservation methods as possible.

Methods of Conservation of Water

There are many methods of water conservation:

Increase the efficiency of irrigation
Protect watersheds and wetlands

Afforestation, reforestation and forest conservation

Preventing losses in:

- Industries
- Leaking in dams and reservoirs
- Leaking in municipal pipelines
- Households

Reusing of waste water by setting up waste water treatment plants

Constructing many small ponds, tanks etc. in low lying areas

Building many check dams at appropriate places in all rivers

Provide drainage facilities in canals and use drainage water for irrigation

Rainwater harvesting in rural and urban areas

Consider water as an economic good and levy reasonable charges for it to discourage wastage

6.4.1 Rainwater Harvesting

Rain is the primary form of water in the hydrological (water) cycle. Rivers, lakes, groundwater are all secondary. However, forgetting the primary source, we depend wholly on the secondary sources. This trend must be changed. The potential of rain is tremendous. It is most unwise to let rainwater go waste and then cry hoarse over water scarcity.

(a) *Water harvesting in villages*: Some of the common water harvesting systems used in villages are as follows.

Tankas: These are common in arid zones. Tankas have been constructed for centuries in Indian households, near temples, schools and public places. Such structures have been reported from many countries like Botswana, Ghana, Yemen, Kenya, Thailand, Indonesia, Sri Lanka etc. A Tanka is a covered underground cistern, a hole 3 to 4 meters in diameter, with plastered sides and a lid of thorns or dried grass. In the Barmer District of Rajasthan alone, more than 10,000 tankas are functioning successfully even now.

Nadis: These are excavated or embanked village ponds, 2 to 12 meters deep. Different linings are used to reduce seepage. To reduce evaporation, rubber or polythene sheets or bamboos are used.

Check dams: These are small dams built with stones or earth and boulders across rivers or streams. Their capacity is 1000 to 50,000 cu.m.

Subsurface water harvesting: This is usually done in riverbeds and is common in India, Kenya, Tanzania etc. The riverbed must be of sand. Bunds are made below the riverbeds. They recharge all the surrounding wells with a large quantity of water.

Floodplain reservoirs: These are ponds dug in the floodplains. They get filled with water during floods and remain full after the flood is over.

Quarries: These are pits left after stones are cut or bricks are made. These are used as storages.

Baolis: These are stepwells. Baolis were very common in Delhi during the Mughal and other periods.

Water harvesting in urban areas: Water harvesting systems are gradually becoming popular in urban areas. The common ones are:

On channel storage: During the rainy season, storm water flows very fast along the sides of the roads and other paths. This water can be stored in the channel itself by constructing regulators at suitable places. Deepening and widening the channels wherever possible will increase the storage capacity.

Off channel storage: Channels cannot store more than 20% of water flowing through them as their capacity is not much. Therefore off channel storage systems (reservoirs away from the channels) can be made, wherever possible.

Lakes and Ponds: In cities, lakes and ponds are misused for dumping solid wastes and sewage disposal. Desilt and widen them.

Rooftop Rainwater Harvesting: Rainwater falling on roofs can be collected and stored for direct use or can be recharged into underground aquifers.

1. *Storage for direct use:* Rainwater is directed from rooftops to storage tanks through drain pipes after filtration. Sand filter must be used and chlorination done if water is to be used for drinking.

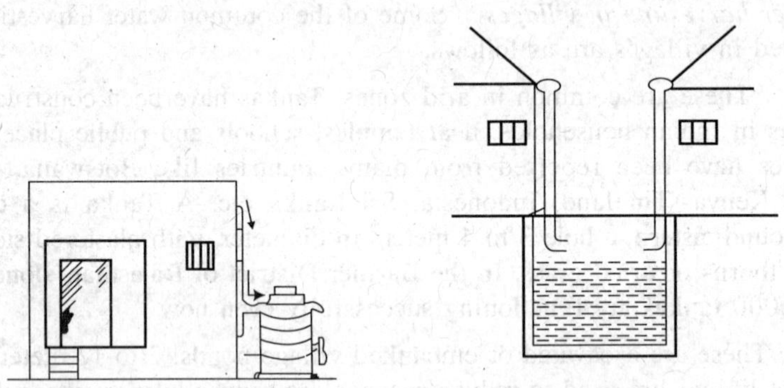

Storage of rainwater in overground containers

Storage of rainwater in underground tanks

Figure 6.1 Storage after simple filtration at the starting point of the drain pipe in overground and underground tanks

Whether the water should be used directly or must be guided underground depends on the rainfall of the place. In Kerala, Mizoram etc, rain falls almost throughout the year. In such places, small domestic sized water tanks can store water. But in dry zones like Delhi, Rajasthan, Gujarat etc, rain falls only for 2 – 3 months a year. To collect water for a whole year, huge containers may be needed and so it is better to recharge aquifers.

2. *Recharging underground aquifers:*

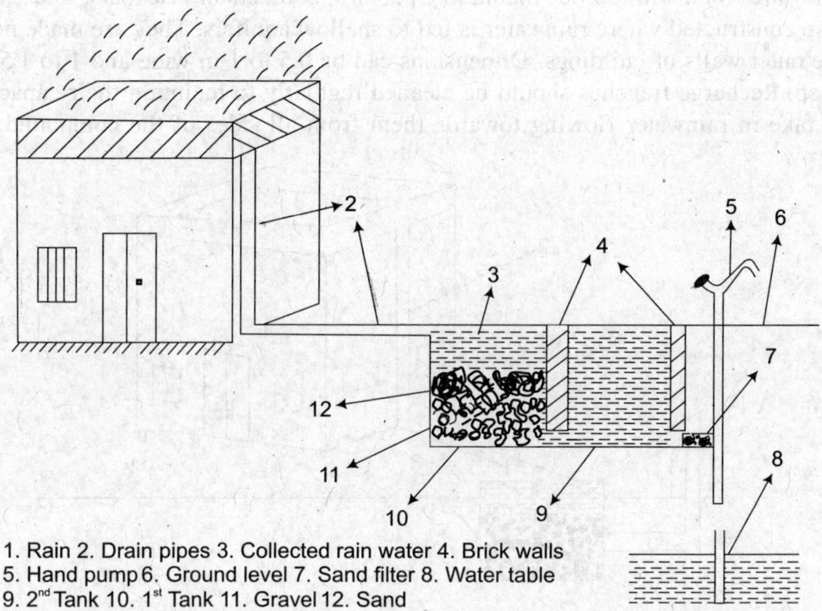

Figure 6.2 Basic plan of roof top rainwater harvesting

Artificial recharge of rainwater into aquifers: There are different methods to do this.

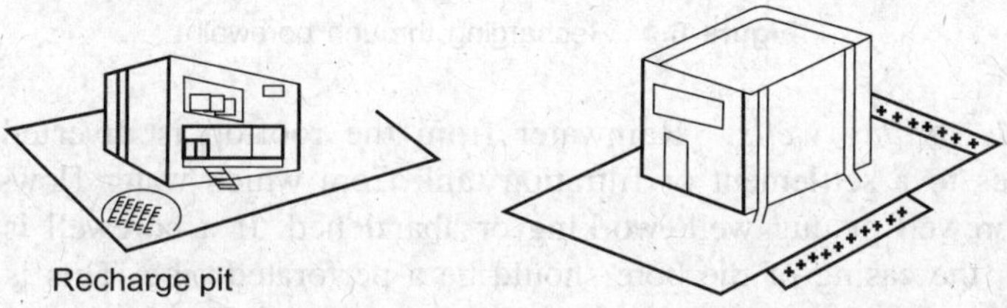

Figure 6.3 Artificial recharge of rainwater into aquifers

Recharge Pits: These are constructed for recharging shallow aquifers. They are excavated pits 1-3 meters wide and 2-3 meters deep. The pit is lined with brick or stone wall with openings called weep holes at regular intervals for sidewise infiltration of water. The top can be covered with a perforated lid. Filter material should be arranged in layer as follows: Pieces of bricks or stones at the bottom, gravel in between and coarse sand on top.

Recharge trenches: These are long continuous trenches dug in the ground. They are filled with porous media like pebbles, boulders or brickbats. These are also constructed where rainwater is fed to shallow aquifers. They are made near the outer walls of buildings. Dimensions can be 0.5 to 1 m wide and 1 to 1.5 m deep. Recharge trenches should be cleaned regularly to maintain their capacity to take in rainwater flowing towards them from all sides of the compound.

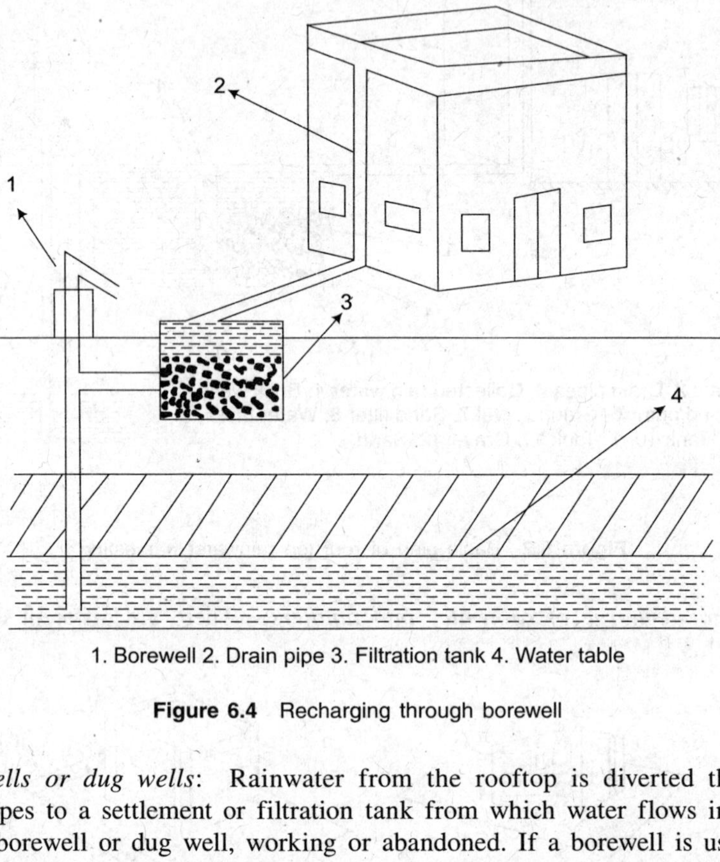

1. Borewell 2. Drain pipe 3. Filtration tank 4. Water table

Figure 6.4 Recharging through borewell

Borewells or dug wells: Rainwater from the rooftop is diverted through drainpipes to a settlement or filtration tank from which water flows into the well—borewell or dug well, working or abandoned. If a borewell is used for recharge, the casing of the bore should be a perforated pipe. This is for side-wise infiltration. Developing the borewell is beneficial. Developing is a process where water or air is forced into the well under pressure to make the soil around the bore loose. This helps to increase infiltration.

If a dug well is used, its lining should have weep holes at regular intervals. The bottom of the dug well should be desilted every year to maintain intake capacity.

Roofs of houses, flat complexes, public buildings like schools, shopping centres, hospitals, companies etc. can be used for rooftop rainwater harvesting. Many schools and some companies in India have installed such systems.

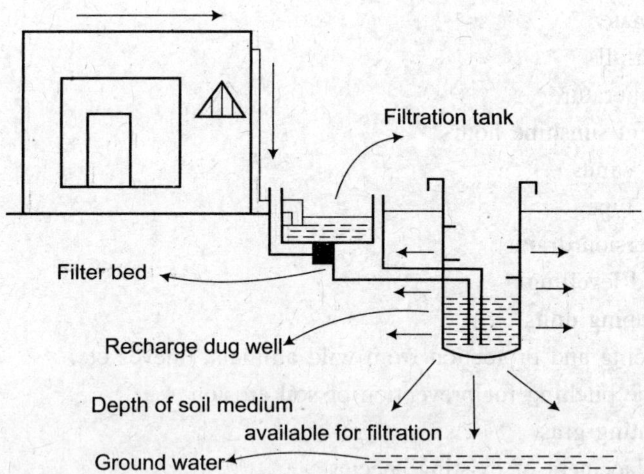

Figure 6.5 Recharging through dug well

Some states and districts have decided to make rainwater harvesting systems mandatory for all future buildings. This is a welcome sign. Community awareness and participation are most essential for any such programme to succeed.

6.4.2 Watershed Management

The deterioration of natural resources in any area can be prevented or at least restricted only by adopting the watershed approach. If a watershed is developed, resources can be properly maintained and developed. A watershed is a manageable unit. It is in fact a drainage basin. In this approach, development is done not only in the area of cultivation, but it covers the whole area, starting from the highest point of the area (ridge line) to the outlet of the *nala* or the natural stream. Thus watershed development includes implementing development works starting from even barren hill slopes, passing through marginal lands, privately owned agricultural lands and badly cut *nalas* to rivers.

Various resources can be developed by watershed management. In addition to water, other resources that are developed include fuel, fodder, livestock etc. Thus, this method is an intensive one and may be expensive in the beginning. However, economic evaluation has shown that long term benefits are many.

A watershed can be of two types. Watersheds, 0.5 lakh hectare or more in area are called macro-watersheds, while those up to 0.5 lakh hectares in area are called micro-watersheds.

For studying watershed management and finding out the ways and means to bring about maximal benefits to the local people and raise their standard of living, many details have to be collected. This can be done under seven sections.

The first section usually includes general information needed to start work on the watershed. These are:

- Climate
- Rainfall
- Temperature
- Bright sunshine hours
- Hot winds
- Soil type
- Diversion drains
- Land levelling
- Pumping units
- Fencing and protection from wild animals, thieves etc.
- Stone pitching for prevention of soil erosion
- Planting grass
- Replacement of existing species
- Effective closure to biotic interference.

In the second section, socio-economic issues are studied. These are:

- Natural vegetation
- Human population
- Cattle population
- Socio-economic conditions
- Land holdings by the local people
- Institutional financing facility
- Means of communication
- Availability of market for disposal of farm produce
- External influence on watershed farming by road building
- Agriculture and horticulture practised in the locality etc.

In section three, farming methods, present land use pattern etc. are observed. It is important to know the farming methods and present land use in detail because it is the local farmers who will be benefiting most from the watershed management measures. For this, information must be collected from the local farmers.

These include:

- Management practices
- Rainfed or irrigated
- Desilting of water bodies present
- Engineering structures (any dam, bridge etc)
- Replacing of uneconomical species by useful trees
- Availability of improved seeds
- Raising fuel and fodder species
- Use of modern implements

- Crop rotation
- Name of the crop
- Area of cultivation
- Preparatory tillage practices—time and method
- Ploughing—whether with tractors or bullocks
- Manures added after ploughing
- Seed availability
- Availability of seed treatment facilities
- Sowing methods—time, method
- Whether transplantation is done or not, if yes, spacing
- Use of manures and fertilizers—time, dose, method
- Plant protection measures—chemicals used and their dose, method of application.
- Average height and diameter of the plants
- Bearing time
- Harvest: time, method
- Yield—yield of primary (main) produce, yield of secondary produce, like hay, mushrooms on hay, compost from plant parts after harvest, fish from farm ponds, if any, farm animals like cattle etc.
- Present land use pattern (cropping single, double or multiple)
- Labour availability
- Market demand
- Total income in rupees per year

Socio-economic issues related to the local people are detailed in fourth section. These may include:

- Human population
- Cattle population
- Land holdings by local people
- Means of communication
- Availability of market for sale of farm produce
- External influence on watershed farming by road building etc.

Problems and needs of the watershed are studied in the fifty section. These include:

- Soil conservation
- Water conservation
- Irrigation
- Lack of electricity
- Lack of proper methods of communication
- Lack of roads nearby
- Biotic interference etc.

Finally, cost-benefit ratio is worked out in the sixth section. This is based on:

- Expenditure on watershed management
- Cost of cultivation
- Income from agriculture, including primary and secondary produce
- Evaluation of social benefits to local populace etc.

Any cost benefit ratio of 1:1.5 is considered good enough. However, with proper watershed management in the area, this ratio can be bettered manifold over.

The seventh and the last section comprises future planning. This section includes:

- Procedure for financing
- Phasing socio-economic benefits
- Future plan with regard to land use, introduction of improved seeds, manures and fertilisers, cultural operations, engineering measures etc.

Future plans could include activities aimed at:

1. Phasing of works:
   - Constructing tanks for drip irrigation by gravity
   - Making greenhouses
   - Buying modern equipment for processing/preservation
2. Financing:
   - Co-operative societies involving local people to get credit from banks
3. Project organisation and institutional structure:
   - Proper market survey
   - Improve shelf-life of produce
   - Community biogas plant
4. Social and economic benefits:
   - Establishing self-help groups
   - Employing more number of local people
   - Other benefits:
     1. Pisciculture
     2. Community policing
     3. Integrated farming
     4. Schools, dispensaries etc.

An ideal watershed development programme involves work in four phases. The first phase is the Recognition Phase, wherein the problems are recognised. The second phase is the Restoration Phase, wherein solutions are suggested. The local people are usually motivated enough by the implementers of the watershed development agency (usually the Government, sometimes any NGO) to initiate the third phase, i.e., the Protection Phase during which the development programmes are implemented successfully. The fourth and the final Improvement phase deals with the evaluation of the overall improvement achieved by the programme.

Thus, watershed development work can help the communities involved and contribute to real development. This is in addition to its capacity to conserve water in the most scientific manner in a short time as well as on a long term basis, solving the problem of water scarcity in an area permanently.

6.5 RESETTLEMENT AND REHABILITATION OF PEOPLE: ITS PROBLEMS AND CONCERNS

Resettlement and rehabilitation of displaced rural masses has always been neglected in many countries, while big projects like major dams are taken up. The land where the original settlers lived belongs to them. Without paying them adequate compensation, they should not be displaced. In fact, the best system would be finding them alternative arrangements and then begin the project. But this is never done. The human rights of the poor people are always sacrificed for the benefit of the powerful.

Many a times, the poor become poorer because of the activities of the rich, like construction of skyscrapers in areas occupied by the poor. Unsuccessful in getting the promised compensation, they are compelled to move to still poorer localities. The gap between the rich and the poor increases further.

Development projects such as irrigation, power, mining etc., contribute to the economic growth and well-being of humans. These projects create employment opportunities and trigger growth in agricultural, industrial and other such sectors. However, one serious drawback of these projects is that many people living in the area are displaced. The social cost of this displacement is not calculated properly. Also, the problem is not paid the attention due to it. In many projects, benefits are highlighted too much and the misery to the local people is not projected correctly. Proper resettlement and rehabilitation of these people has always been neglected in India. The media also does not play its role properly, because the benefits of the project are published everywhere while the condition of the displaced poor villagers, many of them tribal people, does not get the coverage it deserves.

Since 1950, 5 million people have been displaced by various development projects in India. These projects were mainly for dams or mines. Out of this, more than 40% are local tribal people living in remote parts of the forest. They have no access to even basic necessities like drinking water, electricity and primary healthcare. They depend on forest products for their livelihood. They live in isolation. They know only their own language. Their culture, customs and lifestyle are all unique. It is a crime to push these simple people out of the forests, snatching their livelihood from them.

Until 2003, no policy by the central government existed in India for the resettlement and rehabilitation of the displaced people affected by projects in irrigation, power, mining etc. Moreover, there was no common policy. Each case was dealt with by the state government separately. These were faulty. In most cases, the affected lived in abject poverty even after 25 years. While the constitution of India declares right to life loudly, project affected people are often forcibly displaced even before giving them any alternative site for settlement.

6.5.1 National Resettlement & Rehabilitation Policy, 2003

After years of working on a draft policy, Government of India released its National Resettlement and Rehabilitation Policy in 2003. But, the policy has a loophole. Firstly, there is no time-frame for implementation of rehabilitation. Therefore, the policy is not going to stop the present practice of starting project without paying attention to rehabilitation needs of displaced people. Secondly, the policy does not discourage projects that disturb tribals and their livelihoods, as done earlier by some state governments.

In Andhra Pradesh, proposed irrigation projects, such as Polavaram in East Godavari district, are sure to cause huge human displacement. Local tribes will be deprived of their shelter and means of livelihood. Submergence in tribal and forest areas will result in great ecological destruction and affect the cultural identity of ethnic tribes. In Polavaram alone, an estimated 2,50,000 people, of whom at least 50% are tribes, will be displaced from 300 villages.

As usual, this policy also recommends providing alternative land to the land lost by the Project Affected Families (PAFs) belonging to ST community. However, a random study of various projects implemented during 1990 to 2005 has shown that such provisions were never implemented in any irrigation project constructed during this period. Therefore, what really matters is the sincerity of the government.

Since Independence, many irrigation projects were constructed and several others are being constructed in India. Similarly, projects for many multipurpose dams, which aim at hydroelectric power in addition to water storage and irrigation, mines for various minerals etc., have also been cleared. Millions of people were displaced. It is high time that governments become sensitive to the displacement issue.

Though injustice done to displaced population since Independence cannot be undone, following specific steps will help in reducing human misery and at the same time achieving development in future:

- Discourage projects that seriously affect tribal population.
- Consider constructing check dams or minor dams in place of major dams. This will save lakhs of people from displacement.
- Give priority to projects that benefit the tribal communities. This will reduce disparities in development.
- Implement the rehabilitation package before starting work on the project.
- Consult the people facing displacement and get the consent from local Panchayat before planning any project.

- Recognize the sacrifice of the people displaced by the project by displaying their details at the project locations along with other usual details.

6.5.2 Case Studies

6.5.2.1 Sardar Sarovar Project (SSP) and Narmada Bachao Andolan (NBA)

Like all other developmental projects in the country, in the case of the SSP also, people directly involved with the project came to know about its impact only long after work had started to dam River Narmada. The illiterate disorganized farmers and tribals living in isolated villages across hills and forests knew about the dam only when trees began to be felled at the orders of the then chief minister of Gujarat. Soon, media attention was drawn to the large-scale deforestation going on in the area. The World Bank that had started funding the project came under heavy criticism. The outside world considered the forceful evictions of the local tribal people as human right violations. In 1991, the World Bank sent a 2 member team to India to study the rehabilitation aspect of SSP. The report prepared by this team was highlighted by the anti-dam activists, who had started the Narmada Bachao Andolan (NBA). Even after all this, neither the state government nor the central government was ready to reconsider the project. State machinery was brought in full force to crush the resistance to the project.

In 1993, Manibeli in Maharashtra became the first village to be completely submerged under water. The villagers along with NBA activists led by Medha Patkar fought for justice in a non-violent manner. In 1994, heavy monsoons submerged even more villages, which had been declared safe. Protests now spread to Bhopal in Madhya Pradesh. Dharna and fast in Bhopal were meant to stop the continuing construction of the dam to 80 metres, though it had been decided earlier that the dam would not be higher than 69 meters.

By 1995, the government had decided to further raise the height of the dam to 85 meters. NBA filed a case in Supreme Court in 1995 raising several questions regarding the displacement and the economic feasibility of increasing the height of the dam. Though the state government of Gujarat and the central government pleaded with the Supreme Court to increase the height of the dam to 90 meters, the court flatly refused their plea. However the court said that the government could continue with the Research and Development (R&R) activities. The court also called upon the NBA activists to co-operate.

Many interim orders were passed by the court in succeeding years. The most recent award of Supreme Court verdict in the case of Sardar Sarovar Project on rehabilitation practice points out the gap in "policy" and "practice" of resettlement and rehabilitation. The award delivered on 15th March 2005 categorically says that rehabilitation shall be completed before the submergence takes place. This award, in fact, upholds the spirit of Narmada Water Disputes Tribunal Award (1979), which in essence states "rehabilitation before submergence".

6.5.2.2 Polavaram Project in Andhra Pradesh

The judicial intervention in the case of the Sardar Sarovar Project may contribute to the improving of rehabilitation practice in Gujarat, Maharashtra and Madhya Pradesh, which are the affected states. However, how far this court decision influences other projects in several other states is not clear. In most cases, states accept them as awards specific to individual projects and do not extend them logically to other projects in spite of close similarities. For example, a judgment was passed by the High Court of Andhra Pradesh on a PIL filed in 2004 to stop construction activity of Pulichintala Irrigation Project until obtaining environmental clearance from Ministry of Environment and Forest, Government of India. This court order is being obeyed by the state government in case of Pulichintala Project. Yet the Government is proceeding with the construction work on Polavaram Project, which also does not have many clearances regarding resettlement and rehabilitation of the displaced people.

6.5.2.3 Andhra Pradesh Resettlement and Rehabilitation Policy 2005

Different state governments implement resettlement and rehabilitation of people displaced by dams, mines etc. in their own way. The plight of 20,000 people of 12 villages submerged in water during the implementation of Yeleru Irrigation Project in Andhra Pradesh is miserable. The government of Andhra Pradesh had put forward a Policy on Resettlement and Rehabilitation for the Project Affected Families (PAF's) on 8th April 2005. The Policy was released in the form of Government Order (G.O.) No.68 of Irrigation Department. It was claimed by the Government that this was the best resettlement and rehabilitation package in the country. The Policy does have some good elements, but it does not cover the most important area.

The G.O. contains the following decisions:

- There should be no displacement of tribals nor any disturbance of tribal way of life execution of any developmental project.
- Submergence of tribal lands should be avoided as far as possible.
- Plans for rehabilitation should be prepared in association with the affected people.
- It will be ensured that the people are fully compensated and rehabilitated.
- Rehabilitation shall be completed before the execution of the project.
- If sufficient land cannot be provided, employment in the project or in other projects or in government will be provided to at least one member of each displaced family.
- A training programme of entrepreneur development will be conducted for displaced tribals. Follow-up in getting financial support, helping find marketing outlets, supply of raw materials etc., shall also be done at the time of project execution.

It is unfortunate that no effort was made to implement this G.O. in any project during later years. No rehabilitation plan has been implemented. None of the above provisions was implemented in these villages.

The case of dams has been described here in detail. Problems and concerns are the same in the case of mines, townships, building complexes, railway lines, roads, military firing ranges, bridges etc. In all these cases, it is mostly the poor who are affected. How this modern model of development is heaping injustice on the vulnerable is a point to ponder on; corrective action must be taken by the concerned authorities.

6.6 ENVIRONMENTAL ETHICS: ISSUES AND POSSIBLE SOLUTIONS

6.6.1 Issues

Ethics is concerned with:

- what is morally good behaviour;
- what makes a person good; and
- what principles we must use in decision making?

As the great Greek philosopher said, the goal of studying ethics is not to know what is good, but to become good, otherwise there is no meaning in studying it.

Ethics involves a lot of reasoning, therefore it is based on logic. Yet, emotions influence the human mind more than reasoning. Therefore an easier way of motivating people to do anything is to create, encourage and enhance moral emotions in them. This point of view has influenced ethics also because ethical thinkers often argue that if we follow the right principles, it will lead to peace of mind, happiness and a more fulfilling life.

An ethical community comprises three elements—the family, the society and the country. These three factors together make an individual what he/she is. Finally, there are the universal ethical communities also. In the field of environment, there is definitely an increasing international consensus that there is an urgent need to protect the environment. The number of international treaties and agreements regarding various environmental issues is proof of this. All over the world, a general feeling has been expressed that we must respect the rights of the future generations. It is the duty of the present generation to protect the environment with all its biodiversity and indigenous cultures for the generations to come.

The view that nature matters only where human welfare is concerned is wrong. This is a purely western thinking. Some westerners even believe that condemning cruelty to animals is good because it elevates the human race by stimulating compassion! According to this viewpoint, cruel practices towards animals are not wrong in themselves. They become wrong only if humans become less compassionate to other humans. However, as mentioned earlier, ancient scriptures show very clearly how the environment as a whole was treated as one entity in India. We have always believed that man is a part of nature, not that man is master of nature. Our principle has been 'Vasudhaiva kudumbakam' meaning the whole world is one family.

Yet there is need to inculcate feeling of love towards nature in each and everyone on this earth. The earlier this is done the better, because despite of the vast information we have on the dire consequences of human interference on the biosphere, many of our activities continue to harm the environment. We destroy forest covers, drain wetlands and pollute air, water and soil in various ways. The danger that human activities may lead to rising of sea levels and global climatic changes remains true. It is high time therefore that environmental education becomes part of the education process at all levels.

The different ethical issues related to environment are as follows:

Equity Disparity among Countries: There are vast differences between the economies of the developed countries and the developing countries (MDC's or more developed countries and LDC's or less developed countries, as they are referred to these days). Gross national product per person of some countries as per World Bank, World Development Indicators (2000) is given below: (Figures in dollars – value of 1 Dollar was about Rupees 35/- in 1998).

Table 6.1

S. No.	*Name of the country*	*GNP per person, 1998*
1	Japan	32,350
2	United States	29,240
3	Germany	26,570
4	Brazil	4,630
5	South Africa	3,310
6	Russia	2,260
7	China	750
8	Indonesia	640
9	India	440

The figures speak for themselves. Though there are poor people in the rich countries also, most of the poor in the world live in Africa and Asia. World Bank figures show that almost half the world's population survives on about 1 dollar per day.

More than 1 billion people do not have access to clean drinking water. The available water itself is polluted in many areas and this leads to many water-borne diseases. The number of people dying of cholera, typhoid, jaundice, dysentery and such other water-borne diseases is very high in the LDC's. Diseases spread garbage, malnutrition, infant mortality etc. are more in these countries.

(b) *Equity Disparity within Countries*: Within countries also, disparities are very clear. In general, in the rich countries, the difference in income between the rich and the poor is not as much as in the poor countries. In the poor countries, the rural masses are affected most by environmental degradation.

The quantity of firewood collected by the poor from the forests is much less when compared to the huge quantities used by different wood based industries like timber, paper, plywood, rubber, package, etc. Thus, the rich industrialists consume the forest produce the maximum, while the rural poor take the blame for the deforestation and at the same time remain poor. Similarly, the poor fishermen are often blamed for overutilizing the marine resources. The fact is that the rich use trawlers go deep into the sea and collect fish in huge quantities, while the poor fishermen using simple boats cannot penetrate far. It is thus the poor who are always at the receiving end.

As such, more and more people are migrating from the villages to towns and cities. There is neither enough water, food or land for the migrants, nor are enough roads or healthcare systems for them in the towns and cities. Yet, unable to endure the poverty in our villages, many people migrate to urban areas. The final result is that poverty spreads everywhere. If the rich in the rural and urban areas waste resources just because they can afford to waste, it is the poor all over who suffer.

(c) *Inequality in Consumption of Resources*: There is no doubt that the LDC's remain less developed because they have not been able to control their population growth rate to the desired level. When the population is more, demand for everything increases—water, food, firewood, coal and petroleum. There is a big deficit in supply in many of these. Some LDC's do produce enough food and more, but then, these are exported to the MDC's. As a result, the starving poor are not benefited, deforestation, desertification, pollution etc. increase when population increases.

Even with all this, consumption of natural resources is much more in the MDC's. This is because the people in the rich countries demand more from nature. They use much greater quantities of energy resources like petroleum, electricity etc. Their usage of food and water resources is also more. Consumption of resources leads to industrialization and technology, which increase and consume still more resources. Similarly, machines and technology lead to surplus wealth, which is invested in more machines and more technology. In this way, resource consumption becomes a vicious circle in all rich countries. Above all, most of the rich countries are wasteful in nature.

USA has only about 4% of the world's population, but it consumes about 25% of the world's resources. One reason for this high consumption is high wastage.

(d) *The Gender Inequality*: Women who constitute 50% of the population do not get justice in any matter. They work for longer hours than men, for they have to do the household work and in addition work outside the home. They get up earlier and go to bed later than men. They feed the men and children first and then eat whatever is left. Because of this, women are mostly undernourished and often fall prey to anaemia. In areas of water scarcity, women have to walk long distances to bring water for household needs. Again, it is the women who go to the forest to collect firewood. Back home, they have to cook using fuel inefficient stoves. The fuel efficient stoves designed by educational institutions in the country have reached to very few villages. In the field of education, women lag behind men everywhere. Whether the national average or the average of any state in India, truth remains that less women are literate than men. Because of this, they do not know their rights and also are not given any say in matters concerning society. They do not have the freedom even in matters related to themselves.

Environmental degradation affects women the most. Women feel the maximum brunt of water scarcity, deforestation, lack of good alternative energy sources etc. It has always been women whose healthcare is not done properly in society. On the whole, girls and women do not get the share due to them in society. Unless this state of affairs changes, society as a whole can never progress. And only in an enlightened society can there be any environmental awareness.

(e) *Destroying Natural Resources Belonging to Future Generations*: If the present rate of destruction of forests and water bodies and coral reefs continues, what will be left for the future generations? This is a question to ponder. We do not own the earth or its resources. We are only their custodians. We have to pass on these resources intact to our children. In fact, we have not worked hard to get these resources in the first place. They have been handed over to us by our forefathers. So now it becomes our duty to preserve them as best as possible for the benefit of future generations. Living in a clean environment is everyone's right. One generation cannot deny that right to another generation.

When forests are cut down, many animals have nowhere to go and they finally perish. The felling of even a single tree changes the environment. Many birds lose their nests, many squirrels, snakes etc. lose their shelter, many plants living in the shade of the tree die and countless microorganisms and small insects and worms in the soil are deprived of their food, as they eat the leaves fallen from the tree. In the absence of the tree, soil erosion, light intensity, temperature and wind velocity of the place increase, while fertility of the soil and atmospheric humidity decrease. Thus cutting down of one tree destroys one entire ecosystem.

When the temperature of the sea water rises even slightly as a result of our activities, polyps that make the coral reefs die. When mercury and other such toxic chemicals from our factories are released into rivers, innumerable fish die. We drain wetlands, home to a variety of unique animals, only to make homes for ourselves. The list is truly endless. We owe it to our children that they also enjoy the benefits of the vast biodiversity on this planet. We have no right to destroy it.

(f) *Animal Rights*: While thinking or talking about animal rights, the first thing for us to remember is that we are also animals. We are one type of mammals; that is the truth. Agreed, we are more intelligent than all the other mammals, but that is all the more reason for us to understand that they, who depend on us in many ways, deserve our kindness. Cruelty to animals is wrong ethically, just like cruelty to fellow human beings is wrong, ethically. The animals are never cruel to us; they do not harm us for the sake of harming us. Lions and tigers kill their prey only when they are hungry. That is not wrong. Likewise human beings also catch fish or kill animals for food. That is part of the balancing system of nature. But men kill for sport too. This is cruelty. They cage birds; this is cruelty. We must respect the rights of animals, especially as they cannot talk. Finally, understanding that by protecting the animals we are benefiting ourselves too is plain logic. In fact, what is good for nature is good for man as man is part of nature.

(g) *Ethics and morals*: There are positive laws and natural laws in the world. The positive laws are the laws that are enforced by the police and the judiciary. But to enforce the rights of the oppressed or the animals, a different kind of law is necessary. Such laws are called natural laws. These are based on ethics and morals. These are same to all, irrespective of countries, religion, caste, class, creed, age, sex etc.

If human beings have the same rights everywhere, it is because of ethical and moral standards. These are universal in application. Yet, in some countries, where the laws of the land are not being implemented truly, sections of people remain oppressed and there may not be any solace at all as the oppressors naturally have no ethical or moral feelings.

In fact, there is no practical difference between ethics and morals because ethics is the science of morals. It is concerned with human conduct and character. It is a system of morals or a set of rules of behaviour. Morals relate to character or conduct considered good or bad. Moral behaviour is always directed towards right. An individual with high morals is able to differentiate between right and wrong. Moral laws are supported by evidence of reason.

For an individual to feel ethical towards the planet Earth and its resources, environmental education is a must from childhood. If a child grows up admiring the beauty of nature and its infinite variety, he will definitely be concerned about the environment throughout his life. This will lead to the making of successive generations concerned about nature and natural resources.

6.6.2 Possible Solutions

Ethics had been strong in all societies in the earlier times. All over the world, people led simple lives. In those days, natural resources were aplenty. It was believed that these will never end. Gradually, population increased greatly in some countries like India and China. Industrial revolution that spread in the west increased people's desire for luxury and comforts. Everywhere in the world, man's needs increased as his greed grew. This led to a depletion of the natural resources. The situation remains the same today. There is no limit to man's desires and wants. If this tendency is not checked and if we do not take care to replace the resources we are using, the shortages are sure to grow to dangerous proportions.

In India, nature used to be worshipped in the earlier days. The wisdom of the sages and other knowledgeable persons had prevailed upon the common men. Thus, many trees like Peepal, Bargad, Neem, Kadam etc. and small plants like Tulsi used to be worshipped. In tribal areas, there were festivals related to trees like Karam. They also protected Mahua trees, which provided edible flowers and oil from seeds and alcohol.

Many animals were also considered sacred. Many of them have been depicted as vehicles of gods and goddesses or associated to them in different ways in the Hindu mythology. For example, eagle, bull, peacock, swan, mouse, tiger and owl are respected as vehicles of Vishnu, Shiva, Karthik, Saraswathi, Ganesh, Durga and Lakshmi respectively. Animals like cow, snake and monkeys are associated with Krishna, Shiva and Rama respectively. Hanuman was in the form of a monkey, while Jambavant was in the form of a bear. Lord Vishnu's earlier incarnations were in the form of different animals like fish, tortoise, boar etc. Such associations used to motivate people to love and respect these animals and birds.

There were unwritten rules against polluting water bodies. Many rivers in India are considered holy. Some well known examples are the Ganges, Yamuna, Krishna, Cauvery, Godavari, Narmada, Tapti, Gomati, Sarayu, Son, Netravati, Gandaki, Sindhu etc. Mountains like Himalaya, Vindhya, Satpura etc.; hills like Nilgiri, Nallamala, Mahadeo, Parasnath, Rajmahal etc. are all venerated by the people. Forests are thought to be governed by Van Devta's, i.e., Forest gods and goddesses. In this way, plants, animals, forests, mountains, rivers, in fact, all parts of nature were loved and respected in India.

Mahatma Gandhi and many other great men of India preached and practised 'Plain living and high thinking'. His principle of enough ness is very popular: "The earth provides enough for everyone's needs but not everyone's greed". Simple living will automatically reduce over-exploitation of natural resources. Many great men of the past had motivated people to protect the plants and animals. Swami Vivekananda had said, "Not injuring any living being is a virtue; injuring any being is a sin." Though this is not possible fully as many plants and animals are eaten, the words can be followed to avoid cruelty to them and also not to kill or injure them for sport.

It were environmental ethics that motivated Rabindranath Tagore to set up Shanti Niketan. This is a beautiful concept where lessons are imparted to students in the open, under the trees, in natural surroundings. This makes the students nature lovers.

There can be many ethical guidelines to prevent the earth from further degradation. If these guidelines are followed, environment on the earth will improve greatly:

For conservation of vegetation:

- Plant trees wherever possible
- Prevent cutting of trees
- Know the medicinal values of various plants and put them to use
- Pass on the information you have to others so that they can take interest in protecting medicinal plants

- Do not over-water your plants, they may die due to waterlogging
- While going out of station for holidays, place a long hose pipe with holes at intervals along the row of plants in your garden and keep the tap partially open. This is drip irrigation for homes. The plants will remain alive; at the same time, no water will be wasted.
- Paper is made from wood; reduction on use of paper amounts to cutting of lesser number of trees. Therefore, use recycled paper if possible.
- Do not crumple paper, only tear it into pieces before throwing, because crumpling breaks the fibres and reduces the quality of recycled paper.
- Do not waste paper; use both sides.
- Reuse gift papers.
- During functions, give potted plants in place of bouquets to guests.
- If your compound walls are covered by moss, do not scrape them off unless you want to whitewash them for some purpose. Mosses look green and beautiful; they also absorb carbon dioxide from air and release oxygen into air during photosynthesis, like all other green plants.
- Cultivate a love for nature in the young children at home. Tell them all about the usefulness of each part of nature; tell them stories concerned with nature

***For conservation of animals*:**

- Take good care of pet animals and birds.
- Do not throw stones or disturb small animals like lizards, chameleons, birds etc.
- Do not tease animals in the zoos.
- Arrange for water containers outside your homes for wandering animals, if possible.
- Attract birds to your garden by providing water for them.
- Break leftover food into small pieces and throw them to birds.
- Instead of throwing banana skins etc. in the dust bins, give them to cows that are roaming in your locality.
- Do not catch or kill butterflies, moths etc. They alongwith many other insects help in pollination.
- Do not kill earthworms; they improve soil quality.
- Do not catch or kill dragon flies; they eat other harmful insects. Remember that dragon flies are gradually becoming extinct.
- Do not use products made from the skin of snakes, crocodiles, leopards etc. These are endangered animals.
- Do not use ivory products for decorating homes.
- Do not bring small wild animals home for rearing them; most probably they will die; moreover, it is against the law.

***For water conservation*:**

- Conserve water by practising rooftop rainwater harvesting.
- Do not waste water at home.
- Report about leaking pipes in public places to the authorities.
- Do not direct sewage water on to roads, rivers etc.
- Do not keep the tap running while brushing or shaving.
- Do not throw garbage into water bodies.
- For potted plants, reuse the water used for washing vegetables and fruits.
- Water the plants in the morning to reduce evaporation from the soil.
- Use eco-friendly cleaning solutions in place of chemicals.
- Do not cement your courtyard; maintain a lawn instead. The green grass allows water to filter down the soil; it looks beautiful too.

***For waste management*:**

- Use dust bins and do not throw garbage on the streets.
- Keep parks and gardens clean.
- Do not throw used plastics here and there.
- Use paper or cloth bags in place of plastic bags for shopping.
- Make compost manure at home from waste materials like skin of vegetables and fruits, and other biodegradable waste materials.
- Grow plants like banana, ol, kachu etc. near outlets of water from bathrooms, kitchens etc.
- Segregate solid wastes at home so that all degradable wastes go into the compost pit, while only the non-degradable wastes like plastics are deposited in the municipality bins.
- Do not burn plastic wastes at home; the fumes are also harmful.

***For soil conservation*:**

- Do not leave open areas in your garden without any vegetation.
- Provide a cover to the soil by dead leaves and twigs, cut grass etc.
- Raise the edges of the garden space so that no water flows out, carrying precious soil away. Let the water sink in place.
- Use organic manure like cow dung, green manure, compost, vermicompost etc. in place of artificial fertilizers to maintain soil fertility.
- Use biopesticides like neem oilcake in place of chemical pesticides.
- Get involved in community programmes like tree plantation on vacant lands nearby.

***For Conservation of energy*:**

- Reduce electricity consumption by all possible methods like:

  (a) Switching off fans and lights, television, computer etc. while not in use

  (b) Using tube lights in place of ordinary bulbs

(c) Using energy saver bulbs

(d) Turning off electric heaters in the kitchen slightly before the food is cooked as the heat that lingers on will complete the cooking

(e) Placing study tables near windows to use sunlight in daytime

(f) Not keeping hot food in refrigerators as that will consume more electricity

(g) Getting electric metres checked etc.

- While using gas stoves in the kitchen, remember that:

(a) Avoid reheating of food; this is bad for health also

(b) After the curry starts boiling, sim the switch

(c) Soak pulses beforehand so that they cook faster

(d) Use pressure cookers as much as possible

(e) Take out eggs or other cooled items from the fridge beforehand so that extra gas does not burn to bring them to the room temperature first

(f) Use flat bottomed vessels for cooking on flat hot plates, but while cooking on gas, use round bottomed vessels to utilize the energy to the maximum

(g) Cook items covered whenever possible

- Use alternative energy sources like:

(a) Biogas

(b) Solar energy

- Avoid wastage of petrol by:

(a) Walking or cycling as much as possible

(b) Arranging for car pools instead of each person using his own car. If a few neighbours have to go in the same direction for work or college at the same time, car pooling system is ideal

(c) Trying to use public transport like trains and buses

(d) Servicing vehicles regularly and getting emission checks done

(e) Not keeping the car switched on while parked

(f) Stopping fully at traffic signals if the light is red

The list can go on and on. All the practices mentioned here and more are easy to follow. The only thing needed is determination.

The solutions for problems related to environmental ethics can be summarized as follows:

- We should understand that we are part of nature and behave accordingly.
- We should try our best to conserve all the resources that nature has gifted us and not waste any resource.
- Where we have to take out resources, we must see to it that they are replaced.

- We should not degrade the earth's physical, chemical or biological capital.
- We must do Environment Impact Assessment (EIA) studies before starting any project.
- To find out whether our projects will be harmful for the following generations, we can also do a GIA study (Grandchild Impact Analysis) – this is a recent concept.
- We should honour the right of all beings to live on the earth.
- We should try hard to preserve the present biodiversity in nature.
- We should not allow certain human communities also, like forest dwellers, to perish.
- We must be kind to animals.
- We must try to lead simple lives and motivate others to do so.
- We can and must think globally, but we must act locally to solve different problems.

Those who follow these ethical guidelines are True Earth Citizens.

6.7 CLIMATE CHANGE

Climate of any place is influenced by light, temperature, humidity, wind etc. Major changes in climate usually do not take place in nature. They do occur because of volcanoes, rivers changing courses, spontaneous (natural) forest fires etc. When the climate changes in a place, many other changes result, like changes in the species composition, occurrence of floods and droughts etc.

In the earlier days, the climate never changed as a result of human activities. But now-a-days, this has become common. One of the most important changes in climate is increase in the atmospheric temperature and its effects. This is called global warming. Other changes are more rainfall leading to floods in some places, while less than normal rainfall leading to droughts in others, increased occurrence of storms, earthquakes, cyclones, tsunamis etc.

Climatic changes can seriously affect the availability of plant and animal food, drinking water, forest produce etc. They can also bring about changes in the disease causing germs and make them stronger and of more varieties. This can lead to epidemics of known diseases and also newer types of diseases in human beings, animals and plants.

Increased occurrence of diseases coupled with floods, droughts etc, can lead to migration of whole populations from one place to another on the earth. This will cause problems of settlements of refugees and their livelihood.

The climate of a place is the average of its weather over a long period. Rainfall, sunshine, wind, humidity and temperature are the important aspects of climate. Over the last 150 to 200 years, climatic changes have been taking place too rapidly. Many species of plants and animals are not able to adapt themselves to these changed climates. Many have become endangered or even extinct.

Greenhouse gases in the atmosphere have increased to a great extent. Carbon dioxide is the most important greenhouse gas in the atmosphere, mainly because of:

- fossil fuel combustion in industries, and
- increased use of automobiles.

Methane is another greenhouse gas, which comes from:

- Domesticated animals like cows, goats, pigs, buffaloes, camels, horses and sheep.
- Waterlogged paddy fields, where anaerobic bacteria live in abundance
- Waste dumps and landfills
- Leaking pipelines etc.

Nitrous oxides are emitted from fields fertilized extensively by nitrogenous fertilizers.

The natural causes for global climatic changes are continental drifts, volcanoes, ocean currents etc. We cannot do anything to stop these. However, we can do much in the case of the human causes. Climate changes that started with industrialization have been continuing ever since step by step.

- Large scale use of fossil fuels had started during the period of Industrial Revolution in the 19th century.
- The industries created jobs.
- To get these jobs, there was migration from the villages to the towns and cities.
- More and more land was used for houses.
- Natural resources were used extensively for construction, transportation etc.
- Wastes increased tremendously.
- Number of vehicles increased many times.
- New industries came up, which again polluted air, water and soil.

The process has thus become a vicious cycle. It is going on and on. Along with the urban areas, rural areas changed too. These were mainly based on agriculture.

- To feed the increasing population, it became necessary to create more croplands.
- For this, more and more forests were cleared.
- This resulted in erosion, floods and droughts.
- To increase yield, artificial fertilizers were used liberally, emitting greenhouse gases like nitrogen oxides into the air.
- Dams were built for irrigation; this led to further deforestation and flash floods.
- The increased number of power plants released large quantities of greenhouse gases into the air.
- More irrigation led to more salination of soil.

In this way, villages have also contributed to large scale atmospheric changes leading to climate changes.

Our industries, vehicles, and croplands pollute the air and increase the greenhouse gases in the atmosphere. When this happens, climate is bound to change, and this is a global problem. So all the countries in the world must get together to solve this problem. As the global climate change is mainly because of our activities, we can rectify the problem by various methods:

- Cut down the use of fossil fuels
- Reduce consumerism
- Stop deforestation
- Use more environment friendly methods of agriculture
- Reduce demand for energy
- Use alternative sources of energy
- Preserve wetlands
- Tax the polluting industries to compensate for cost of pollution control
- Use non-polluting substitutes for raw materials in polluting industries
- Create public awareness about climate changes and its effects etc.

In short, climatic changes due to human interference in the working of nature do pose a serious threat to the present generation. The future generations will have to face these problems on a larger scale if we do not do something about it now.

6.7.1 Global Warming

Our environment provides us with a variety of natural resources. The atmosphere is one of them. It forms a protective shell over the earth. Under normal conditions with a 0.03% by volume of carbon dioxide (CO_2) the temperature at the surface of the earth is maintained by the energy balance between the sun rays that strike the earth and the heat that is radiated back into the atmosphere. However, when there is an increase in CO_2 concentration (the major sources of which are industries, automobiles and burning of fossil fuels), it prevents heat from being re-radiated out. The thick layer of CO_2 in the atmosphere functions like the gas panels of a greenhouse, allowing the sunlight to pass through, but preventing the heat from re-radiating out. This is called the greenhouse effect.

The major sources of increased concentrations of carbon dioxide in the atmosphere are industries, automobiles and burning of fossil fuels. Most heat is absorbed by CO_2 and water vapour in the atmosphere, which adds to the heat already present. The net result is the heating up of the earth's atmosphere. This is called global warming.

Some amount of CO_2 in the atmosphere is absorbed by the oceans. But as industrialization accelerated and the number of automobiles increased, rate of CO_2 release has become more than the rate of absorption by the oceans.

All gases which contribute to the greenhouse effect are known as Greenhouse Gases or GhGs. These are:

- CO_2 (Carbon dioxide)
- N_2O (Nitrous oxide)
- CH_4 (methane)
- CFCs (Chloro Fluoro Carbons)
- SO_2 (Sulphur dioxide)
- HFC's (Hydro Fluoro Carbons)
- About 25 other gases

These are discharged into the atmosphere mainly by industries and partly by other human activities, leading to an increase in surface and atmospheric temperature. The pattern of this rise has been studied by different agencies. Though the results have not been the same, all agree that there has been a gradual increase in global temperature.

The overall increase in temperature over the past one hundred years has been estimated to be about 0.5 to 0.7°C. According to one study, average air temperature started increasing in the Northern Hemisphere first because of rapid industrialization in the developed countries, while the increase started in the Southern Hemisphere comparatively late.

In India, during 1997-1998, a rise of 2°C in the temperature from the normal was recorded in the Indian Ocean. This had resulted in the death of 70% of the corals in the Andaman & Nicobar and Lakshadweep islands.

There have been predictions that if the present amount of carbon dioxide is doubled, the temperature of the earth's surface will increase by 2.9°C. In such a situation, continued global warming may have grave consequences like melting of the polar ice caps leading to a rise in sea level resulting in flooding of low lying cities like London, Dhaka, Kolkata etc.

Realising the seriousness of the issue, UNEP (United Nations Environment Programme) chose the slogan "Global Warming: Global Warning" to alert the people on World Environment Day, June 5th, 1989.

Soon, the UNCED (United Nations Conference on Environment and Development), commonly called the Earth Summit, which was held at Rio de Janeiro, capital of Brazil in 1992. It was a historic assembly of world leaders from 170 countries. This marked the setting of a global agenda in international relations and economic development.

The Kyoto Protocol was in fact a follow-up of the Earth Summit. This treaty is called so because it was negotiated in Kyoto, Japan. It was held in 1997. As of April 2006, a total of 163 countries have signed the agreement. These countries are bound to reduce their emissions of greenhouse gases.

Advocates of the Kyoto Protocol claim that reducing these emissions is crucially important; carbon dioxide, they believe, is causing the earth's atmosphere to heat up. Most prominent among the advocates of Kyoto have been the European Union and many environmentalist organizations. The United

Nations and some individual nations have also issued reports favoring the Kyoto Protocol.

India, China and Brazil have the greatest potential for the new Clean Development Mechanism (CDM) proposed by the protocol. There are many opportunities in the non-conventional (alternative) energy sector and other sectors based on increasing energy efficiency, utilisation of wastes etc. To cite one example, using fly ash to make cement can be a CDM project. For each tonne of fly ash used in this manner, the amount of CO_2 is released into the atmosphere is reduced by one tonne.

If countries do not heed to the warnings by scientists, global warming is sure to bring about drastic climatic changes on the earth. Rise in temperature will not be the only result, rainfall patterns are sure to change, leading to more floods and droughts, more cyclones and storms. The recent erratic rainfall in India, according to some, is an effect of global warming. Therefore, if the countries that signed the Kyoto Protocol do act as promised, and more countries sign the Protocol, the rate of global warming at present will surely fall.

A counter argument has been put forward by some. According to these arguments, global warming is not a very serious problem, moreover, nature has the acpacity to maintain its equilibrium. But not many believe in this theory. Majority believe that global warming is a reality and it will have serious consequences if not controlled by the present generation.

6.7.2 Acid Rain

When fossil fuels like coal, petrol, natural gas etc. are burnt, large quantities of carbon dioxide, sulphur dioxide and nitrogen oxides are formed and released into the atmosphere. Exhausts from automobiles also contain some of these gases. These gases dissolve in the water droplets in the atmosphere and form acids like sulphuric acid, nitric acid etc. These fall down on earth along with rain. This is called acid rain. Sometimes, they remain in the atmosphere in the form of clouds or fogs.

Acid rain is more common in the developed countries because they burn more fuel for their industries and the number of automobiles there is much higher than in other countries. They also have more power stations. In fact, burning of fossil fuels for power generation contributes to almost 60 – 70% of the sulphur dioxide released into the atmosphere in the world annually.

The oxides may travel for very long distances in the air. So other countries are at a risk too. Moreover, during these travels, the oxides may be converted into more hazardous chemicals which may remain in the atmosphere.

Acid rains cause many problems. These are:

- Formation of acidic soils
- Acidification of freshwater bodies like ponds, lakes, streams etc. by direct rain and also through runoff from land
- Reduction in yield of different crops
- Reduction in number of aquatic plants and animals, mainly fish

- Human health problems
- Corrosion of buildings, bridges, monuments, fences, statues etc.
- Leaching (moving downwards) of nutrients from the top layers of the soil
- Death of many nitrogen fixing bacteria and blue green algae
- Loss of fertility of soils due to washing away of nutrients
- Damage to leaves leading to reduced photosynthesis (making of food by green cells)
- Destruction of wax covering of trees leading to more attacks by pests; such trees cannot tolerate cold too
- Toxic chemicals like cadmium and mercury dissolve in the acid rain and are washed down from the acidic soils to water bodies
- Lowering of pH of water leads to death of many planktons (microscopic organisms) which are food for insects, small fish etc. This affects the whole food chain
- Damage to monuments like the Taj Mahal
- Fish caught from acidic waters are not fit for human use
- Acid rain of nitric acid leads to excess nitrogen levels in the soil. This can help some plants to grow better at the expense of others, thus disturbing the pattern of vegetation
- Aluminium ions are released from the soil, which damage the roots of trees
- Aluminium ions from the soil, when washed into water bodies, kill fish by increasing the production of mucus that blocks their gills
- A very interesting chain effect of acid rain has been suggested by some Dutch scientists. Of late, in Dutch forests, there has been reduction in the numbers of song birds. The chain effect being cited is:
- Acid rain leaches out calcium from the soil
- As a result, the population of snails falls
- Song birds now get fewer snails to eat, so they eat other food
- Due to calcium deficiency, the shells of the song birds' eggs are thinner and porous
- Chances of the eggs breaking and drying are more
- Most of the chicks that are hatched have poorly developed bones; such defective birds die faster

Just see how balanced the whole ecosystem is in nature. What right do we have to disturb this perfect balance?

Solutions to the problem of acid rain:

1. Use less fossil fuels
2. Use more environment friendly fuels
3. Use natural gas in place of coal and petrol
4. Use coal with lesser sulphur content

5. Do research on safe automobiles. Cars running on hydrogen fuel have already been found in Germany and Japan. These will hit the market shortly for sale. Still, more research is required as these cars are costly.
6. Reduce the number of automobiles used. For this,
   (a) adopt car pooling system instead of each person driving a car
   (b) use bicycles in place of scooters and motor cycles whenever possible
   (c) walk short distances; this is excellent for health too
7. Install scrubbers in industries to prevent the harmful gases from entering into the atmosphere
8. To neutralize the acidity of soils damaged by acid rain, add gypsum (calcium sulphate) to the soil

Acid rain is not a very serious problem in LDC's (Less Developed Countries) as of now. However, it is assumed that acid rain will be more in these countries in the near future as they are following the path of development adopted by the western countries. At the same time, the West has realized its folly and is trying hard to correct the mistake. Therefore, instead of blindly copying the West, we must adopt more eco-friendly methods of living.

6.7.3 Ozone Layer Depletion

The atmosphere that forms a protective covering around our earth comprises many layers. The lowermost layer of the atmosphere in which man and other organisms live is called the troposphere. It is about 20 km thick. It is a mixture of several gases. Above this is the stratosphere, about 30 km thick. In this layer, ozone is formed by a photochemical reaction. Solar energy splits the oxygen molecule to form atomic oxygen. When this combines with molecular oxygen, ozone is formed. It is also formed through chemical reactions involving pollutants like sulphur dioxide, nitrogen oxides etc. with ultraviolet radiations. Above the stratosphere is the mesosphere, about 40 km thick. Ozone layer is present in the lower part of the mesosphere also. Thermosphere follows mesosphere for about 400 km and this is followed by the exosphere or outer space, which extends for up to about 32,200 km from the earth.

The ozone layer in the stratosphere protects us from the harmful effects of the ultraviolet rays radiated by the sun. Its depletion by human activities may have serious implications. For example, pollutants like CFC's (chloro fluoro carbons) emitted by air-conditioners, refrigerators etc. entering the stratosphere remain there for years, react with ozone and deplete it, forming what is called the 'ozone hole' in the ozone layer. The thinning of the ozone layer leads to many serious disturbances in the natural phenomena.

***Effects of thinning of ozone layer on the environment*:**

- Temperature changes
- Rainfall failure

- Increase in the filtration of UV radiation to the lower layer of the atmosphere leading to:
  - Decrease in the chlorophyll content of plants
  - Increase in harmful mutations in all organisms
  - Reduction in fish productivity
  - Increase in skin cancer among humans etc.

A thick ozone umbrella is thus essential for the harmony and balance of our ecosystems.

Ozone depletion was the first global environmental problem recognised by the world community in mid-1980's. In 1987, an international conference was held at Montreal on substances that deplete the ozone layer. Came to be known as the Montreal Protocol, it was signed by 175 parties. The countries agreed that they would phase out CFC's, HFC's (hydro fluoro carbons), halons, methyl bromide etc. which are ozone depleting chemicals. Under the treaty, countries were required to gradually stop the production of such chemicals. Developing countries were given more time. CFC's are supposed to be phased out completely by 2010. Similar phase out time tables exist for the other ozone-depleting substances covered under the treaty. Scientists believe that if all countries comply fully with the Protocol, the ozone shield will gradually begin to heal within the next few years, with a full recovery to pre-1980 levels expected by about 2050.

Most countries have met their obligations to the Montreal Protocol in a timely fashion. By 1997, global CFC production was down by 85% from its 1986 level as the treaty began to encourage the development of alternatives to these harmful chemicals and the illegal CFC trade violations in the US and Europe declined.

***Some reasons for illegal trade in CFC*:**

1. Regional refrigerant management programmes were poorly organized
2. Alternatives were perceived to be too costly
3. Capacity to detect CFC release was inadequate
4. Officials were untrained in identifying the harmful chemicals etc.

Anyway, enforcement in the west is improving. Moreover, the World Bank and several other international donors have stepped in with millions of dollars as funding to countries that lagged behind, like Russia, Azerbaijan, Bulgaria etc. Most of the developing countries are also well on their track to meet their commitments.

The good news is that the ozone layer has started healing. The World Meteorological Organization and the United Nations Environment Programme announced this good news in 2002. In August 2006, two UN agencies confirmed the news. The main reason for the healing is that shorter lived harmful gases like methyl chloroform and methyl bromide have been removed. The report will be published any time next year. As per this report, by 2050, large areas over Europe, North America, Asia, Southern Africa, Latin America and Asia will have their ozone layers repaired. Yet the time is not ripe for any celebration. For, unexpected delays are also occurring in the process of repair. This is because some types of CFC's are still in use. Also, the substitutes for CFC's that cause some depletion will be used more in future.

There is thus no relaxation. We just cannot rest until assured that all is well. In 2005, the Ministry of Environment and Forests, Govt. of India had found that many companies are still using ozone unfriendly substances in India. The technologies used here are different from those being used in developed countries. Awareness among the erring companies is therefore necessary in our country. As individuals, what we all can do is plant more trees in our surroundings. This helps to heal the ozone hole faster, as plants use large quantities of CO_2 for photosynthesis, thus reducing its level in the atmosphere.

6.7.4 Nuclear Accidents and Holocausts

Radioactive isotopes escape frequently from nuclear power reactors. Wastes add such isotopes to water reservoirs, rivers etc. Nuclear tests emit large quantities of isotopes into the air. These and other such incidents are collectively called nuclear accidents. They are nothing less than a holocaust, causing misery and disease to a large number of people, animals and plants in a vast area.

In the tundras of Lapland, the food chain is very short. It starts with lichens, which grow slowly getting their nutrients from air and water. These lichens are eaten by reindeer. The reindeer are killed and eaten by the Lapps themselves. The Lapps eat up to 2 kgs of reindeer meat per day. Thus the Lapps are the most affected by radioactive substances left in the environment when nuclear tests are conducted in the Arctics. Extensive nuclear tests were conducted in the Arctics in 1961 and 1962. The fall out from these tests contained Caesium-137, Polonium-210 and Iron-59. Caesium was the most abundant. Even if tests are conducted far away, people are affected, at least slightly. (More details in Unit 5, under Nuclear Hazards)

6.7.5 Case Studies

6.7.5.1 Bombing of Hiroshima and Nagasaki

A powerful atom bomb containing Uranium- 235 was dropped on Hiroshima, Japan in 1945. This was the greatest man-made disaster in the history of the world. The Second World War had begun in 1941 and was continuing without any sign of ending. It was then that the American forces dropped the first ever atom bomb used against civilians on the city of Hiroshima on 6th August 1945. 70,000 people died on the spot. The radius of total destruction was 1.6 km. Resulting fires engulfed 1.4 km^2. Ninety per cent of Hiroshima's buildings were destroyed. By Dec. 1945, thousands more had died from injuries and radiation poisoning.

Just 3 days after the bombing of Hiroshima, the American forces dropped a second atom bomb on Japan—this time in another city, Nagasaki, on 9th August 1945. The radioactive isotope used was Plutonium – 239. Forty thousand civilians died on the spot; 60,000 were injured. On account of radiation poisoning, thousands died in the following months.

The number of deformed children in Japan after these two bombings also ran into thousands. Even in 2005, 266,000 exposed people were living in Japan.

The world war ended after these two bombings, but what happened was a crime against humanity. Atom bombs or fission bombs (nuclear fission is the splitting of the nucleus of the atom) generate enormous amounts of energy in terms of air pressure and heat. In addition to this is the significant amount of γ radiations, which kill man and mutate genes.

6.7.5.2 Chernobyl disaster

Forty one years later, on 25th April 1986, a series of explosions took place in a nuclear power plant in Chernobyl, Ukraine, then Soviet Union. As a result of the explosion, the heavy roof of the reactor building was blown off. Radioactive waste materials and radioactive dust containing Caesium-137 flew high into the atmosphere. These spread gradually over the entire earth. The reasons for this horrible disaster were both poor design of the reactor and human error. The after effects were terrible.

- Thirty-one people died on the spot from massive radiation exposure. Two hundred and fifty nine people were hospitalized due to acute radiation sickness.
- After the accident, 65,000 people were employed to build a concrete shell around the destroyed reactor. They were also exposed to dangerous levels of radiation. Thousands who worked on this project died.
- Thousands were affected with cancer. Four million people are estimated to be suffering from the effect of constant low levels of radiation. Due to weakening of immune systems, exposed people get various infections.
- All plants in the neighbouring areas were poisoned. Thousand of cows and goats had to be shot because they had eaten radioactive grass. Poisoned cats and dogs were abandoned when people were evacuated. 1,35,000 people were evacuated few days after the disaster. However, in 1991, when the authorities realized that this was not enough, 1,25,000 more were evacuated. The severity of the disaster had not been fully understood.
- Twenty months after the Chernobyl disaster, radioactive snow fell in Antarctica.
- Lichens, the main food of reindeer were contaminated. Thousands of reindeer of Lapland had to be killed.
- Even 8 years after the disaster, some lambs could not be sold in 1994 as they were too radioactive.

6.8 WASTELAND RECLAMATION

Land where plant growth is sparse is called wasteland. This may be due to physical, chemical or biological reasons. For example, the soil in a particular wasteland may be acidic, alkaline, saline, sandy, or even waterlogged. Soil may be too hot or too cold. Water may be actually lacking as in deserts or may be present, yet unavailable to plants due to high mineral concentration in the soil etc.

Reclamation means claiming the soil back to fertility, i.e., making the wasteland fit for plant growth. Thus bringing degraded land back to normalcy is called land reclamation. There are many methods of wasteland reclamation. These may be mechanical, chemical or biological.

6.8.1 Mechanical Methods

- This method can be used to reclaim alkaline and saline soils. A network of deep cuts is made in the land to lower the water table. Salts leach deep down. Then irrigation is done by good water. Alkali on the surface is removed by scrapers. Salt or alkali tolerant species are planted in the beginning.

6.8.2 Chemical Methods

- Alkali soils can be reclaimed by the addition of gypsum (calcium sulphate)
- Acidic soils can be reclaimed by adding lime.

6.8.3 Biological Methods

- Reforestation can be done through micropropagation. For this, trees species are selected that are adapted to grow in degraded land. They must also be highly productive. Multiplication in thousands can be done through tissue culture. Multiplication of 10,000 to 1,00,000 trees per year is possible for many species plant on the degraded land. For this, clonal propagation of superior genetic stocks rather than seeds of unknown genetic quality is recommended.
- To fix sand-dunes and prevent desertification, *Casuarina* trees can be planted. They grow fast and improve soil fertility by nitrogen fixation. It can be used for tropical reforestation on wastelands. *Casuarina* trees give firewood with high calorific value. It can also be converted into excellent charcoal.
- Varieties of many plant species tolerant to salts have been found by plant breeding and genetic engineering. Plant such varieties in the land degraded by excess salts.
- On degraded land, only a brief period is favourable for growth. Within this period, if saplings do not establish, no vegetation cover is possible. Certain fungi are available which when injected into the roots of

saplings, promote growth. Such fungi form symbiotic relationships with the roots of the cultivated plants. These associations are called mycorrhizae. They can improve seedling survival and growth by increasing the uptake of nutrients, particularly phosphorus and by lengthening root life and protecting the plants against disease causing germs, i.e., pathogens. A list of such fungi has been prepared. They can be inoculated into the roots of saplings.

- Microbes can be used for improving the soil fertility of degraded land. Trees that have nitrogen fixing bacteria in their roots, like Shisham (*Dalbergia sisoo*), Siris (*Albizzia lebbek*) etc. can be planted on degraded land. Nitrogen fixing bacteria like *Rhizobium* can also be used to induce nodule formation in many species, both leguminous and non-leguminous. These can be utilised to improve the fertility of degraded land.
- Plants resistant to salinity, acidity etc. have been found. Research is going on to understand the molecular basis of salt tolerance and to isolate genes responsible for this, so that better varieties of tolerant plants can be developed through genetic engineering. Examples of species resistant to acidity are chilli (*Capsicum annum*), carrot (*Daucus carota*), barley (*Hordeum vulgare*), tobacco (*Nicotiana tabaccum*), rice (*Oryza sativa*), potato (*Solanum tuberosum*) etc. Saline resistant plants are *Prosopis spiciglea*, palash (*Butea monosperma*), *Terminalia bellerica* etc.
- *In vitro* selection for tolerance to aluminium toxicity has been successful in tomato (*Lycopersicum esculentum*), soy bean (*Glycene max*), rice, barley and wheat (*Triticum* sps.) These plants can be cultivated in wastelands degraded by aluminium.
- Soils are contaminated with heavy metals when water containing these from industries is used for irrigation. Microbes engineered genetically are used to reclaim such soils.
- Triticale (a hybrid of *Triticum*, i.e., wheat and *Secale*, i.e., rye) is suitable for growing in acidic soils in Poland, Kenya, Ethiopia, India, Mexico etc.
- Triticale can also grow on dry and sandy soils of savannas of Brazil and some places in Asia.
- In Australia, Triticale grows on mineral deficient soils containing high amounts of boron.
- In China, many new varieties of rice, wheat and tobacco have been developed through anther culture. Some of them are stress resistant.
- Microbes are available, which can grow on land polluted by toxic heavy metals released by the industries. These can reclaim the land degraded by industrial pollution.

- Blue green algae can be used to reclaim alkaline and waterlogged land. They are grown in summer and are removed later. Afterwards, crops like rice can be grown in these areas.
- On making vermicompost in the degraded soil, earthworms make the soil fertile.
- Afforestation programme should be held on land degraded by mining. Suitable species must be selected. Care must be taken to plant as many species as possible for biodiversity and protect them well. Water holes must be made for animals.

6.9 CONSUMERISM AND WASTE PRODUCTS

Consumerism is on the rise in all countries, especially the rich. A very good example of increasing consumption and its associated impacts is the increase in consumption by children. Children's markets are enormous and there are many products and foods made attractive to the children. Parents are facing a hard time raising children; at the same time, children are being increasingly influenced by commercialism.

This is just one outcome of the increasing trend of consumerism. In fact, as science and technology advance, more and more finished products are being produced every year by the industry for the comfort and luxury of man. Part of the wastes produced during their manufacture may be blown away by the wind or carried away down the streams. Steps may not be taken to dispose them of safely. If the by-products or side products are utilized in any way, throwing away of these can be avoided, which will also help in keeping the environment clean.

High level of urbanization has resulted in the mushrooming of towns and cities all over the world. While in the MDC's (More Developed Countries) the disparity between the rich and the poor is not very high, in the LDC's (Less Developed Countries), this disparity is very high. More and more people believe that more money would improve their life. So people dream of a life of wealth. All their efforts are directed towards maximization of wealth. In this rat race, no one has the time to stop and think of the conditions of the poor. The poor are ignored. They remain poor without getting opportunities.

Most of the poor, especially in LDC's like India are uneducated. They have no idea of hygiene. They therefore spoil the environment by throwing away the unwanted things just outside their huts. The slums in cities thus become breeding grounds of diseases. The environment gets choked by used tins, plastics, rubber tyres etc.

In India, the educated are no better, in many cases. They also throw away plastic bags, plastic glasses, and many other unwanted things carelessly around, in a most unethical manner. The recent great flood in Mumbai was caused by the choking of drains by plastics. After a heavy rain for 18 hours on 26th July 2005, the city of Mumbai came to a complete standstill for the next 48 hours. People travelling in trains throw away the used plastic glasses and thermocol dishes just out of the windows. Nobody bothers to think of the effect of these on the environment. There are no proper waste management systems in trains or for that matter in most other places in India.

As people are getting wealthier, the throw-away concept is also catching up. Earlier, people used to buy refills for their ball point pens. But now-a-days, nobody bothers. As soon as the refill is over, the old pen is thrown away and a new pen bought. The concept of repair and reuse is almost nil now.

So what is happening to the environment? Unwanted things are accumulating in the ecosystems. Plastics are the main culprits. In India, we see cattle and buffaloes roaming around the streets. These animals eat the plastic bags in the absence of good green grass and develop tumours in their stomachs. In Lucknow, recently, cows have been operated on only to take out even ten or fifteen kilograms of plastics from their stomachs.

Plastics of today are the electronic wastes of tomorrow. There is no limit to the electronic equipment like televisions, computers, mobile phones, calculators etc. thrown away by people. Earlier, this was a problem for the MDC's alone, but now this is a serious problem for LDC's like India also.

With growing computerization, the quantity of the e-waste is increasing day by day. For example, over 85 million cell phones were in use, in 2006. This number is expected to go up to 275 million or more by 2010. Now 7 Indians out of 1000 own a computer. By 2008, this is going to double. Along with these are the other electronic equipment like calculators, music i-pods etc. Televisions are more in number than the land line phones now. The IT industry is growing in India, and such changes are bound to happen. All this is undoubtedly good for the economic boost of India. The question is how we are going to dispose of this scrap. There is no safe method yet. The old computers of the USA and Europe land in Africa, India or China. There, they are manually dismantled, releasing the toxic materials like lead, mercury, cadmium etc., all highly toxic, along with copper. The LDC's are becoming the waste bins of the MDC's.

Most of the industrialized nations are thus high-waste or throw-away societies. They try to sustain their economic growth by increased input of raw materials and energy. Utilization of fossil and other forms of energy is also maximum in developed countries. The wastes finally end up in low waste societies. This is absolutely against all ethics.

It is however a welcome change that 'recycling' is catching up everywhere. Many companies are also eager to become 'zero waste companies' by recycling or reusing its waste materials in various suitable ways. These business houses recycle their by products and do not send them to landfills, as was being done earlier. New revenues are also generated in the process. This has led to cost reduction and resulting increased profits. Sanyo, Canon, Toyota etc. have reported that they have achieved zero waste production. Such activities will help in reducing the number of throw-away societies in general.

Increase in consumerism and the trend to waste anything and everything in the present times is leading to a more and more unsustainable society. Many aspects of nature itself, in a broad sense, can be classified as sustainable and non-sustainable. Resources, economy, development, society etc. can all be sustainable or otherwise.

The harmful effects of consumerism and wastefulness can be countered in different ways:

- By rational exploitation and optimal utilization of resources
- By adopting eco-friendly technologies for various processes
- By popularizing environmental education
- By ensuring continued yield of resources for the present and future generations.
- By changing personal attitudes towards resources and environment
- By motivating every individual not to waste things
- By linking development with conservation

Due to the wide disparities in the economies of the developed and the developing countries, it is difficult to form policies for establishing sustainable earth societies. Developing countries like India accounts for a very large population; in fact every sixth world citizen is an Indian. The developed countries are pointing fingers at such countries (mainly India and China) and accusing them of depleting natural resources much faster than they can regenerate by natural processes. On the other hand, the industrialized west is utilizing a major portion of the earth's resources, in spite of having low population levels. Accusing each other in this manner is not going to solve the problems. The rich and the poor nations together should formulate policies for establishing a sustainable earth society. Every individual will then follow the ethical principles involved in such a society.

6.10 ENVIRONMENTAL LAWS

All countries have various laws for environment protection. Even in olden times, laws were formulated by governments to protect the environment. In India also, several Acts were in force even before independence. Some examples are Indian Fisheries Act (1897), Bengal Smoke Nuisance Act (1905), The Motor Vehicles Act (1935) etc.

However, environment became a big issue only after the United Nations Conference on Human Environment was held at Stockholm, Sweden, in 1972. In this conference, Prime Minister Indira Gandhi had spoken strongly and effectively for India and other developing countries. It was after the Stockholm Conference that the world woke up to the seriousness of environmental issues.

6.10.1 The Environment (Protection) Act of 1986

Union Carbide, an American company, had set up a chemical plant in Bhopal. On the midnight of 3rd December 1984, forty tonnes of highly toxic methyl isocyanate (MIC) stored in the company leaked into the atmosphere and spread throughout the city. Though according to the Govt. of India, death toll on that night was 2500, according to environmental activists, 8000 people had died on that night itself. Another 8000 people had died later due to toxic effects. In addition to this, 2,00,000 people were injured due to the gas leak. Thus the Bhopal Gas Tragedy was the worst industrial accident in history.

The exact number of deaths and injuries that night will never be known, as there was total confusion after the tragedy. However, so many American lawyers came to Bhopal immediately. Seeing this and to speed up the payment of compensation to the relatives of the dead and the affected, Govt. of India passed the Bhopal Act in March 1985.

It was with the Bhopal Act in mind that the Govt. of India passed the Environment (Protection) Act in 1986. This is called EPA 1986 for short. The Act was passed under Article 253 of the Constitution.

***The purposes of the Act are*:**

1. To implement the decisions of the United Nations Conference on the Human Environment of 1972
2. To protect and improve the human environment
3. To prevent hazards to human beings, other living animals, plants and property
4. To provide a framework for the Central Government to co-ordinate activities of the Central and all State authorities established under previous laws.
5. To give wide powers to the Executive to enable bureaucrats to frame necessary rules and regulations.

The scope of the Act: The Act was passed to provide protection to the environment and to improve the environment. The act includes 26 sections in 4 chapters. The Act is applicable to the whole of India.

The scope of EPA 1986 is quite broad. Section 2 of the Act defines the following terms thus:

- "Environment" includes water, air and land and the interrelationship that exists among and between water, air and land, and human beings and other living creatures, plants, microorganisms and property.
- "Environmental pollutant" means any solid, liquid or gaseous substance present in such concentration as may be, or tends to be, injurious to environment.
- "Environmental pollution" means the presence in the environment of any environmental pollutant.
- "Hazardous substance" means any substance or preparation, which, by reason of its chemical or physico-chemical properties or handling, is liable to cause harm to human beings, other living creatures, plants, microorganisms, property or environment.

As per Section 3 (1) of the Act, the Central Government can "take all such measures as it deems necessary for the purpose of protecting and improving the quality of the environment and preventing, controlling and abating environmental pollution."

Based on the Act, the Centre can:

- Set new national standards for the quality of the environment (ambient standards)
- Set standards for controlling emissions and effluent discharges
- Regulate industrial locations
- Prescribe procedures for managing hazardous substances
- Establish safeguards for preventing accidents
- Collect and disseminate information regarding environmental pollution

As per Section 23 of the Act, the *Central Government can delegate its functions and powers to:*

- Any officer
- State Government, or,
- Any other authority

According to Section 24 of the Environment (Protection) Act, the provisions and the orders based on the Act will over ride any other law. Under Sections 6 and 25, the Centre has the power to make more rules.

Chapter 3 of the act is on environmental pollution. Almost all the subjects listed in Section 3 are about industrial pollution.

All this shows how powerful the Act is. It also shows how vast the powers of the Central Government are and through it, the state governments are controlling pollution and protecting the environment. The Centre has rightly used the powers the Act has given it in passing several rules and regulations regarding pollution and toxic substances. For example, the Centre has extended protection to ecologically sensitive areas like the sea coast. It has also introduced a system of labelling products with 'Ecomark'.

The details of the Act in different areas are:

1. *Pollution control*: Section 7 of EPA prohibits discharge or emission of pollutants in excess of the standards set. There are 3 types of standards:
   - Source standards – The polluter must reduce at the source of emission and discharge of pollutants.
   - Product standards – New manufactured products must follow pollution norms
   - Ambient standards – These are standards set for maximum pollutants in the air so that the concerned people live a healthy life.

   In Schedule 1, 89 industries have been listed. These industries must follow the standards fixed for each of them regarding disposal of wastes. In the case of other industries not listed in Schedule 1, general standards must be followed. The general standards are also called Minimum National Standards. These have to be met by every industry.

The Pollution Control Boards can set stricter standards where necessary. Their duties are:

- Advise the industries to set up waste water treatment plants
- Encourage industries to follow the 3 R's – i.e. Reduce, Reuse and Recycle wastes
- Motivate industries to set up biogas plants and use other alternative energy sources
- See the effluent carrying capacity of the medium into which it will be discharged etc.

New motor vehicles must meet emission and noise limits. Air-conditioners, refrigerators, air coolers and some types of construction equipment manufactured after 31st December 1993 must meet the noise levels set for them.

Levels of ambient air quality have been fixed for various air pollutants and noise, for industrial, residential and commercial areas. Some zones like areas near hospitals must be 'Silent Zones'.

2. *Hazardous substance regulation*: Many rules have been formulated based on EPA. According to these rules:
   - A person generating hazardous waste materials must do proper handling, storage and disposal of these wastes.
   - Hazardous industries must identify major accident hazards and submit a safety report to the authorities stating the preventive measures being taken.
   - An importer of hazardous chemicals must give complete information on product safety to the authorities. It must transport the chemicals according to the rule set.
   - Biomedical wastes must be disposed of properly by hospitals and clinics.
   - Manufacture, import, export, storage and use of hazardous microorganisms and genetically engineered cells have to get proper licence for such activities.
3. *Environment Impact Assessment*: An assessment of the possible effects of an industry, construction work etc. on environment is called Environment Impact Assessment (EIA). It was in 1994 that our Government decided to make EIA reports compulsory for industries. Only after the Impact Assessment Agency clears the project can any industry be started. In some cases, further review by experts will be necessary before the clearance can be given. The agency is the Ministry of Environment and Forests.
4. *Coastal Regulations and Protection of Special Areas*: Special regulations have been made in ecologically sensitive areas. The Coastal

Zone Regulations were issued in Feb. 1997. As per these regulations, developmental activities and tourism cannot be done within 500 metres from the sea shore. This rule is applicable for the entire coast of India. Setting up of new industries cannot be done within this range. Commercial activities are restricted. Building activity is regulated according to the level of urbanization and the ecological sensitivity of the coastal region.

Some horticultural belts and the Himalayan foothills in some places, parts of the Aravalli range etc. are also ecologically sensitive areas. These regions face special environmental risks. All such areas are also covered by this rule.

5. *The Ecomark scheme*: Manufacturers of consumer goods can use the label 'Ecomark' if the products follow the rules for environment-friendly products set by the Government. This scheme was announced in 1991. Its aim is to encourage manufacturers to make environment-friendly products. It will also help consumers to make choices between environment friendly and other products. However, till 2003, no product carrying Ecomark had reached the market.

   Another rule brought recently, in 1999, prohibits packing of food stuff in recycled plastic bags. They must be virgin bags, or they must be at least 20 microns thick. This ensures that thin bags made of plastic recycled over and over again cannot be used to food packing.

Violations and Penalties under the Act: The EPA was the first Act to give the Government sweeping powers for environmental protection. It can issue direct orders to:

- Close, prohibit or regulate any industry or operation or process
- Stop or regulate the supply of water, electricity or any other service

Officers of the Central Government assigned with the specific duty can enter the industry for:

- examination;
- testing of equipment; and
- taking samples of air, water, soil or any other substance from any place for analysis.

Discharging of pollutants in excess of the standard permitted is unlawful. Persons who break EPA must prevent pollution and report the discharge to the authorities.

Penalties:

- For each failure to follow the rules, a prison term of up to 5 years or a fine of Rs. 100,000/- or both
- For every day of continued violation of the law, fine will be Rs. 5,000/-
- If violation is for more than 1 year, prison term to be up to 7 years

Just an outline of EPA is given here. It is an elaborate Act, with many sections and sub-sections. It is essential to create awareness regarding this Act among the public. It must be followed by one and all. Punishments should be given to all offenders.

6.11 AIR (PREVENTION AND CONTROL OF POLLUTION) ACT OF 1981

The Air (Prevention and Control of Pollution) Act was passed by the parliament under Article 253 of the constitution. This was done to implement the decisions taken at the United Nations Conference held at Stockholm in 1972.

The Act displays the 'detrimental effects (of air pollution) on the health of the people as also on animal life, vegetation and property.'

The main provisions of the act are:

- Providing for prevention, control and abatement of pollution
- Providing for the establishment of Central and State Pollution Control Boards
- Give powers to the Boards to implement the Act

The Air Act is similar to the Water Act which was passed in 1974. The Central Pollution Control Board (CPCB) and the State Pollution Control Board (SPCB) were already having certain powers under the Water Act. The Air Act expanded the authority of CPCB and SPCB's to include air pollution control.

All sources of air pollution come under this Act, such as:

- Industries
- Power plants
- Vehicles

With the help of Air Act, the Pollution Control Boards can check the emission of all types of air pollutants like:

- Carbon dioxide
- Sulphur dioxide
- Carbon monoxide
- Nitrogen oxides
- Lead
- Smoke
- Volatile organic compounds

Under the Air Act, all industries operating within such areas must get a permit from the State Boards.

The states must prescribe emission standards for different types of industries and automobiles after consultation with the Central Board. While doing so, the states must note the ambient air quality standards of the industries.

Before 1987, the Air Act was not very strict. It used to give mild penalties to the law breakers. In 1987, there was an amendment which made the authorities more powerful and the penalties more stiff. Now the PCB's can stop water supply or electricity supply to a defaulting industrial unit or even close it down. The PCB's can also apply to the court directly to restrict the industries to keep emissions or effluents under the permitted limit. Also, noise pollution was included in the Air Act. This was a good step as it is air that gets polluted with noise and so noise pollution is definitely one type of air pollution.

The Air Act of 1981, as amended in 1987, became more powerful in certain areas:

- The amendment gave permission to state governments to designate certain areas as pollution control areas. All industries operating within such areas must get a permit from the State Board. Under the Air Act, all industries operating within such areas must get a permit from the State Boards.
- Magistrates were given power to restrain air polluters from discharging emissions
- Citizens can file suits; they can also ask the PCB's to give them the necessary emission data so that they can file suits.
- The State Boards were given the power to notify standards independently

Violations and Penalties: Anyone disobeying any order or breaking the Air Act can be put in jail for 3 months or can be fined for Rs. 10,000/- for each day of disobeying. Both imprisonment and fine can also be given to the industries, based on the seriousness of the offence.

Air Act Cases: There have been many Air Act cases in India. Many are pending in different courts too. Some famous examples are:

- Dwarka Cement Works case
- Stone quarrying case of Dehradun
- The Taj Trapezium (The Taj Mahal) case
- Motor Vehicles Case etc.

6.12 WATER (PREVENTION AND CONTROL OF POLLUTION) ACT OF 1977

Discussions went on between the Centre and the states for more than ten years before the Water (Prevention and Control of Pollution) Act was passed in 1977. It was passed under Article 252 (1) of the Constitution. This Article empowers the parliament to enact laws on subjects under state control if the state legislatures of 2 or more than 2 states demand such laws. In this case, twelve states wanted the Act and so it was passed in the parliament.

Since 1982, the Central Board has been attached to the Central Government's Department of Environment, Forests and Wildlife. The Water Act covers different types of water bodies like:

- Streams
- Inland waters
- Underground waters
- Sea water
- Tidal waters
- Wells
- Sewers etc.

The Act provides for a permit system to prevent and control water pollution. The Act prohibits discharging polluting matter into streams, wells and sewers or on land in excess of the standards established by the State Boards. A person must obtain permission from the State Board before starting any industry, operation or process, any treatment or disposal system or any extension or addition to any such system.

As per the Water Act, the CPCB (Central Pollution Control Board) can:

- Advise the Central Government on water pollution issues
- Fund investigation and research related to water pollution
- Develop an all inclusive plan for control of water pollution

Important features of the Water Act are as follows:

- The Water Act suggests that state governments can make water pollution laws.
- The Act gave power to the state boards to regulate pollution
- The State Boards will encourage, conduct and participate in research and investigations on water pollution.
- They will develop reliable and economical methods of treatment of sewage and effluents from industry and other sources.
- The Central Board will co-ordinate the activities of the State Boards.

Powers of the State Boards with regard to the Water Act:

- The State Boards can establish and enforce standards of effluents discharged into water bodies by factories.
- They can control sewage discharges by industries.
- State Boards can approve, reject or modify applications submitted by factories in any matter concerned with effluent or sewage discharge.
- To minimize water pollution, they can advise state governments which sites would be best suited for starting new industries.
- The State Boards can give conditional permission to industries, setting restrictions on location, construction and use of outlets and nature and composition of the discharges.
- They can start necessary construction, if even after 30 days the defaulter has not started it, and charge the amount on the defaulter.
- The State Boards can enter and inspect the industries or operations when necessary.

Violations and Penalties: Before the amendment in 1988, the Water Act was not very powerful. Enforcement was done by the following methods prior to 1988:

1. Boards could start criminal proceedings against the erring industries
2. Applications could be submitted to magistrates to restrain polluters.

After the amendment, State Pollution Control Boards have more power. They can:

1. Withdraw power supply of the industry at fault
2. Withdraw its water supply
3. Close a defaulting industrial plant
4. Put severe punishment on the defaulters

***Water Act Cases*:**

The first and second cases mentioned were against the Central Board for Prevention and Control of Water Pollution, while the third was versus State of Rajasthan.

- Pondicherry Papers Ltd. Case
- Delhi Bottling Co. Pvt. Ltd. Case
- Agarwal Textile Industries Case

The Water (Prevention and Control of Pollution) Cess Act was passed in 1977 to help meet the expenses of the Central and State Boards. Cess means tax. As per this Act, a differential taxing system is followed—higher rates for water are to be paid by the defaulting units. This revenue is used to implement the Water Act.

6.13 WILDLIFE PROTECTION ACT OF 1972

Based on the resolutions passed by eleven states under Article 252 (1), the Parliament enacted the Wildlife (Protection) Act in 1972. It has the following features:

- Wildlife advisory boards can be set up and wildlife wardens and other staff can be appointed to implement the Act.
- Schedule 1 animals and birds cannot be hunted at all.
- For hunting animals and birds of Schedules 2, 3 and 4, special permission is needed.
- State governments can declare any area of ecological importance as a sanctuary or a national park.
- In both these, public entry is restricted.
- No wildlife or habitat can be destroyed here.
- Cattle and other animals can be grazed in sanctuaries, but this cannot be done in national parks.
- Regulates trade in wild animals, animal products and trophies.
- Imposes penalties for violating the Act.

Species have been listed in different Schedules by the Act. These are as follows:

Endangered species	Schedule 1
Species requiring special protection	Schedule 2
Big game	Schedule 3
Small game	Schedule 4
Some species classified as vermin	Schedule 5

Bringing any harm to the endangered species is prohibited throughout our country. There are regulations for hunting other species. Species belonging to Schedule 5 can be hunted without any permission. It is the responsibility of the wildlife wardens and staff to see that the Act is followed by all.

The Act was amended in 1982. As per this amendment, wild animals can be captured and taken to other places for scientific management of their populations. For example, *ex situ* conservation of wild animals is done in zoos. (See Unit 8) This beneficial procedure should not suffer due to the Act. That was why such an amendment was brought in 1982 for the Wild Life (Protection) Act of 1972.

In 1991, the Act was amended again. In this amendment, three major additions were made:

1. Special chapters were included to protect specific plants and animals.
2. Functioning of zoos was also regulated.
3. The needs of tribal people and forest dwellers were recognized and suitable changes were made keeping their welfare in mind.

The Wildlife Act adopts two policies for protection of wildlife. Accordingly:

- Some species (the endangered species) are protected everywhere.
- All species are protected in special areas like sanctuaries and national parks.

The Wildlife Act was passed in parliament under Article 252, after eleven state legislatures passed the necessary resolutions. The Forty-Second Constitutional Amendment shifted the subject of wildlife to the Concurrent List. By this, the parliament got the power to bring laws on wildlife more easily.

***Violations and Penalties*:**

- Any violation of the Act attracts imprisonment and fines.
- Citizens, after giving a notice of 60 days, can prosecute the Government.

There are some pitfalls in the Act. For example:

- The Tehri Dam will destroy 12 rare and endangered species. Is this hunting? If so, can the construction of the dam be stopped?
- For hunting animals and illegal wildlife trade, punishment is very mild in Jammu and Kashmir, while it is severe in all other states. Are such discriminations right?

- Why is there a provision for supplying certificates of ownership to those who possess skins, horns etc. of rare animals?

Some cases:

- The Sariska Case—Tarun Bharat Sangh, an NGO wanted to know-how the Government of Rajasthan could allow mining in a protected forest.
- Jamnagar Marine Park case—Gujarat Navodaya Mandal challenged the permit given to Reliance Petroleum Ltd. by the Chief Wildlife Warden for laying a pipeline through a sanctuary.
- The most recent cases are against some celebrities for hunting endangered animals.

6.14 THE FOREST (CONSERVATION) ACT

The Indian Forest Act of 1927 used to be followed by the forest departments of individual states for many years. When India got independence, forests were placed under the State List of the Constitution. At the time of independence, forest cover in India was 25%. This went on decreasing and touched an all time low of 10% according to some environmentalists in the 70's. However, due to large-scale reforestation programmes, there has been some improvement. The forest cover in India stood at 18% in 1997. Even this is much less than the stipulated percentage of 30 to 33% for ecological stability.

The harmful effects of felling trees indiscriminately were becoming very prominent in India. At the same time, specialists in different fields debated how to balance the different demands of the nation as forests had to be tapped for various purposes. It was then that the Government of India passed the Forest (Conservation) Act in 1980. It was later amended in 1988. A revised National Forest Policy was also adopted in 1988.

Under the Forest Conservation Act, an Advisory Committee has been constituted. This committee will advise the Centre on funding and the various other issues. As per the amended Act, states have to get approval of the Central Government for:

- De-reserving a Reserved Forest
- Using forest land for non-forest purposes like mining, building dams for irrigation, hydroelectric power etc.
- Assigning forest land to a person or corporation
- For clearing forest land for reforestation etc.

At the same time, the State Governments have the power to:

- Use the forest for forestry purposes
- Construct water holes, trenches, checkposts, fences, pipelines etc.
- Stop any undesirable activity in the forest as soon as observed

The 1980 Forest Act was passed by the Centre in alarm. Until then, an average of 1 million hectares of forest land was being cleared for various purposes in India. After the Forest (Conservation) Act was passed, deforestation rates did come down to some extent. But this was not enough. Therefore some amendments were made in the Act again in 1992. Accordingly:

- Cultivation of crops like tea, coffee, spices, rubber etc. is non-forest activity.
- Cultivation of oil yielding and medicinal plants and fruit trees need prior permission. No permission is needed if the crop is a native species.
- Taking stones and boulders from forests is a non-forestry activity.
- Tusser culture in the forests is a tribal activity to make silk from insects. It is the livelihood of tribals. So this is allowed. But local host plants must be used. Cultivating mulberry bushes in the forest for rearing silk worms (sericulture) is not allowed.
- Any proposal for any non-forest activity must provide a cost-benefit analysis first. An EIA study (Environment Impact Assessment study) must also be done to know the possible effects of the proposed activity on the surrounding environment as a whole.

***Violations and Penalties*:**

The Act prohibits the cutting of trees without the prior approval of the Central Government. Forest land cannot be used for non-forest purposes. Disobeying the Act will lead to up to 15 days in jail.

***Some Cases*:**

- The Dehradun Valley litigation—This was the first case where the Supreme Court had to balance ecological integrity against industrial needs. Thousands of acres of Mussorie Hill Range of the Himalayas were blasted by dynamite for limestone mining.
- ARC Cement case—This company operated a cement factory in the Dehradun valley violating the Forest Act. The company was ordered to shift elsewhere.

6.15 ISSUES INVOLVED IN ENFORCEMENT OF ENVIRONMENTAL LEGISLATION

It took long years for man to know that he was degrading his own environment; it took longer years for him to realize that there should be environmental legislations to prevent environmental degradation. Now that legislations are there, and it is our duty to obey the rules and regulations. And it is the duty of the Governments to see that we do obey them, and if we do not, to punish us.

Bringing up issues in the parliament, pass them and make them into Acts is all very good. Yet enforcement of legislations on everyone, irrespective of class, creed and every other divisive factor in society, is as important as enacting them in the parliament.

For enforcement of legislation, an effective agency would be needed. Such an agency has three important functions:

1. Collect all necessary data

2. Process the collected data in detail
3. Pass on the data and the analysis of the data to the law enforcing agency

Environmental legislations may be made by states, countries or international agencies and organizations. Internationally, breaking any environmental law is considered an environmental crime. Under international law, countries that have ratified (signed or agreed to) treaties are responsible for carrying them out. But governments cannot do this all by themselves. Governments must take the support of business houses, media and ordinary citizens to ensure that the treaties are followed properly.

In 1920, there were only about ten international treaties. By 2000, the number had gone up to 240. This shows how the world is getting more serious about the environment.

6.15.1 Some Important International Treaties

- 1971 – Ramsar Convention on Wetlands of International Importance – 123 parties (signed by 123 countries)
- 1972 – London Convention on the Prevention of Marine Pollution – 78 parties
- 1973 – Washington Convention on International Trade in Endangered Species of Wild Fauna (animals) and Flora (plants) – CITES – 152 parties
- 1987 – Montreal Protocol on Substances that Deplete the Ozone Layer – 175 parties
- 1992 – Convention on Biological Diversity – CBD – 178 parties
- 1997 – Kyoto Protocol on Climate Change – 30 parties

These agreements cover many issues like:

- Conservation of wetlands
- Biodiversity
- International fisheries
- Ocean pollution
- Atmospheric pollution etc.

Enforcement of international environmental treaties is done by many institutions and officials. Secretariats are staffed by international civil servants and are often managed by UN Agencies like United Nations Environment Programme (UNEP). Governments of the countries that signed the treaties are in general very co-operative and provide the necessary data.

For example, a network of national organizations called TRAFFIC was created in 1976. It monitors international wildlife trade. It sees how different nations are enforcing the legislation on wildlife trade, finds out the routes of smuggling of wild animals and plants and passes on the information to CITES. International trade in more than 900 endangered species, like great apes, sea turtles, crocodiles, some orchids etc. is monitored by CITES.

In India, many NGO's like Bombay Natural History Society (BNHS) help the affected people to take matters to court. Responsible citizens often file public interest litigations (PIL's) in court. In many towns and cities, there are citizens' forums which pursue matters of pollution, conservation etc. and take the offenders to court. We also have consumer courts for dealing with cases in adulteration. In villages also, where at least some people are educated and are aware of human rights regarding environment, they form co-operatives and file cases. Small environmental crimes can be dealt with at local levels also by the village elders, Panchayats, Lok Ayukts etc. This shows that environmental awareness of the public is very important in preventing environmental crimes.

6.15.2 Environment Impact Assessment (EIA)

Wherever there is development, some damage to the environment is done. The population is increasing in our country every minute, demand on the space available is becoming more and more. The land area remains the same, but the number of people who have to share this area is increasing rapidly. Just to build more homes, imagine how many trees will have to be cut. To put up industries, to make dams, to construct roads and railway lines, environment has to be disturbed.

However, development cannot be stopped to protect the environment. The solution to this problem lies in a procedure called Environment Impact Assessment (EIA). By this procedure, all possible effects on all aspects of the environment will become clear to the persons or industrialists or agencies that are planning a particular activity in a particular locality. After seeing the report, the proposed activity can be started as it is or in a modified form or the locality can be changed. If the activity is very harmful to the environment, it will have to be dropped altogether and some other proposal will have to be thought of.

The Ministry of Environment and Forests has made EIA compulsory before the permission is given to start any project, private or public. After the enactment of EPA in 1986, EIA was made compulsory for any project.

Many organizations and agencies have been given the power to do EIA studies. The person or persons who propose the project have to give an application to the SPCB first. After going through the project carefully, the SPCB will instruct the proposers to get an EIA done. The next step is for the proposers to approach a certified organization or agency to prepare an EIA report.

Preparation of an EIA report will take many months. This is because all the effects on all the aspects of the environment will have to be studied thoroughly. Mostly, studies are done on the effect of the industry on the air, water and soil of the locality. But in fact, this is not enough. The locality may have great biodiversity. By setting up of the industry or building complex there, this may be lost. This is especially true of large projects like dams, roads or railway lines through forests etc. If the organization does not have competent staff in all areas of environmental studies, the EIA report will be sketchy and not complete.

As it is the proposers who pay for the reports, the reports may be incomplete too. The agency may not list all the expected problems so that the proposer's project will be cleared by the Ministry.

In spite of these drawbacks, the rule that EIA report is a must for new as well as old industries planning for expansion is a very good procedure to help in environmental protection. The report finally goes to the Ministry of Environment and Forests, which is the impact assessment authority.

In 1997, the Ministry announced that a public hearing is also essential. The SPCB will advertise in 1 or more local newspapers the date, time and venue of the public hearing. Local people are supposed to attend the meeting and express their opinions. They will also be shown the EIA report.

This system works very well in developed countries, but in our country, most of the villagers are not aware of the fine details mentioned in the report. Therefore, they cannot identify the seriousness of the impact of the proposed project on their locality. This drawback can be rectified only by increasing the level of education and environmental awareness in the common people.

The damage to the environment may be small or big, temporary or permanent, restricted to some areas within the proposed area or spread everywhere. In some cases, the damage can be partly undone over a few years. All these must be clearly mentioned in a good EIA report. The effect of the project on the local people must be given importance too. Especially, when big dams and other such structures are built, thousands of people are displaced, and this creates big rehabilitation problems. In case of industries dealing with toxic chemicals or radioactive substances, special study is necessary to judge the long term effects of the emissions and effluents on human health.

As water is a major natural resource and is in great demand, exact information on how much water will be needed for the industry and how much of it will be from the underground aquifers must be looked into. The local people must not face water shortage due to the setting up of the industry. It has been made mandatory in many states that unless an industry has made provision for rainwater harvesting, permission will not be given to it for starting the enterprise. Similarly, any new industry must have a fool proof system of waste water management so that effluents are not discharged into water bodies nearby. The maximum noise that will be produced by the proposed project has to be seen too as noise pollution has serious consequences on human health.

Only some aspects have been discussed here. Plans made by the proposers for minimizing all types of pollution and conservation methods to be adopted will be studied in detail by the organization or agency conducting the EIA.

Many organizations have been certified as competent by the Government for conducting EIA studies. One such institute is the National Environmental Engineering Research Institute (NEERI), Nagpur. As per the Annual Report of NEERI, twenty-three EIA reports were prepared by the Institute in 2004-05.

To explain how the procedure works, one example is given here. The study was Impact Assessment of diverting irrigation water from Pakadiguddam tank Project (PTP) to an industry. The State of Maharashtra wanted to divert water from PTP in Rajura Taluka, Chandrapur District, Maharashtra, to cement and power plants of M/S Maratha Cement Works. Based on writ petitions filed, the Nagpur Bench of the High Court, Bombay directed NEERI to submit a report on the impacts of agricultural operations and suggest mitigation measures. NEERI scientists studied the water conducting system, PTP site and nearby areas, discussed the issues with the farmers, State Government officials and the industry.

The study showed that management by the Irrigation Department was not satisfactory and that water supply to the industry had caused depletion of irrigation. Suggestions were given as to how maximum benefit could be obtained for both the industry and the farmers.

This shows that EIA study is done not to obstruct industrial or any other type of development. It is meant to give suggestions to all parties as to how to utilize natural resources fully without causing harm to the environment.

6.16 PUBLIC AWARENESS

Pollution of air, water and soil, and mindless utilization of natural resources have raised environmental degradation to dangerous heights across the globe. One of the very best methods to reverse this trend, undoubtedly, is creating public awareness in environmental issues.

Methods of creating public awareness have been discussed at national and international levels. The goal is to develop a world population that is aware of and concerned about environment and its associated problems, and committed to work individually and collectively towards solving the current problems and preventing new ones. To achieve this target, an action plan is needed at all levels of implementation—global, national and local.

6.16.1 At the International Level

There are many environmental organisations and agencies at the international level active in different fields of work. Some examples are:

- *United Nations Educational, Scientific and Cultural Organisation (UNESCO)*: This UN agency was founded in 1945. Among other activities, it also supports activities related to environmental quality, human settlements, training to environmental engineers and other socio-cultural programmes related to environment.
- *International Union for Conservation of Nature and natural resources (IUCN)*: It is an autonomous body founded in 1948. IUCN always involves the local people in any project, thus spreading awareness.

- *United Nations Environment Programme (UNEP)*: This is a UN agency established in 1972. It acts in collaboration with different governments for environmental monitoring and protection.
- *Man and Biosphere Programme(MAB)*: This programme, launched by UNESCO in 1971, has 14 projects. India is also a member country. Indian National Committee on MAB is under the Dept. of Science and Technology in New Delhi. Latest activities of MAB include sustainable tourism, eco-job training for young people etc. Such activities attract the youth, which is a sure way of spreading awareness.

  Earthscan: This is an agency founded by UNEP in 1976. It collects original articles on environmental matters and sells them to newspapers and magazines, especially in developing countries, creating public awareness.

  Earthwalks: These are designed to focus international attention on environmental issues. Children are taken out by seniors for programmes like "Meet a Tree", "Leaf Slides" etc. Such nature studies create environmental awareness in young children.

6.16.2 At the National Level

All countries have different governmental and non-governmental organisations to promote public awareness in environmental issues.

(a) *Governmental Organisations in India*: The Government naturally yields the maximum power and does most of the environment related work.

- *Ministry of Environment and Forests (MoEF) of the Central Govt.*: It was set up in 1980. From 1986 onwards, MoEF has been organising National Environment Awareness Campaign (NEAC) every year. The objective is to create public awareness at the national level. Proposals are invited from NGO's, educational institutions etc. for conducting various activities like seminars, workshops, camps, padayatras, competitions, festivals, training courses, rallies, public meetings, exhibitions, demonstrations, preparation of educational resource materials, folk songs and dances, street theatres, science fairs, preparation or use of CD-ROMS and other multimedia tools etc.

  Many institutes work under MoEF. One of them is the Centre for Environmental Education (CEE), Ahmedabad. CEE is a resource agency for National Green Corps (NGC) scheme of the Govt. of India. In this scheme, NGC promotes formation of eco-clubs in schools. Each school is given Rs, 1,000/- to start an eco-club. Thousands of schools have already made use of this facility. To create awareness in the youth of the country, CEE introduces courses in colleges and taps youth energy through volunteer activities. CEE has published books in 15 languages on environment related issues. It uses media to raise awareness of the general public. It works with the industrial sector also to convince them of the need for waste minimization. CEE involves students and communities in water quality monitoring with people's participation in local environmental issues.

- *Autonomous institutions like Indian and National Institute of Technology's (IIT's and NIT's)*: These institutions do different types of research works on environmental issues and popularise their findings. It is due to the result of such activities that many villages in our country now have gobar gas plants, smokeless chulhas, solar cookers etc.
- *The University Grants Commission, Universities and Colleges:* UGC has recently introduced a syllabus on environment for undergraduates in all sections—Arts, Science and Commerce. (This book is meant for this particular course). Earlier, Environmental Science was taught only to Science students. Research work done in many universities is published for public awareness. In many colleges, eco-clubs have been established which conduct various types of public awareness programmes.
- *State Governments*: Governments of all states have ministries of Environment and different Developmental Authorities. There are the Corporations, Municipalities and Panchayats also. All these implement various programmes related to environment in India. These help to create awareness from the top rung to the grassroots level. Celebration of Van Mahotsav twice every year is one such event. Ever since Rabindranath Tagore presented to the nation concepts and practice of "Vriksharopana" in 1927, tree planting has become a common practice in India. Slogans highlighting importance of trees can be seen everywhere. Area under plantation forests has increased greatly in recent years.

Spreading awareness about the need for family planning and the need to increase literacy can never be over-emphasised. Non-formal methods of education like Adult Education, Total Literacy Mission etc. have been taken up by both Central and State Governments and various NGO's in India. These have been partially successful. The fact that maximum environmental awareness is shown by Keralites proves how important the role of education is.

(b) *Governmental or government supported agencies in some other countries*: Many such institutions are active in all countries. A few of them are briefly described here.

- *Environmental Protection Agency (EPA)*: The best known of all is this independent Federal Agency of the U.S. Govt. established in 1970. EPA believes that for work on environmental awareness, partnership is needed among the states, tribal and local governments, research communities, NGO's and industry.
- *Texas Environmental Awareness Network (TEAN)*: This is an association of state agencies and environmental and educational organisations dedicated to creating public awareness in environmental issues.

- *Universities*: Some established universities, especially of the developed countries, make use of latest technologies like remote sensing, DNA tracing etc. to identify problems and solve them. In one recent application, scientists from the University of Auckland in New Zealand confirmed with DNA tracing that whales are being sold in fish markets of Japan. Wide publicity was given to this finding and soon corrective action was taken and the culprits punished.
- *Governmental Ministries and Departments*: In Denmark, the Govt. has promoted wind energy since the adoption of the national energy plan of 1976 and has established wind cooperatives. There are subsidies for wind and taxes on polluting fossil fuels in Denmark.

6.16.3 At the Local Level

Though governments have great powers, they look to NGO's, businessmen, media and citizens for help.

- *Non-Governmental Organisations (NGO's)*: The role of NGO's in creating public awareness is most significant. Kerala Sastra Sahitya Parishad (KSSP), Bombay Natural History Society (BNHS), World Wildlife Fund (WWF) India etc. do valuable work in this field. The number of NGO's all over the world has gone up greatly.
- *Business houses*: These have an important role too. Consumers are increasingly getting interested in green products and are willing to pay more for them. As a result, organic industry has blossomed globally. Waste reduction is another common greening business strategy. Such actions by business houses create awareness in society.
- *Media*: Mass media like radio, TV, newspapers, documentary films, hoarding boards etc. have a great influence on the people. We regularly get information on population growth and disappearing forests from the media. They also help in giving the public information on how to construct rainwater harvesting systems, where to get eco-friendly products etc.
- *Individuals*: Efforts by individuals are no less commendable in creating awareness among larger sections of the society. Rajendra Singh, the Magsaysay Award winner and Medha Patkar who leads "Narmada Bachao Andolan" are well known. Much less known are people like Simon Oraon, the 71-year-old uneducated tribal in a village near Ranchi, who persuaded his villagers to construct check dams to conserve water, to practise rainwater harvesting and to follow multiple cropping for better utilisation of soil. Sarah Jewitt, who did her doctorate from Cambridge University, stayed with Simon for 9 months to learn his methods. A copy of her thesis is Simon's prized possession though he cannot read it. Setting practical examples like this goes a long way in creating public awareness.

Environmental Calendar: The Environmental calendar can be used for organizing appropriate activities. This is done in all countries to create public awareness in environmental issues. Some days that can be celebrated for motivating the society are also included in the following list.

1. January 12 – National Youth Day
2. February 2 – World Wetland Day
3. February 28 – National Science Day
4. March 8 – International Women's Day
5. March 21 – World Forestry Day
6. March 23 – World Meteorological Day
7. April 7 – World Health Day
8. April 18 – World Heritage Day
9. April 22 – Earth Day
10. May 31 – Anti-tobacco Day
11. June 5 – World Environment Day
12. June 11 – World Population Day
13. August 6 – Hiroshima Day
14. August 9 – Nagasaki Day
15. September 8 – World Literacy Day
16. September 16 – World Ozone Day
17. September 27 – World Tourism Day
18. October 1 to 7 – Wild Life Week
19. October 4 - World Animal Welfare Day
20. October 16 – World Food Day
21. December 1 – World AIDS Day
22. December 10 – Human Rights Day

Various activities like seminars, debates, quiz, elocution, essay, drawing, painting, slogan, poster, exhibitions etc. can be organized indoors by students in schools and colleges on these days. Outdoor activities can be street plays, rallies, public speeches, folk songs and dance etc. All such and similar activities will help create public awareness on various environmental issues.

QUESTIONS

A. Very short answer type:

(a) Multiple choice type – Tick the right answer:

1. In the world's population, India accounts for:

(a) 10% (b) 5%

(c) 25% (d) 16%

2. In 1992, the Earth Summit was held in:
   (a) Geneva (b) Montreal
   (c) Rio (d) New York
3. Coal, petroleum and natural gas are:
   (a) Fossil fuels (b) Renewable fuels
   (c) Alternative energy sources (d) None of these
4. Silviculture is cultivation of:
   (a) Herbs (b) Medicinal plants
   (c) Vegetables (d) Trees
5. EPA is an agency of the Government of:
   (a) India (b) China
   (c) USA (d) Japan
6. One example of alternative energy is:
   (a) Coal (b) Petroleum
   (c) Wood (d) Solar energy
7. Pisciculture is rearing:
   (a) Fish (b) Aquatic plants
   (c) Pearls (d) None of these
8. To neutralize acidity of soils damaged by acid rain, add:
   (a) Potassium nitrate (b) Urea
   (c) Gypsum (d) Superphosphate
9. World Environment Day is celebrated on:
   (a) 5th June (b) 11th June
   (c) 12th January (d) 8th September
10. Radioactive pollution affects
    (a) Air (b) Water
    (c) Soil (d) All of the above

(b) Define:
1. Sustainable development
2. Conservation
3. Silviculture
4. EIA
5. Unsustainable development
6. Greenhouse effect
7. Climate
8. Watershed

(c) Explain briefly:
1. Environmental problems created by unsustainable developmental processes
2. Comparison between sustainable and unsustainable development
3. Agroforestry
4. Social forestry
5. Sand mining from rivers

(d) Name the following:

1. 3 alternative energy sources
2. 3 aspects of forest management
3. 3 meaningful methods of water conservation
4. 2 energy sources that pollute the environment
5. 3 emissions that can be checked by Pollution Control Boards
6. 3 methods by which deforestation can be checked
7. Reasons for cutting down the use of fossil fuels
8. 3 greenhouse gases
9. Important International Treaties
10. The person who leads 'Narmada Bachao Andolan'

(e) Briefly explain the following in 2 or 3 sentences:

1. Sustained yield concept
2. Greenhouse gases
3. Methods to prevent over-fishing
4. Provisions of Air Act 1981
5. Kyoto Protocol
6. How burning fossil fuels causes atmospheric pollution
7. Disadvantages of major dams
8. Acid rain
9. Sources of carbon dioxide in the atmosphere
10. Global warming
11. Consumerism
12. Ozone hole
13. The scope of EPA 1986
14. Hazardous substances regulation
15. Environment Impact Assessment (EIA)
16. Coastal zone regulation
17. Ecomark
18. Centre for Environmental Education, Ahmedabad
19. Environmental calendar
20. Reforestation

(f) When and where was the Earth Summit held?

(g) What can we do as individuals to fill the ozone hole? Give reasons for your answer.

B. Short Answer type:

Write short notes on:

1. Earth Summit
2. Fossil fuels

3. Conservation of water bodies
4. Interlinking of rivers
5. Powers of the State Pollution Control Boards with regard to Water (Prevention and Control of Pollution) Act, 1977
6. Energy conservation
7. Central and State Pollution Boards
8. Effects of climate change
9. Ozone hole and its effects
10. Global warming and its effects
11. Harmful effects of acid rain
12. Solutions to the problem of acid rain

C. Long answer type:

(a) Give a detailed account of climate change and its effects on environment.

(b) What do you understand by 'acid rain'? Explain in detail.

(c) Write notes on any two:

(i) Wasteland reclamation

(ii) Air (Prevention and Control of Pollution) Act

(iii) Watershed management

(iv) Problems concerning resettlement and rehabilitation of people

(d) Write an essay on Ozone layer depletion.

(e) Give an account of EPA 1986.

(f) Explain in detail methods to create public awareness on environmental issues.

(g) Write long notes on any two:

(i) Water Act 1977

(ii) Wildlife protection Act, 1972

(iii) Forest Conservation Act, 1980

(iv) Enforcement of Environmental legislation

(h) Give an account of the various methods of rainwater harvesting.

(i) What is Environmental Ethics? Discuss the issues related with it and possible solutions.

(j) Discuss in detail what can be done to conserve our natural resources.

(k) What is consumerism? What are the bad effects of consumerism and wastage.

Unit 7

Human Population and the Environment

7.1 INTRODUCTION

At the World Food Conference held in Rome in 1974, Henry Kissinger, then U. S. Secretary of State, declared that by 1984, no man, woman or child would go to bed hungry. Thirty-six years have passed since; not just ten as Kissinger had wished; more than a billion people in the world do go to bed hungry even today. One main reason for this is definitely the uncontrolled population growth in many parts of the world.

7.2 POPULATION GROWTH, VARIATIONS AMONG NATIONS

As per the 'pro-population school' of thought, most of the ills that the present generation is suffering are because of the extremely fast rate of population growth. They also predict that the condition of the future generations will be worse, if this trend is not reversed.

In fact, the rate of population growth was less in the earlier centuries than compared to later centuries, though there has been some downward trend of late. Let us examine some figures (last three figures estimates by UN, 1978).

Table 7.1 World population in different years

Year (A.D.)	*1800*	*1850*	*1900*	*1930*	*1950*	*1970*	*1980*	*1990*	*2000*
World population in millions	919	1,091	1,571	2,070	2,513	3,678	4,415	5,275	6,199

From these figures, it is clear that population growth was comparatively slow till 1950. It was about 0.8%. From 1950 to 1970, growth rate was about 2%, while the decline started slowly after that. Between 1970 and 1990, the growth rate fell to about 1.9%, while it fell further to about 1.6% from 1990 to 2000.

Population growth is not same in all countries of the world. It is very much uneven. Populations have been growing very fast in the developing countries (Less Developed Countries or LDC's). In the developed countries (More Developed Countries or the MDC's), populations have been growing very slowly. The difference is so much that world figures show high population growths, as is clear from the table.

Table 7.2 Uneven distribution of world population (in millions)

Year	*World*	*MDC's*	*LDC's*
1950	2,513	832	1,681
1960	3,027	945	2,082
1970	3,678	1,050	2,628
1980	4,415	1,131	3,284
1990	5,275	1,206	4,069
2000	6,199	1,272	4,926

The Table also shows that LDC's accounted for about 70% of the world's population in 1950. This increased to 79% by 2000. In the MDC's, on the other hand, rate of population growth has declined since 1970.

Within both developed and developing regions, population has been increasing fast in the urban areas. Increase in urban population has led to the formation of slums, pollution and environmental degradation. Increasing urbanization also results in economic imbalance, which leads to social problems also.

Table 7.3 Size of urban population in millions

	1950 A. D.	*1980 A. D.*	*2000 A. D.*
World	724	1807	3208
Developed Regions	449	834	1092
Developing Regions	275	972	2116

Overpopulation results in economic poverty as the available resources are shared by too many people. Moreover, poor countries with large populations are selling their natural resources like forest produce to the rich countries. Because of this, the rate of deforestation has increased greatly in the tropical countries. Twenty million hectares of forests are lost per year in these countries.

Brazil is one of the LDC's. Its economic condition is so poor that it is forced to export huge quantities of timber from its rain forests to MDC's. The dense Amazon forests are losing millions of hectares of rain forest every year. The fast depleting natural resources of Africa, Latin America, South and South East Asia have changed vast tracts of natural vegetation into wastelands.

Exploitation of natural resources to support the ever increasing human population has many adverse effects on the environment like:

- reduction in forest cover due to unmindful felling of trees;
- excavation of land for mining purposes;
- reduction in cultivable land;
- water shortage due to excess withdrawal of groundwater without recharge; and
- pollution of air, water and soil and all the after effects of pollution.

Social effects of population growth have also begun to show in different countries and different regions within the same country, like:

- illiteracy and unemployment;
- migration from rural to urban areas;
- increase in violence and crime;
- deterioration in hygiene and health etc.

7.3 POPULATION EXPLOSION

Human population in the world is not just growing; we can say that it is exploding. The six billion mark was reached on 12th October, 1999. World population has doubled since 1960.

More than 95% of the population growth is in the developing countries. Meanwhile, population growth has slowed down in Europe, North America and Japan. The Population Division of the UN Department of Economic and Social Affairs has estimated that world population will grow to 8.9 billion by 2050.

The poorest countries often have the highest population growth rates. In 62 countries in Africa, Asia and Latin America, over 40% of the population is under 15 years of age. Population is exploding fastest in the poorest countries. And within countries, it is the poorest families that mostly tend to be the largest ones. The share of the global population of the poor countries has been increasing steadily for more than 40 years now.

Population explosion results in:

- Internal instability
- Natural disasters
- Poverty
- Social disruption
- Energy crisis
- Global climatic changes
- Shortage of water
- Water pollution
- Increase in water-borne diseases
- Reduction in fish resources
- Soils becoming infertile due to intensive farming

- Increased air pollution
- Desertification due to overgrazing
- Destruction of coastal ecosystems, the most delicate ecosystems
- Marine pollution leading to destruction of coral reefs
- Encroachments into forests
- More floods caused by mining of sand from rivers for building construction
- Increase in the number of accidents as traffic becomes unmanageable
- Inflation due to shortage of essential articles
- Increase in crime due to unemployment and frustration
- Reduction in discipline and other such values
- Increased waste generation and resulting spread of diseases etc.

An estimated 13 million refugees have fled their own countries to escape from harassment and discrimination, armed conflict or violence. A considerable number has been forced to leave due to social or environmental reasons though they are not refugees. Millions are displaced within their own countries. Most of them have added to the number of the urban poor.

Table 7.4 Continent-wise population (in million) in the year 2000

Asia	*Africa*	*Europe*	*Latin America & Caribbean*	*N. America*	*Oceania*	*Total*
3,684	800	728	518	306	31	6067

As is clear from the following Table, the population of the world has exploded so much that we have started to refer to the numbers in billions!

Table 7.5 Taken from Population Division of UN Economic & Social affairs Dept., 1998

	Africa	*China*	*India*	*Rest of Asia*	*Latin America & Caribbean*	*Europe*	*USA & Canada*
Population in 1995 In billion	0.7	1.2	0.9	1.3	0.477	0.728	0.297
Population in 2150 In billion (Projected)	2.8	1.6	1.7	2.8	0.916	0.595	0.424

Let us have a look at the present situation. Many figures have been given by the Population Reference Bureau. Some selected figures from this 2006 World Population Datasheet are given here, just for us to realize how urgent it is to put a stop to the population explosion.

Table 7.6 World Population Clock, 2006

	World	*More Developed Countries*	*Less Developed Countries*
Population	6,555,336,000	1,216,198,000	5,339,138,000
Births per year	137,035,288	13,496,866	123,538,422
Births per minute	261	26	235
Deaths per minute	107	23	84
Infant deaths per minute	13	0.2	12.8

The statistics are very clear. Differences between the MDC's and the LDC's are too contrasting not just in the total population, but also in the number of births per year. The number of births per minute has been compared here for the readers to understand the gravity of the situation. Population is so high in the LDC's that poverty is more, resulting in poor healthcare. This is shown by the contrasting figures in deaths per minute. 78.5% of the total deaths per minute in the world take place in the LDC's. In the case of infant deaths per minute, the difference is still more. 98.5% of the infants who die are in the LDC's.

Even after so many infant deaths, the total population is very high in the LDC's. This shows how extremely high birth rates in these countries are. Let us analyze some statistics.

The Indian subcontinent is adding about 21 million people a year, the equivalent of another Australia. India's population has nearly tripled over the last 50 years. It has grown from 350 million in 1950 to 1 billion in 2000. According to the projections by the UN, India will add 515 million more people by 2050. That is almost twice the population of the US at present.

Pakistan is expected to double its population in the next 50 years by adding 189 million people by 2050.

Bangladesh is expected to add 83 million people by 2050.

The subcontinent is thus expected to add another 787 million people by the middle of this century.

From the above account, it is clear that for removing poverty, it is essential to curb the population explosion going on in the less developed countries of the world. For this, practising family planning is a must. Moreover, it is necessary to improve the overall health of the entire family. In healthy families, infant mortality will be minimized and life expectancy will be more. All these are included in what is known as 'Family Welfare Programmes'.

When people are healthier, they will be stronger, with more stamina and this will lead to more efficiency. This will lead to overall increase in the prosperity of the nation. There will be fewer persons to share the resources. Healthy people will be happier too. For all this to happen, it is imperative to practise and popularise family welfare programmes.

7.3.1 Family Welfare Programmes

To curb the present population explosion, especially in the developing countries, it has become imperative to adopt family planning measures. Along with family planning, other health related activities are also a must. All these together are called 'Family Welfare'. It is good to see that the developing countries have started working seriously in this direction.

The most essential part of any family welfare programme is family planning. As the words indicate, family planning is planning to keep the family small. Small families are in general happier and healthier.

(a) *Family planning*: By family planning, couples can choose how many children to have and when to have them. Family planning programmes are conducted mainly by governments and NGO's. The central as well as state governments in India have Health and Family Welfare ministries. In addition to birth control, these ministries look into birth spacing, breast feeding, antenatal (before delivery) care of pregnant women, post-natal (after delivery) care of mothers, parental care of infants, immunization programmes for infants etc.

There are many methods of birth control.

1. *Sterilization*: This is a minor operation done under local anesthesia. It can be done in both males and females. There are no post-operative problems. Both are permanent methods of birth control. However, in case the need arises later to have babies, these can be revoked by a second operation.
   (a) *Vasectomy*: Sterilization operation in males is called vasectomy. Here, the tubes that carry sperms are tied so that sperm is not able to go out. There is a common misunderstanding in men that this operation is injurious to health and that men lose their strength and stamina after vasectomy. The truth is that vasectomy does not affect men's sexual ability or general physical or mental strength. Sperms are still produced in the testis, but they are absorbed back into the cytoplasm of the cells. The fluid that comes out during intercourse is just a mixture of some other secretions, and does not contain sperms. Making Indian men aware of this truth will lead to more men going for this sure method of birth control.
   (b) *Tubectomy*: Sterilization operation in females is called tubectomy. In this operation, the fallopian tubes that carry the egg (ovum) to the uterus are tied with the help of surgical threads. Women who have undergone tubectomy produce ova as usual, but they die within the fallopian tubes after 3 or 4 days. The ovum can be fertilized only if it reaches the uterus. Now-a-days, tubectomy has become easier after the invention of the laproscopic method.
2. *Condoms*: These are rubber sheaths used by males during intercourse. Sperms are not able to enter into the vagina, thereby preventing pregnancy. One condom can be used only once. The use of condoms is the most widely used method of birth control in India. Condoms made by

different companies are available in the market. The trade name of condoms made by the Govt. of India is 'Nirodh'. It is provided free of cost, and is part of the family planning programme of the Government of India.

3. *Intrauterine devices* (*IUD*): There are a few types of IUD's, but Copper T's are the commonest. A Copper T is a small T shaped object which is inserted into the uterus by a doctor. This prevents the fixing of the ovum on the wall of the uterus, even if it gets fertilized. The fertilized ovum can grow further only if it is attached to the uterine wall. It is essential to change the Copper T every 3 years. Copper T does not cause any discomfort and the woman can do all her works as usual. However, rarely, the woman may have excess bleeding during menstrual periods or may develop backache. In such cases, she should consult her doctor. If her doctor advises her to discontinue using IUD, she can do so and go for other birth control measures. However, this happens only rarely. Most of the women who use Copper T are very comfortable with it. Other IUD's like loop, spiral ring etc. are also available, but Copper T's are the most popular.
4. *Oral pills*: Many brands of oral contraceptive pills for women are available in the market, like Mala D, Saheli etc. Such pills prevent the formation of ova in women. While using these, care should be taken not to miss even a single dose. Before starting pills, a doctor should be consulted. Oral pills may have side effects like high blood pressure, irregular bleeding, nausea etc. Yet many women do use them. Pills to be taken daily and those to be taken weekly—both types are available.
5. *Following the safe period*: In the monthly biological cycle of women, there is what is commonly called 'safe period' and 'danger period'. In one month, a woman produces only one ovum. If this ovum is not fertilized in time, the woman will not become pregnant in that month. The ovum is active and ready for fertilization during the middle of the monthly cycle. During these days, intercourse should not be done. Starting from the 1st day of menstrual period, the first ten or eleven days are 'safe'; the last ten or eleven days are also 'safe'. The days in the middle is called the 'danger period', as there is the risk of pregnancy if intercourse is done on these days. This is actually the fertility period. The ovum is usually formed on the 14th day, counting from the first day of menstruation. It is to be safe that 3 or 4 days are left on both before and after this day, because there may be slight variations in the day of formation of the ovum. This method works well with women who have regular menstrual cycles. For those whose periods are not regular, this method is not advisable.
6. *Use of spermicides*: Spermicides are chemicals that kill sperms. These are available in the form of pastes, jellies etc. under different trade names. The paste is to be applied in the vagina before intercourse.

These are usually harmless, yet women should see that they are not allergic to these formulations.

7. *Abstinence*: This is the only method which is 100% sure of results. Abstinence means abstaining from, i.e., avoiding intercourse. This is not possible always, yet is adopted by some couples at times.

(b) *Abortion*: All the methods described above are meant to prevent pregnancy. In case a woman does conceive even after following any of these methods, she can go for abortion. Abortion is medical termination of pregnancy. It should be done in early pregnancy, before the foetus reaches a maximum of 60 days (preferably 45 days). The age of the foetus is calculated from the first day of the latest menstrual cycle. Some couples go for abortion as a birth control measure, but it is always better to use contraceptives or go for sterilisation. Abortion raises many ethical issues; it also affects women's health, especially in the long run.

The methods described above can be used to space the children also. The ideal difference of age between two successive children is three or four years. The idea is that it is better to have a second baby when the first baby starts to go to school. Spacing of babies is essential for the health of the mother and the child.

In India, quite often we see that parents' craving for male babies is the main reason for their not adopting family planning measures. This attitude is deep rooted in many societies, but times have changed. Like many other social evils, it is high time this attitude is changed too. Whatever the sex of the first two children, no third baby should be given birth to. If every Indian follows this rule, we can at least stabilize our population. It is the duty of every educated person to practise family planning in his or her own life and also to convince others to do the same.

Family planning measures have been very effective in China, Indonesia and Brazil. But in India and Africa, it has had moderate or poor results. This is because family planning services are not readily available in many rural areas. According to UN estimates, one half of births in LDC's are unplanned and one fourth is unwanted. If services reach every corner of the country, thousands of abortions can be prevented.

Some countries use the incentive-disincentive strategy to promote family planning. This is a sound method by which families who follow the national policy are rewarded and those who do not are not rewarded or even punished in some manner. For example, the first two children will get fee concessions, while the third child will have to pay extra fees. Or, government houses allotted to families with two children can be bigger than houses allotted to families with more children. Such measures are often effective.

To reduçe births, it is essential to empower women. In the LDC's especially, most women are fully dependent on men and do not have any say in any matter, including the number of children they should bear. This situation will change only if women become educated and financially independent. All right thinking men are helping them achieve this goal.

In general, family planning measures are more concentrated on women. An opinion is gaining ground that more research is needed to find out pills and other methods for men.

As mentioned earlier, family welfare includes not only family planning, but also other healthcare practices. Pregnant women must take nutritious food and the exercises advised by the doctor. They must have peace of mind also. Babies born to such women will be healthier. For delivery, it is best for women to be admitted to nearby hospitals. After delivery, women must again have good food and the prescribed dose of iron tablets.

Babies must be given all immunization as required. These can be given either at hospitals or at family welfare clinics. The first vaccines to be given to a newborn baby are BCG and the 1st doses of oral polio vaccine and hepatitis B vaccine. Further doses of polio and hepatitis B are to be given at required intervals up till the age of 5. Other vaccines to be given in time are DPT, measles and tetanus. Till the child is 15 to 16 years of age, the vaccinations will continue at fixed intervals. Immunization cards given from the hospitals/ family welfare clinics will have all details. Parents must get their children immunized as per the instructions given in the cards.

It goes without saying that healthy children grow into healthy adults. In making this possible, contribution of both the parents is essential. Fathers must not think that immunizing children is the duty of mothers alone. As an old saying goes, 'Health is wealth'.

7.4 ENVIRONMENT AND HUMAN HEALTH

Human health is closely related to the environment in many ways. Environment has both physical and mental influence on us.

Physically, there are many diseases caused by unhygienic living conditions. These afflict mostly the poor and the uneducated people. One main reason for the spread of such diseases in our country is the overcrowding due to the population explosion. Too many people live in crowded localities without proper sanitation. People throw garbage here and there and there is no proper garbage removal system. There is no waste water management either and dirty water accumulates in patches here and there. Cattle and other animals are also reared by some, and there is no drainage system from their sheds. Ponds in which people take bath are also used to bathe animals. Ponds and other water bodies are polluted by the throwing of garbage. Lack of toilets is another serious problem in such localities. Smoke from 'chulhas' pollutes the air. In short, soil, water and air are polluted in overcrowded localities.

The main reason for such unhygienic conditions seen in villages and slums in cities is lack of education. It is essential to include lessons on hygiene in school textbooks. The role of family welfare workers and NGO's also helps a lot in making people realize the importance of hygiene.

Many diseases are caused by the lack of sanitation and the resulting environmental pollution. The most important of these are:

- *Water- and food-borne diseases*: Diarrhoea, dysentery, cholera, amoebiasis, jaundice etc. are the common gastrointestinal diseases found mostly in the LDC's. These diseases are caused by drinking unclean water or by eating food without washing hands. Thus, these diseases are clearly associated with hygiene. Millions of children die due to diarrhoea every year. In India, as per one estimate, one out of every five deaths of children is due to diarrhoea. The Lack of education compounds the situation because most of the deaths in all these cases are due to dehydration. The uneducated do not understand the importance of supplementing water intake to compensate water loss from the body through loose motion and vomiting. Poliomyelitis and viral hepatitis are also water-borne diseases. Typhoid spreads by drinking water or eating food containing the disease causing bacteria.
- *Insect-borne diseases*: Malaria, filaria, encephalitis (brain malaria), dengue, chikungunia etc. are spread through mosquito bite. Mosquitoes breed in accumulated water. This is a direct result of poor water management. Moreover, mosquito eradication schemes are not very successful in the LDC's due to overpopulation and lack of awareness and education of the poor. Moreover, when insecticides like DDT are banned in many places due to its non-biodegradable toxic nature, other effective alternatives have not been made available.
- *Diseases caused by parasitic worms*: Children of poor families often play in the mud and then eat food without washing hands properly with water and soap. As a result, they ingest eggs of disease causing worms like round worm, thread worm etc. along with food. Hookworms enter into the bodies of people who walk barefooted on infected soils. Tapeworms infect those people who eat pork that is not well cooked. All such cases are much more common in the LDC's as they are all concerned with poverty and lack of awareness.
- *Plague*: This is also a disease associated with sanitation. Plague is a primary disease of rats. Fleas that bite infected rats get the disease. Human beings get plague if they are bitten by the fleas.
- *Kalazar*: This disease is spread by sandfly. It is seen in rural areas. It occurs when cattle sheds and poultry farms are built near houses.
- *Respiratory diseases*: The commonest respiratory disease is asthma. Other examples are bronchitis, lung tuberculosis etc. These diseases are either caused by or are aggravated by air pollution. Inhaling smoke from chulhas, dust from the air, exhaust fumes from automobiles, gases released from industries etc. all lead to various respiratory diseases. It is essential for people to know the importance of ventilation in houses. As far as possible, fresh air must be inhaled. Lung cancer is caused mainly by tobacco smoke. The smokers themselves (active smokers), and also others closely in contact with them (passive smokers) may get cancer.

- *Diseases caused by noise pollution*: Very loud noise is harmful to human health. Loud noises are common in factories, during festivals, on busy roads, near aerodromes etc. In addition, some people have the habit of keeping the TV or radio on high volume. All this loud noise damages the ears, and cause deafness. Other problems are sleeplessness, nausea, ulcer, lack of concentration, irritability, high blood pressure etc.
- *Diseases caused by pesticides*: Farmers in general sprinkle too much pesticides on their crops to protect them from various diseases. But they do not realize that in the process, they are exposing themselves to various disorders. Inhalation of pesticide fumes affect hormonal balance and even cancer. In addition to this, the fruits and vegetables that we buy from shops almost always have pesticides on their surfaces. Therefore, it is most important to wash them well before use.
- *Diseases caused by herbicides*: Herbicides are chemicals used to kill small plants (herbs), which are weeds growing with crop plants. These cause skin diseases in human beings.
- *Cancer due to exposure to radioactive substances*: People working in factories or projects dealing with radioactive materials are at a risk of getting radiation induced diseases. There is a maximum level of radioactive substances, which can be tolerated by humans. If these levels are crossed unknowingly or by neglect on the part of the authorities, workers get diseased. Cancer is one of them. Mutations may be produced in the sperms and eggs of workers. This will affect future generations.
- *Diseases due to exposure to hazardous chemicals*: Toxic chemicals are manufactured in many factories. Most of these factories are in the developing countries. This is because the developed countries know that such factories pollute the environment. In the developing countries, environmental laws are not strictly followed; offenders often go unpunished; corruption is also more. Therefore, the MDC's try to set up their polluting units in the LDC's. They try to send their old ships containing tonnes of hazardous waste to countries like India for breaking down. Labour is cheap in India. Population and poverty are so high that people are prepared to work in unhealthy situations, that too for lesser wages. Constant exposure to toxic chemicals causes different diseases, including cancer, tuberculosis, skin diseases etc.

7.4.1 Detection of New Diseases and Reappearance of Controlled Diseases

As and when man detects diseases, research works start and in most cases, drugs are found to cure those diseases and in many cases, to prevent them too. These drugs act on the disease causing organisms (parasites). However, as time passes, some of these germs become immune to the drugs by changing their own structure by mutation. Many of these diseases spread fast in the slums and overcrowded localities and sometimes become epidemics.

The great flood in Mumbai in 2005, it was found soon, was caused mainly by drains clogged by plastic. People had to wade in water to go to work for even 3 or 4 days. Many got infected with a strange disease, which was diagnosed as 'Spirillosis'. It was caused by a certain type of spirillum (a bacterium).

7.4.2 Diseases due to Climate Changes

There is no doubt that global climate is changing due to human activities. Temperatures are rising, rains have become unpredictable, incidence of floods and droughts, heat waves and cold waves have increased. Human health is deteriorating due to all these. As more regions become warmer, malaria, which is a tropical disease, will certainly spread more north and south of the equator. The *El Nino,* which causes periodic warming, may affect rats and this may increase the diseases in humans.

Death and destruction due to climate changes brought about by human interference in the world environment will be discussed under Disaster management.

7.4.3 Efforts to Prevent Environment Related Diseases

Efforts in this direction should be taken first by the government and secondly by NGO's and other social welfare organizations. Citizens also have a duty to live clean lives and try to prevent the diseases from spreading.

- Provide clean drinking water to all
- Provide separate localities for rearing domestic animals
- Set up many more public lavatories in all localities
- Run adult literacy programmes sincerely
- Prepare textbooks for adult education including chapters on health and hygiene in detail
- Make education for children compulsory
- Change patterns of agriculture in such a way that dependence on chemical pesticides is highly reduced
- Popularise biopesticides like neem, garlic, 'hing' etc.
- Do not allow industries to come up near residential areas
- Control traffic
- Be strict with automobile servicing and checking rules
- Arrange for tree plantation drive etc. so that air will be purified
- Make rainwater harvesting compulsory for upcoming structures to solve water problem
- Punish people polluting water bodies
- Encourage the use of alternative energy sources like solar energy
- Make family planning programmes stricter

- Set up health clinics in all villages and towns based on the number of residents in each area
- Make mobile clinics available to the people
- Construct roads in villages; if roads come up, all other facilities are sure to follow. This will remove the disinterest shown by qualified doctors at present to practise in villages
- Give importance to educating women and empowering them; health of the whole family depends a lot on the women of the house
- Set up family welfare clinics with qualified doctors, nurses and midwives in all villages. Many women in India die due to complications related to childbirth. Clinics are helpful in immunization programmes of children too.
- Punish polluting industries and projects severely
- Arrange frequent cleanliness drives in villages and also slums in cities
- Make industries using wood as raw material plant trees
- Conduct various programmes to create environmental awareness among the common people
- Award a certain percentage of marks to school and college students who spread environmental awareness, literacy etc.
- Enforce all environmental legislation with full force
- Poverty is closely related to health; implement poverty eradication schemes.

This is not a comprehensive list at all. Central and state governments, non-governmental organizations, self-help groups, social service and family welfare agencies etc. must all pitch in with full vigour and enthusiasm to prevent environment related diseases in society.

7.5 HUMAN RIGHTS

It is an established fact that Vedic India had a strong tradition of respect for human rights. The principle was 'constant desire of giving to every man what is due to him'. The king was duty bound to enforce the law of the land and also the rights and claims of his subjects. Later, Kautilya asserted in the *Arthasasthra*, "In the happiness of the subjects, lays the happiness of the king". *Mathsyanyaaya* (the law of the fish -- i.e. big fish eating small fish) was never allowed in the state. In other words, the strong and the powerful were not allowed to trouble the weak. Traveller historian Megasthanese wrote, "Kings employed spies to detect violators of human rights."

In the modern times, it was in 1948 that the Universal Declaration of Human Rights was adopted by the UN General Assembly. This was the result of an emerging view that the manner in which countries treat their own citizens must be subject to international standards. There was a long list; some are reproduced here:

- Equality of rights without discrimination
- Equal protection of the law
- Social security
- Work under favourable conditions
- Food, clothing and housing
- Healthcare and social services
- Special protection for children
- Education etc.

Few governments deny the fact that these are legitimate rights of any citizen. They agree that good life means suitable means of sustenance, good health, clean environment etc. As all human beings deserve dignity, their claims to these and other basic necessities of life are legitimate.

All over the world, the belief is growing that man is part of nature (this was the basis of all environmental studies in ancient India; we were much ahead of others in this respect, as in many others). Therefore, nature must be preserved for man's well-being. This led to the opinion that there must be a global effort to protect the environment. Many international agreements were drafted for this purpose. Most of the countries of the world ratified these agreements. Human right to equity in different countries, in one's own country and society was widely accepted. So were the rights of future generations to a clean environment, rights to biodiversity, rights to protection of indigenous cultures etc.

There is however a doubt that globalizing environmental issues will not be beneficial. Though the causes for environmental degradation are indeed global, corrective measures are to be taken locally. Hence the slogan, "Think globally, act locally".

Some workers use the term, "Ecological rights" to denote human and environmental rights, which include the following relationships:

- Relationships of human beings of the same country with one another
- Relationships of human beings of the rich and poor countries with one another
- Relationship of all mankind with natural environment

Many environmental issues are closely associated with human rights. One of them is equal distribution among the people of:

- wealth;
- resources, and
- energy.

In all the three, we see that disparities are very high between:

- the rich and the poor;
- men and women, and
- present and future generations

7.5.1 Disparities between the Rich and the Poor Countries

As is clear, in each category, there is an advantaged group and a disadvantaged group. The burden of polluted air, water and land and dwindling natural resources always falls on the poor. Often, it is the poor who are compelled to encroach into a forest just to survive, or keep their surroundings dirty as they cannot afford to keep them clean.

The challenge of bringing in an era of equity is very large indeed. To make this happen, there must be sincere efforts at the national level and also at the global level. Responsibility for the present position of the planet has to be shared between the rich and the poor countries. Therefore, for finding out solutions and implementing them, the rich and the poor countries must work together. For this, trade talks and flow of money is not enough. Social and political ties between governments and civil societies must be strengthened.

7.5.2 Disparities Between the Rich and the Poor Within Countries

In fact, the disparities are more in India than America. Some waste resources and live lavishly while other have almost nothing to eat. Resources belong to all, irrespective of class. But when rich contractors are given permission to fell trees, it is the poor 'vanvasi' who suffers. In addition, the poor forest dwelling tribal people are accused of destroying the forests. Same is the case with poor fishermen. The ocean does not belong to anybody. But when mechanized trawlers catch fish, the poor fishermen suffer losses and even go hungry. The poor fishermen have as much right to the fish in the sea as the rich and powerful trawler owners have.

Rehabilitation of displaced rural masses has always been neglected in many countries while big projects like major dams are taken up. The land where the original settlers lived belongs to them. Without paying them adequate compensation, they should not be displaced. In fact, the best system would be finding them alternative arrangements and then begin the project. But this is never done. The human rights of the poor people are always sacrificed for the benefit of the powerful.

7.5.3 Disparities Between Men and Women

Unequal status of women is seen in many cultures. As per the Universal Declaration on Human Rights, human rights apply to all people equally, "without distinction of any kind such as race, colour, sex, language....or any other status". But the truth is that women are sufferers, doubly. They, like men, are victims of human right violations; in addition, they are discriminated against as women. Women of all countries, races, classes, cultures, religions and nationalities are treated unequally, though in some cases, the treatment is worse.

The Convention on the Elimination of all forms of Discrimination Against Women (CEDAW) or the International Women's Human Rights Treaty was adopted by UN in 1979. CEDAW was the first document to address women's rights.

Most of the world's displaced people and refugees are women, most of the world's poor are women. Still, as they have been given a subordinate status in society, their human rights violations are continuing, as always.

Nutritious food and health are human rights. Yet it is unimaginable how many women are discriminated against in these fundamental rights. In many Indian households, it is the women who take food last, after the men have finished food. In low income households, mostly, the quantity left is not enough for the women. Regarding health issues also, in many houses, the women do not get what is their due. Studies have shown that in India, it is the women and the girl children who are most anaemic.

7.5.4 Disparities Between the Present and the Future Generations

We are depleting the natural resources belonging to the future generations. Of all the creatures of nature, only human beings are capable of planning change. However, we have been using this extra mental capability to change the course of our own development! This is because human beings have a natural instinct to overuse resources and very little incentive to conserve them. Human activities are thus destroying forests, draining rivers dry and raising the level of the seven seas!

It is important that we think of the future generations and not degrade the Earth's life support systems. It will give them options for dealing with the problems they will face. If our ancestors had left us much less of what we have today, what would have been our position? We would not have been able to enjoy life as we are doing now. So also, it is our duty to leave the earth in good shape to our descendants. Our descendants have a human right to a clean and rich environment. On the other hand, we have no right to destroy what is their property. The land, the water, the air, the variety of plants and animals – they all belong to them as much as they belong to us now. We must hand over these natural resources as they are to them. We cannot be greedy and use them all up.

7.6 VALUE EDUCATION

Values are principles based on which we decide what is right and what is wrong. Just being literate does not make one educated. Even one who has a degree, for that matter, need not be educated in the right sense of the word, unless he has been taught the values of life on earth. If an educated person is intolerant, selfish, greedy and dishonest, what is his worth in society? An educated person must be able to differentiate between right and wrong, only then can we say that he or she has had value education.

The search for harmony, knowledge and truth had inspired the ancient Indian minds more than their counterparts in other parts of the world. More than 5,000 years ago, ancient Indian philosophers and thinkers had put forward a theory of morals and values. They were humanists and moralists. They evolved a system of legal theory based on higher values and ideals. Every aspect of life was regulated by *Dharma*, i.e., righteousness. Kings and subjects alike were bound by the law of *Dharma*. It was the concept of values helping mankind to live in harmony with each other and with nature, free from conflicts, exploitations and misery. This system was in vogue for thousands of years.

It is true that values declined gradually in Indian society. Therefore, it has become all the more important to spread value education in today's India.

There are different types of values. They help us live fulfilling lives. By thinking of the welfare of others, we feel happy ourselves. By expanding this theory a little, the coming generations can be included in the 'others'. And then, we will stop exploiting nature. This will automatically result in sustainable development.

7.6.1 Environmental Values

These are important values as environment is all that surrounds us. How we should treat nature and how we should pass it on to our descendants is what environmental values teach us.

Environmental values were etched strongly in the minds of ancient Indians. People knew and felt that environment, as the provider of all, must be conserved. They felt one with nature. They never thought of themselves separate from nature or as superior to other creatures of nature. All were environmentalists. There were great thinkers and astronomers among them. The ancient Indians therefore knew not only about our earth, but also about the universe in general. In many Vedic prayers, we see that peace is sought for the whole universe. Plants and animals were given a lot of importance in the Vedic period. The concept of 'Pancha bhoothaas' was prevalent. The 'Pancha bhoothaas' or the five elements are prithvi (earth), jal (water), agni (fire or energy), vaayu (air) and aakaasham (space). These were known to make the earth and therefore were respected. All 'chara' (movable) and 'achara' (immovable) objects were given due respect. In short, the Vedic period was the golden period of man-environment understanding.

On the other hand, western cultures always considered man the master of the universe. The belief was that God created everything in nature for the sake of man. This approach gave no importance to other living or non-living things. Man could exploit nature to the maximum as per this concept. And this he did, for years, in the west. Now also, the belief remains the same, but people in the West have begun to realize that natural resources are limited and hence he must conserve them. So the West is trying to teach the new doctrine to itself, though the wasteful style of living continues there, especially in the affluent industrialized West.

In general, all over the world, seeing the results of man induced destruction of the earth, people have become worried and hence cautious. This has resulted in the branch of studies called 'Environmental Education'. The aim of this study is to become aware of all environment related issues oneself and also to create awareness among the general public.

- *Man as part of nature*: To realize the environmental values, one has to just go out of the house and look around. The sky, the soil, the plants and animals—they are all part of nature, just as we are. There is oneness in everything. When this feeling comes, it becomes easy to conserve nature and difficult to even think of destroying it. This is the first environmental value.
- *Realizing the harmful effects of our actions*: It is not only that we must not cut trees. We will have to cut some, for our various needs, But we should replace the cut ones. We must not allow soil to be washed away—soil takes years to form. We must not throw garbage here and there—it stinks, it causes diseases; it also makes the beautiful nature look ugly. We will not be able to appreciate the beauty of nature if we ourselves dirty our surroundings. In short, while we need not and cannot stop the development, we must take care that the environment is not degraded.
- *Appreciating the positive effects of our actions*: Suppose we stop wasting energy. We can be happy that we are saving precious energy, brought to our house by the efforts of many. While wasting water, let us remember the village woman walking for 10 kilometres a day to fetch a pot of water. While caging a bird, let us ask ourselves, "How would I have felt if I was locked in a tiny room day and night?" We won't feel like wasting energy or water or cage a free bird. These are small things indeed, but when the message spreads and more and more people begin following such environmental values in life, the overall change will be big.
- *Enjoying the aesthetic values of each part of nature we see*: We go about our work the whole day. We get up in the morning and begin brushing. Do we listen to early morning birds chirping? Can we identify birds with their chirps? (Except the cawing of the crows, of course!) Do we notice the dew drops on the grasses? How many times do we look at the changing shapes of the clouds or the leaves of some trees closing in the evenings? How many times do we stop to take in the scent of opening flowers? Let us enjoy the sights and sounds and scents of nature. We feel one with them; we also experience tremendous peace of mind.
- *Valuing our roots*: One, who values his place of birth or the place where he spent his childhood, will feel like protecting that piece of earth to perfection. This is because he is emotionally attached to the place and cannot see it destroyed.
- *Valuing the holistic character of nature*: Nature is one whole entity. It is not 'a total of the sum of its parts'. We should realize this even while learning of its parts individually. Every part of nature depends on other parts and other parts depend on it. To know and appreciate nature, this realization is essential that all its parts are connected to each other and interact with each other.

- *Valuing all cultures*: All cultures have the right to exist. And they have a right to live life their own way. We should not interfere in their way of living. For example, some try to change the lifestyle of tribal people in an attempt to bring them into the main stream of life. This should not be done. What they need is education and healthcare. Without disturbing their culture, these advantages of modern society must be made available to them.

  Understanding how the earth works: If we understand how the earth works and sustains itself, we can use such earth wisdom to guide our lives. With the same wisdom, we can live peacefully on earth.

  Value of Social Justice: The underprivileged should not be neglected. It is the duty of the privileged to see that everyone is taken along in the path of the forward march of humankind. If this does not happen, there may be revolts in the long run, which will be harmful to all.
- *Learning how to evaluate and resist advertisements*: In the modern world, millions are spent on advertisements. These are meant to tempt us to buy things that we may or may not need. If we learn to analyze the advertisements and resist buying what we do not need, a lot of resources can be saved.
- *Valuing heritages*: We have a rich history. It is our duty to preserve it for our own sake and also for the sake of the generations to come. They are priceless symbols of the past and cannot be reconstructed if damaged. This is why it is essential to preserve the Taj Mahal, our ancient temples and other such heritages.
- *Valuing ecosystems*: Rivers, streams, forests, grasslands, deserts, oceans, croplands etc. are all ecosystems. It is the duty of all of us to protect them. Ecosystems are interdependent and are connected in space and time. Destroying one of them will have a negative effect on others. Therefore we should not do anything that will degrade any ecosystem.
- *Valuing fresh air*: Air is everywhere. We do not need a penny to breathe air. However, if it gets polluted, how we suffer health-wise and monetarily. If we do not want others to pollute the air we breathe, we have no right to pollute it ourselves too. Let us keep the air fresh.
- *Wishing sincerely for a better world*: Value education should not stop at learning the values of nature and all it contains. Its purpose will be served only when every student decides to do something about conserving nature.

7.7 HIV/AIDS

The word "disease" can be split into two – 'dis' and 'ease', i.e., not at ease. Human beings are susceptible to different types of diseases. Most of them are caused by microbes like bacteria and viruses. Fungal infections also lead to some diseases. Some diseases are due to deficiencies of minerals. These are called deficiency diseases or physiological diseases.

Examples of viral diseases are common cold, influenza, measles, mumps, polio, rabies, chickenpox, smallpox etc. and the most serious of them all, with no documented medication found yet, the dreaded AIDS.

Viruses are neither fully living nor fully non-living. They thus form a connecting link between the living and the non-living world. Different types of viruses cause diseases in plants, animals and human beings. All viral diseases are infectious.

AIDS is the short form for Acquired Immune Deficiency Syndrome. It is the progressive destruction of the immune system of the body by the Human Immunodeficiency Virus (HIV). It was first diagnosed in 1981 in a small population of homosexual men. Soon it spread to heterosexual males and females with multiple sex partners, habitual intravenous drug users, infants and small children.

The HIV virus has been recovered in the saliva of infected persons, but its concentration in the saliva is very low and so does not cause infection. Insect vectors have not been found yet, which can spread the disease.

The virus binds to T lymphocytes, also called T_4 cells, and destroys them. This results in gradual non-functioning of the immune system. The result is that the host becomes susceptible to all disease causing microbes like viruses, bacteria, fungi and other parasites. The host thus becomes more and more vulnerable to infection.

HIV can also be seen in certain cells of the intestine and bone marrow. It can also cross the blood-brain barrier, infecting the cells of the central nervous system, such as the brain and spinal cord.

AIDS runs a slow course in many stages. This course is very unpredictable. There are no symptoms immediately after infection. This period of the 1st stage lasts for 6 weeks to 1 year. Only after this period can the virus be seen in the blood. At this 2nd stage, the patient may not show any symptom or may feel weak, or may have fever, swollen glands or rashes. These symptoms disappear within a few weeks. The symptom of swollen lymph glands however lasts for 3 to 5 years.

In the 3rd stage also serious symptoms are not seen, but tests show that the immune system is gradually breaking down.

About one and a half years after this stage, many unrelated symptoms develop. Together, these are called ARC, i.e., AIDS Related Complex. It includes unexpected fevers, night sweats, chronic diarrhoea etc. Disorders of the central nervous system may develop. Cancer of skin, tongue etc. and pneumonia has been reported. Throughout this period, the patient is susceptible to many infections that are normally controlled by the immune system of the body. But now as the immune system has broken down, the patient becomes susceptible to all types of infections. This results in death.

As HIV controls all the functions of the host cells, drugs that destroy the viruses kill the host cells also. This is why it has become so difficult to find out any cure. No effective vaccine is available.

As of now, education is the only way to check the growing number of AIDS cases. Early diagnosis and medical care may delay the complications. It also helps to change the patient's behaviour and prevent transmission.

Some cheap drugs have arrived in the LDC's, but a terrible side effect has been noticed in many cases. The antiviral drugs cause the immune system to recover, and then old hidden infections of leprosy make an appearance. Patients develop painful ulcers on their faces and their fingers become numb. The first such case was reported in 2003 in London. This has proved to be a major drawback of the antiviral drugs discovered.

7.7.1 Some Important Facts about AIDS

(a) It is not exactly known whether everyone infected by HIV develops AIDS. Ten to 30% of infected persons develop AIDS within 5 years. The percentage increases with time. HIV infection is believed to be life long.

(b) HIV-infected people who do not have AIDS can spread the virus. They are the healthy carriers. They can be infected for years without knowing it and can transmit the disease for many years.

(c) *HIV is transmitted mainly by 4 methods*:

1. Sexual intercourse
2. Transfusion of HIV contaminated blood
3. Through injection needles, skin piercing instruments like ear piercing or tattoing needles etc.

A mother can infect her child in the following ways:

1. During pregnancy
2. During birth
3. Through breast feeding

(d) *HIV is not spread by:*

- Casual contact in schools, colleges, at offices, markets, public transports, telephone booths etc.
- Taking bath in swimming pools
- Sharing toilets
- Handshakes
- Hugs
- Kissing on the face
- Eating from the same dish
- Drinking from the same glass
- Massaging each other's bodies
- Mosquitoes or other insects
- Sitting next to infected persons
- Coughing or sneezing

(e) *One can protect oneself in different ways:*

1. Do not have multiple sex partners
2. Use condom—it does not guarantee protection, but reduces the risk greatly
3. Do not share items that could have blood on it, like
   - Razors
   - Tooth brushes
   - Any skin piercing instrument

(f) Persons who are already infected or think that they may be infected should seek medical help immediately. They should inform their partner that they have contracted AIDS. They should not donate blood or sperm or body organs to anyone.

In the year 2000, there were 24 million people all over the world, who had AIDS. Of all the places in the world, Africa has the highest incidence of AIDS. Even by 2000, HIV infection rates had become 20% in South Africa, 25% in Zimbabwe and 36% in Botswana. These are countries with limited government health services.

The role of governmental programmes is very important to curb AIDS. For example, in 1990, both Thailand and South Africa had less than 1% of pregnant women infected with HIV. By 2000, the percentage increased to 20% in South Africa while it was just 2% in Thailand. This was because of extensive public education programmes and a 100% condom campaign. The results were quick to see—condom use rose to 90%, and there was a sharp decline in premarital and extramarital sex and visits to prostitutes.

In India, the first case of AIDS was reported in 1986. Since then there has been a gradual increase in HIV infection. Of the total AIDS cases in India, 89% are in the age group of 15 to 44 years. Seventy-nine per cent of the HIV positive persons are males, while 21% are females. Heterosexual contact is the commonest method of spread, accounting for 74% of the transmissions.

Central and state governments and various NGO's in India are trying to spread awareness about AIDS among the public. Inclusion of this topic in the UGC curriculum for college students is itself a proof if this.

7.8 WOMEN AND CHILD WELFARE

Women and children are at the receiving end in many cases, especially as they are not physically as strong as men. Many injustices are meted out to them in different parts of the world.

7.8.1 Plight of Women in Many Countries of the World

Women account for half of the population in the world. They do almost all the domestic work and childcare. Rural women also help in agriculture and bringing firewood and drinking water from distant places. They have no holidays and no pay for the work they do at home. After all this, it is the women who are treated badly in most cases.

In villages and lower income families of the towns and cities, women eat last, after serving food to all other members of the household. Many a times, there is not enough food left by that time. They get up earlier than the other members of the household every morning and sleep last. While the men folk get holidays to relax and rest, the women work 24 hours a day, 7 days a week. In poor urban families, women remain inside poorly ventilated tiny houses or shanties the whole day and night and inhale the smoke coming from chulhas. They get no chance to rest or relax. In families where family planning is not done, women give birth frequently.

All this has a major effect on their health. Women suffer more malnutrition and get less medical care than men. They become anaemic and get infections faster. About two million women, 95% of them living in LDC's, die every year from pregnancy related diseases, which could have been easily prevented. In many poor households, men do not work or work less. So the burden of looking after the entire family falls on the women. On top of this, men come home drunk in the evenings and beat and ill-treat their wives. In short, poor women have to bear the brunt of every possible inequality.

The inequality is seen in all spheres. Women account for two-thirds of the illiterate persons in the world. In many LDC's, the percentage of literate women is only half of that of men. Adult female illiteracy rates are as high as 57% in India. Women do not have legal rights to own land or borrow money in some societies. The social evils of child marriage, dowry system, the non-acceptance of widow remarriage etc. are practices that go against the welfare of women.

7.8.2 Plight of Children

Population levels and poverty are so high in many LDC's that children are forced to work. There are laws banning child labour in many countries. In India also there is the law, but its implementation has always been poor. Recently, the Centre has decided to enforce the law strictly. All children below the age of fourteen are being taken off from their work places and sent to their homes.

But how far this entire exercise will have the desired result is debatable. There is no guarantee that the parents will take them back. They had to go out in search of work because of poverty. Without poverty being removed, how the parents can suddenly support the children and give them education is not exactly known. It has been proposed that children, whom their parents refuse to accept, will be handed over to NGO's. The desirability of this scheme is also controversial. The problem is that there are too many problems in the LDC's and they all are interrelated. To solve one, the others have to be solved first. Much thought therefore needs to be put while trying to come to a conclusion on the priority list of things to be done.

Inclusion of the Directive Principles of State Policy in the Constitution of India was aimed at protecting the basic rights of individuals. Those who framed our constitution also wanted it to become an effective instrument for social revolution. Article 37 declares that the Directive Principles are "fundamental in the governance of the country" and that "it shall be the duty of the State to apply these principles in making laws". It is thus clear that the Directive Principles must be treated as the human rights provisions of the Indian Constitution.

However, the Directive Principles did not give any legal rights to individuals in case of their violation. This was because the availability of resources and the readiness of the society to accept the changes had to be taken into consideration.

Anyway, even after independence in 1947, some of the provisions of the Directive Principles remain unfulfilled. For example, Article 39 says that the state shall direct its policy in such a manner that all men and women have the right to an adequate means to livelihood, the health and strength of workers, men and women, and the tender aged children are not abused; childhood and youth are protected against exploitation etc. Article 45 relates to the free and compulsory education of all children up to the age of 14.

Education of children is a luxury for many parents even now. Due to poverty, boys are compelled to help their fathers in grazing cattle or work in the field. Sometimes it is due to the absence of schools nearby. In many cases, boys are sent to schools while girls are not. Even if they are sent, girls are taken out of schools soon for helping mothers at home in house work or looking after younger siblings. Daughters also get less food and rest than sons. Percentage of anaemic girls in the LDC's is alarming.

Tens of thousands of children die annually in India due to various diseases and malnutrition. Diarrhoea, measles, malaria, pneumonia etc. are the main causes of death. Many children become victims of breathing trouble because of polluted air. Contaminated water and food causes stomach and intestinal diseases more in children.

Blindness in children is mostly due to malnutrition. In addition to not getting enough quantity of food, intake of food items low in nutrient value also results in malnutrition. In the case of infants, stopping breast feeding early causes malnutrition. The period between stopping breast feeding and starting normal food is more risky for infants. They do not get proper nutrition at this time if mothers do not take extra care. But, for this, mothers need to be educated and families must be able to afford nutritious food.

In fact, unhygienic environment, poverty and lack of awareness are the main causes of many diseases, especially in women and children. This clearly shows how urgently we need reforms in India to correct these imbalances. Steps have been taken in this regard. Ministries of Women and Child Welfare have been set up in almost all the states. Various NGO's are taking active part in organizing awareness programmes like the Breastfeeding Week and the like.

It has been estimated that a minor increase in breast feeding can prevent up to 10% of all deaths in children under five years of age. This is due to 2 reasons: 1. Mother's milk is the purest form of milk and it contains all the nutrients needed for infants till about 6 months of age. 2. When breast feeding is stopped early, infants are given outside food and water which are easily contaminated. This leads to diarrhoea or dysentery and the infants die due to dehydration. Therefore infants should be breastfed as late as possible. In later stages, additional food must of course be given, but not before 6 months. For nursing mothers to produce sufficient milk, they should also know what to eat. Methi seeds are known to increase milk production in nursing women.

Educating the girl child is one major step towards women empowerment. Another is formation of Self Help Groups (SHG's) for women. Over 5.4 million women have organized themselves into 380,000 SHG's in the state of Andhra Pradesh (A.P.). The model of A. P. is being highly appreciated all over India. Most of the other Indian states also have SHG's.

Self-help groups of women in India have been recognized as an effective strategy for women empowerment. These are operative in rural as well as urban areas. They bring together women from all walks of life. The aim is to fight for their rights or a cause. Many issues like health, nutrition, agriculture, forestry etc. are dealt with through self-help groups. The system is based on micro credit, i.e., women deposit small amounts daily or monthly (the amount may be as small as one rupee) to office bearers selected from themselves and take loans from the collected amount when in need. The money is used in income generating activities. The women thus make profit; more than that, their self-esteem and self-confidence increase to a great extent.

The Nobel Prize for Peace of 2006 was won by Muhammed Yunus, Bangladeshi Economics Professor and micro-credit pioneer, who started the Grameen Bank more than thirty tears ago. It provides credit to the poorest of the poor in Bangladesh. This type of credit is called micro-credit. At present, the bank has 1092 branches in 36,000 villages. The bank has 5 million borrowers, of which 96% are women. This silent revolution in Bangladesh is sure to inspire many other countries.

7.9 ROLE OF INFORMATION TECHNOLOGY IN ENVIRONMENT AND HUMAN HEALTH

Information Technology (IT) is concerned with the use of technology in managing and processing information.

IT especially deals with the use of computers and computer software to:

- convert;
- store;
- protect;
- process;
- transmit, and
- retrieve information.

Information technology is also called Information Services (IS) or Management Information Services (MIS). Its new name is Information and Communication Technology (ICT).

Information Technology has been a revolution, just like electricity. It brings about changes in economies and societies all over the world.

IT also helps in environmental issues in different ways, like:

- changing the way how goods are produced and consumed, and
- finding out new means to address new changes.

Computers have become common now even in villages. In many developed states in India, a large quantity of data regarding ownership of land, agricultural practices, details of institutes and agencies which sell seeds, fertilizers etc. and various government communiqués regarding subsidies etc. are stored in computers. The villagers get all information regarding their crops from the internet. Weather forecasts are very important for farmers. These are also available on the net.

Information brings about:

- Equality
- New work environment
- New work ethics
- New work standards
- Objectivity
- Analytical capabilities to decision making and development

Information technology constantly focuses on performance. In the present system of information revolution, the Indian system cannot and should not remain isolated. Indian institutions should be connected to international networks and have access to international data bases and information on new technologies. Such an Information system has a few components like:

- Management Information System
- Project information system
- Personnel Information system
- Financial information system
- Library Information system.

A lot of research is going on in different laboratories and educational institutions on environment and health. The best way to pass on information regarding the results of these experiments is through the Information Communication Technology system.

For example, farmers must know the latest developments in tissue culture, genetic engineering, biotechnology etc. as early as possible for procuring new seeds etc. and for adopting new methods of cultivation. IT helps the farmers in this. The information is put up on the websites of the laboratories and institutions as soon as any breakthrough is made. Computer literate farmers surf the net daily. Thus, they get all the information they need immediately.

Similarly, environmentalists around the world are in touch with each other through the IT system and exchange all types of new knowledge.

Early warning system of disasters can work best along with IT. Predictions of floods, droughts, storms, cyclones etc. and meteorological information on the forecast of weather will be very helpful to farmers and the common man alike.

The World Wide Web (WWW) is the most powerful tool. Any information on any subject is available at the click of a mouse. Therefore, being computer literate helps a lot. As far as India is concerned, many villagers have learnt how to use Information Technology and gather information from the websites. The craze for computers is catching up fast.

Satellite technology has advanced greatly. Of late, meteorological satellites are being used more and more to forecast weather in India also. Use of computers has played a major role in improving the accuracy of predictions. This is because complex calculations can be done very fast using computers. Weather supercomputers can provide an enormous quantity of weather related data to the forecasters.

Mobile phones have become very common. They have penetrated to even villages with very little development. Information on current market prices, weather and other information is passed on through mobile phones (cell phones). In Kerala, owners of fishing boats are constantly using this facility. Sitting at home, they find out every day how the catch was and decide the price of fish for that day. Internet kiosks have come up in states like Tamilnadu. These include internet services, telephones and other related equipment. Each kiosk connects 300 to 400 villages.

IT has been of great help in the day-to-day running of hospitals and other healthcare centres. For example, data regarding birth and death rates, immunization and sanitation programmes etc. can be maintained more accurately than before in computers at health centres. It helps the doctors to monitor the health of the people living in that particular area. Information regarding outbreak of epidemics from remote areas can be sent more quickly to the district information to take corrective action. Patients can seek the help of specialist doctors living far away. Many hospitals have started taking on-line help from expert doctors to provide better treatment to their patients. All this is possible only because of advances in information technology.

There are many websites on health and medical information these days. Anyone can go to these sites with the help of search engines like Google, Yahoo etc. People in far-flung places can get up-to-date information on any health problem from these medical sites. For example, medexplorer.com is one of the several websites for medical information on the Internet. Who.org is the website of WHO, i.e., World Health Organization, one of the important wings of the UN (United Nations).

In short, awareness about IT (ICT) has spread far and wide rapidly in the present times, and people at all levels have come to realize that this technology can bring innumerable benefits to them and change their lifestyles for the better.

7.10 CASE STUDIES

Cases of Information Technology spreading far and wide in India are so many. In the West, it had spread a few decades earlier, no doubt, but India is moving forward very fast. Some case studies in this regard and from other topics in this Unit are given here. Some are from other countries.

1. In Hyderabad, two NGO's, SKS and SHARE are helping women borrowers use electronic 'Smart Cards' to increase savings and find out details about micro-credit services that were available only to the rich clients earlier.
2. In Bangladesh, the Grameen Phone Project gave cellular connectivity to 1,90,000 subscribers, covering 55% of the country by 2002. Poor women entrepreneurs started earning $700/- per annum, which is more than double the per capita income of the country.
3. A website www.ewg.org is keeping track of all subsidies given by the US Government to farmers in the US. The amount was $170 billiion in 2002. The site is getting millions of hits. The system gives information to the citizens while at the same time gives transparency and accountability to the Government.
4. M. S. Swaminathan Research Foundation has created a multimedia database on ecological farm families in India. This database contains many details of farms, like management practices, new technologies, farm economics etc.
5. India's population has grown to 110 crores due to the population explosion. Sixty-five per cent of India's population is farmers. The farm population is increasing by 1.84% per year. The average farm size is becoming smaller each year. Marketing infrastructure is poor, especially in perishable produce. As a result of improved agricultural practices, yield of all crop produce has increased tremendously in India; however, there is no good storage facility. For example, while Indian farmers are suffering huge losses by the rotting of tomatoes, some sauce and ketchup companies in India are importing fresh tomatoes from China. This is most unfortunate. A recent NSSO survey has revealed that 40% of the farmers would like to quit farming, if they have the option to do so.
6. Study in Italy has shown that diets high in refined cereals, especially white bread, is associated with kidney cancer. The study was done by Francesca Bravi and his associates. Seven hundred and sixty-seven cases were studied for the purpose.
7. For every 4 malnourished adults in the world, 5 more are overweight, according to a study by the World Health Organisation (WHO). In other words, out of the 6 billion people in the world, 1 billion are overweight while 80 million do not have enough to eat.

8. Government investments in education and healthcare have increased greatly in some countries. The results are beginning to show. For example, the share of children in the secondary schools of Brazil went up from 46% in 1980 to 66% in 2001. During the same period, the proportion of illiterate women in Brazil has fallen from 27% to 16%.
9. There is a very large health gap between the developed and the developing countries of the world. Polluted water is one of the main reasons for this. Although infant mortality rates have dropped from 25 to 50% in many countries from 1990 to 2000, 43 infants die per thousand in China and 70 per thousand in India as per the 2000 estimates. Most of these deaths are due to water borne diseases.

QUESTIONS

A Very short answer type:

(a) Multiple choice type – Tick the right answer:

1. Present world population is:
   (a) Less than 6 billion (b) More than 6 billion
   (c) More than 7 billion (d) 6 billion
2. One of the diseases caused by the lack of hygiene is:
   (a) Dysentery (b) Chicken pox
   (c) Pneumonia (d) Cancer
3. A disease which is spread by mosquitoes is:
   (a) Influenza (b) Typhiod
   (c) Jaundice (d) Malaria
4. Kalazar is spread by:
   (a) Tapeworms (b) Rats
   (c) Mosquitoes (d) Sandflies
5. Universal Declaration of Human Rights was adopted by UN General Assembly in:
   (a) 1984 (b) 1992
   (c) 1948 (d) 1990
6. The virus that causes AIDS is:
   (a) HIV (b) TMV
   (c) HMV (d) None of these
7. IT helps in providing information on:
   (a) Tissue culture (b) Health and medicine
   (c) Disaster warning (d) All of the above

(b) Define:

1. Values
2. AIDS
3. Information technology (IT)
4. Micro-credit
5. Values

(c) Explain briefly:

1. Why Amazon forests are losing millions of hectares of forest every year
2. Population explosion in India
3. Family welfare
4. Functions of Family Welfare ministries
5. Insect-borne diseases
6. Diseases caused by worms
7. Plague
8. Muhammad Yunus and Grameen Bank
9. Diseases caused by pesticides and herbicides
10. Components of Information System
11. How IT can help farmers
12. Radioactive substances and cancer

(d) Name the following:

1. Three harmful social effects of population growth
2. Five basic human rights

(e) What does AIDS stand for?

(f) In which year did world population cross the 6 billion mark?

(g) What is the full form of HIV?

B. Short answer type:

Write short notes on:

1. Impact of population explosion
2. Sterilisation
3. Types of pollution in overcrowded localities
4. Water- and food-borne diseases
5. Respiratory diseases
6. Cancer
7. Methods to prevent environment related diseases
8. Disparities between men and women in society
9. Environmental rights of future generations
10. Symptoms of AIDS
11. Self-help groups
12. Unhygienic living conditions

C Long answer type:

(a) Describe in detail family welfare programmes.

(b) What are the common methods of birth control?

(c) Which are the diseases caused by environmental pollution? Elaborate.

(d) Give an account of human rights and the various types of their violations.

(e) Write an essay on the different types of 'values'.

(f) Describe in detail what you know about AIDS.

(g) Give an account of the plight of poor children in the developing countries.

(h) What are the various ways in which women are discriminated against?

(i) Discuss in detail the role of IT in environment and human health.

Unit 8

Fieldwork

8.1 INTRODUCTION

Fieldwork is an important part of any scientific study. It makes the subject more interesting and opens up the minds of the learners. They come across many unexpected phenomena and learn much by observation. The learner's observation power increases manifold.

The presentation of the fieldwork in the form of a field report is done in a systematic manner. Usually, one or more teachers take a group of students out for the field study. Sometimes, students go out in groups unaccompanied also. Whatever the case may be, each student must carry a note-book and a pen to note down the observations then and there. Some students must carry a camera too. If they plan to study plants or insects, they must carry a hand lens; if they would like to test water, they may carry a water testing kit, with which some characters of water can be tested then and there; if soil is to be studied, a spade can be used to dig out some soil, which can be brought to the laboratory for testing etc. These are not compulsory; a field report may be made just by noting down the observations and drawing one's own conclusions.

The field report may be hand written or typed. The format of an ideal Field Report is given below:

- **The Cover Page**—This will be a beautifully decorated, colourful page. Here the name of the project must be written in large, beautiful letters at the top or middle of the page. Eg:

  FIELD VISIT TO A POND
  in ...., ...
  TO STUDY POND ECOSYSTEM

  Below this, after a gap, write:

  FIELD REPORT SUBMITTED TO
  THE DEPARTMENT OF.................,
  .... COLLEGE
  IN THE YEAR ....

At the bottom of the page, on the right side, university roll no. may be written in one line and the batch no. in the line below. For example,

11BAHIS0143
SESSION – 2008-2011

- **Certification**—Here, the teacher who took the students will certify the work of the student. It may run like this:

CERTIFICATION

This is to certify that this Field Report has been prepared by Roll No. .... of Batch .... after actual study done in the field on ....

Date .... Signature ....

Name of the teacher ....

- **Acknowledgements**—On this page, the student thanks all people who helped him/her during the work done. He/she will put her signature at the right hand bottom corner of the page and write his/her name below the signature. Date is to be mentioned on the left side. A sample is given below:

ACKNOWLEDGEMENTS

I would like to thank ...., our Principal, for giving us permission to go on this study visit. My sincere thanks are due to ...., our teacher who took us to .... and explained to us all relevant matters in detail. I would also like to thank all other staff members of my department for the co-operation rendered to me. I am grateful to my parents for their wholehearted support in the preparation of this field report.

Date............... ....

(..............................)

- **Contents** - A sample is given below.

CONTENTS

After this page, each chapter must be written in detail. Writing will be done only on one side of the paper, preferably on the right side. On the left side, related photographs can be pasted, or drawings done. Any graph or chart etc. can also be drawn or pasted on this page. If there is nothing related to the matter on the right side, just leave it blank. In this way, continue till Conclusion.

The sheets thus prepared neatly can be given for spiral binding. Channel binding is easier, and can be done by the student. A hard bound readymade file can also be used, with one side ruled papers. If loose sheets are used, they can be punched and inserted in a file. Files must be covered properly and labelled. The report of the fieldwork is now ready for submission.

Suggested Topics: For all topics, the general format of the fieldreport will be the same. Only the names of the chapters except 'Introduction' and 'Conclusion' will change as per the requirement. Under 'Introduction', how the planning for the visit was done, why this particular site was selected, how the trip was made, date of the trip, time when the journey began and when it ended, who all went etc. can be written in the beginning. In another paragraph, a general idea about the site can be given. Under the last chapter 'Conclusion', how the field visit was beneficial, a summary of what was learnt from it, the need of arranging field visits for students in general etc. can be dealt with. In the last chapter namely References, give the names of the books or journals that you have referred to for preparing the report. Names of the authors must be given in alphabetical order, mentioning the titles first and using them for the order. This will be followed by the year of publication, the name of the book or research paper, (if research paper, volume and number), name of the publisher and finally page number/s. In the case of book, page numbers need not be given if you are referring to different pages here and there. In such cases, you can just mention the number of the 1st and the last page, or only the total number of pages. In the case of research publications, it is essential to give the page numbers at the end of each reference. There is no need to number the names of the authors. Some sample formats are given below:

Goel, P.K., 1996: Water Pollution Causes, Effects and Control. New Age International

Publishers, New Delhi

Sharma, P.D., 2007: Ecology and Environment. Rastogi Publications, Meerut.

In the main chapters concerning the study part, various suggestions are given below, under different topics given in the syllabus. All or some of these suggestions may be followed; new ideas can be added too. In the part where observations are being recorded, make sure that past tense or past participle tense is used throughout. For example, write, "The water was greenish in colour and slightly oily" or "A crucible was weighed and some soil was put in it. The weight of the crucible with the soil was taken. The difference on weight gave the weight of the soil. The crucible with the soil was placed in a hot air oven….."

8.2 VISIT TO A LOCAL AREA TO DOCUMENT ENVIRONMENTAL ASSETS

8.2.1 A River

Chapter 1 – Introduction: …………..

Chapter 2 – Location: Find out where the particular river originates from and where it empties into which sea or ocean. This can be done either by searching the internet, by taking help from the teachers or students of Geography Department etc. The latitude, longitude and the altitude of the place can be mentioned. Presence of islands, disturbances in the flow due to any natural or man-made barrier etc. are to be mentioned. Nearest city, distance of the spot from this city, names of nearby villages, names of crops cultivated in the vicinity etc. can all be included in this chapter.

Chapter 3 – Nature of water: Note the colour of the water if any. Try to find out why the water is of this colour. For example, see if any coloured effluent from any nearby factory is falling into the river at this spot, colouring the water. Note if the water has any smell. If you have brought a thermometer, measure the water temperature and the air temperature near the river. If you have brought pH paper, find out the pH of the water. Observe details about the wave motion in the river. If you have got a Secchi Disc, you can measure how clear the water is till what depth.

Chapter 4 – Plants and animals – Try to find out the names of the plants growing in the water. You may see that some are floating near the bank while slightly into the river, some are submerged under water. Both these types of plants may be rooted or may be rootless. Observe all about all the pants, their size, size and shape and arrangement of their leaves, colour, size etc. of flowers if any, and other such facts. You will observe that it is vegetative propagation that is taking place commonly in the large water plants like *Hydrilla, Vallisneria,* water lily (*Nymphaea*), lotus (*Nelumbium*) etc. See what fauna is living in the water. Different types of fish will surely be there. Other than that, at the edges, there may be different kinds of worms, insects etc. Dragon flies may fly over the water. Also observe the birds that catch fish from water.

Chapter 5 – Soil at the periphery: Observe whether water at the edge is clear so that pebbles etc. on the river bed can be seen. Note the colour and nature of the soil. It may be sandy. If so, see whether the sand particles are big or small. It may be clayey and sticky in some parts. Note down the names of the plants and insects/worms on the river bank. Take the local people's help for this. If scientific names cannot be found out, just write common names of the plants and animals wherever necessary.

Chapter 6 – Food chains observed: In this section, mention the food chains observed both in water and on land. Bring some water to the laboratory and study the microscopic plants and animals. Planktons form the base of the food chain. At higher levels, you get various insect larvae, insects, crustaceans, small fish, big fish etc. Also mention the storks or kingfishers or whatever birds you see near the river in the food chain in the water. On the land, a detritus food chain can be seen. Describe that.

Chapter 7 – Human activities: Observe people catching fish, taking bath, washing clothes, bathing animals, boating (for recreation or for carrying people with farm produce across the river or school children crossing the river in a boat to attend school on the other side etc.), village women collecting water plants for food or for sale, or whatever else is happening at the river being studied. Part of the river may be used for immersing ashes and bones of the dead, sewage or effluents from factories may be emptied into the river in some parts. People may be using the banks for defaecation. Dead bodies may be burning on the banks. Washermen may be drying their clothes in some parts. All activities of human beings must be observed and described here in detail.

Chapter 8 – Benefits to local people: Rivers are useful to us in many ways. It gives us fish and other aquatic animals like prawns, and some water plants for food. Water from the river is used for cultivation of vegetables or other crops in the fields nearby. A major or minor dam or a check dam may be present. A hydroelectric project may be present. Sporting or adventurous activities may be there. Water from the river may be pumped into a dam far away for drinking water supply. Local people may be using pumps to withdraw water for cultivation. It may be a picnic spot. In this way, make proper observations and also ask the local people to get all details regarding the benefits of the river to them. (Add points from the preceding paragraph also, briefly).

Chapter 9 – Maintenance: See whether any stone pitching is done at the banks to prevent the soil from the edges falling into water. Ask the local people whether any measures are taken by the panchayat or the locals themselves to clean the bank regularly. Ask whether materials for pujas etc. remain on the banks for days together after immersion of idols, after festivals etc. People may throw plastic covers, plates and glasses after picnics, and these may not be cleaned regularly. In this way, find out all about the maintenance of the river and its banks.

Chapter 10 – Future planning: Find out from the local people whether there are any plans by the local authorities or state or central governments regarding the part of the river studied, like building dams, developing the spot for tourism, diverting the water into canals, start water sports etc. Ask them whether they themselves have any plans for this stretch of river, like increasing fish catch, cleaning the river, protesting against water pollution at panchayats or municipalities etc, if there are frequent floods, how are they planned to be managed etc.

Chapter 11 – Conclusion: ....

Chapter 12 – References: ....

8.2.2 A Forest

Chapter 1 – Introduction: ....

Chapter 2 – Location and General Information: The exact location of the part of the forest being studied must be mentioned – in which state, district, village, tehsil the forest is situated. Find out the latitude, longitude and altitude of the place. The usual temperature pattern, which months show maximum and minimum temperatures, rainfall pattern throughout the year etc. also may be included in this section. For this, the Forest Department officials or the District Head Office can be contacted. The atmosphere in the forest may be warm and humid or dry. Presence of mining activities, dams etc. must be noted down. A rough map can be drawn by hand showing the flora and fauna and streams etc.

Chapter 3 – Uses of forests: Local people, especially women can be asked about the various purposes they use the forest. They may be collecting for their own use and also for selling forest products like fruits, flowers, leaves, roots, nuts, fuelwood, *daatun* (small twigs for brushing teeth), honey, bamboos, medicinal plants, timber, gum etc. The forests may also be used for grazing the local people's cattle, water from forest streams may be used for various purposes by them.

Chapter 4 – Plants and animals in the forest: See whether it is a natural forest with great biodiversity or a plantation forest with only 1 or 2 types of trees, planted at regular intervals. Discuss the advantages of natural forests over plantation forests. If a natural forest, observe whether it is evergreen, with mango, tamarind, teak etc. or deciduous with sal, arjun etc. or temperate with various types of conifers like pines. Bamboos may grow in clusters. Epiphytes may grow on the branches of trees. Know that presence of epiphytes and lichens indicate that the air is not polluted. The forest may be closed (dense) or open, with open spaces here and there. Footprints of different animals may be seen, especially if the soil is soft. Wild berries may be hanging from the trees; some may be poisonous. So without consulting the forest dwellers or the village folk, no berry should be eaten. Monkeys and squirrels may abound. Various birds make nests. Study one or more types of nests. Grasshoppers, praying mantis, different insects and worms may be there on the forest floor.

Chapter 5 – Forest floor: Grasses may form a felt-like layer on the forest floor along with ferns, mosses, lichens etc. Or the forest floor may be clean. The soil under your feet may be soft and warm, indicating that humus is formed well. Or the leaves will be retaining their shape etc.; this shows decomposition has just started. Identify the insects and worms on the forest floor. Discuss the detritus food chain starting from the dead leaves etc. on the forest floor.

Chapter 6 – Over-exploitation of the forest: Find out in what all ways the forest is being exploited by the local people and also outsiders. Using forest produce is different from exploitation. When the local people take out fruits or leaves etc., they do so only for a living. They do not plunder the forests with the aim of amassing wealth. Search for signs of soil erosion, cutting of huge trees, formation of wastelands, overgrazing etc.

Chapter 7 – Awareness of local people: Find out from the local people whether they know of the role of forests in making rainfall possible, preventing erosion, purifying the air, preventing floods, preventing pollution etc. If they do not know these, discuss with them about these and such other points and try to increase their awareness level so that they will take care of the forests in a better manner. Also, see if there are any forest protection group, vigilance group, self-help groups of women etc. in the area. Give them your own ideas.

Chapter 8 - Conclusion: …………

Chapter 9 – References: …………..

8.2.3 Grassland

It is difficult to find true grassland in a tropical country like ours. They are the characteristics of temperate climates. However, true grasslands are seen in Himalayan areas, semi-arid areas, shoals etc. Common grazing lands, forest clearings etc. can also be studied as grasslands.

Chapter 1 – Introduction: ……….

Chapter 2 – General information: Size, location (state, district, village etc.), altitude, especially on hilly regions, latitude, longitude, presence of shrubs or trees here and there, any river flowing nearby and the like.

Chapter 3 – Uses of the grassland: The first and foremost use is for grazing cattle. It is also used for cutting fodder, collecting fuelwood from the twigs of the small plants and the dried grass itself etc. Observe that a huge quantity of grass is needed to support cattle. Relate this with carrying capacity, i.e., the maximum number of herbivorous animals that this particular grassland can support without harming itself. You can ask the villagers how many animals they feed on this grassland. The quantity of milk, and milk products that the villagers use for themselves and sell for a living, quantity of eggs, meat, wool etc. obtained directly or indirectly etc. are all included in the uses of a grassland.

Chapter 4 – Mapping and documentation: Prepare a map of the grassland showing the places where the animals graze, where they drink water, where the local people collect fuelwood, in what parts the grassland is showing signs of degradation, how far is the nearest river or pond or stream etc. Document all information collected along with the source from which these were collected. You can even mention the names of the villagers who gave you this information, and what the village head thinks of the truthfulness of these data.

Chapter 5 – Grassland as an ecosystem: Here, you can give in detail all aspects of the abiotic and biotic components of the grassland you are studying. Give the names of only the plants and animals that you see there, like tall and short grasses, herbs like *Phyllanthus* (Bhoomi amla), *Euphorbia* (Doodhi) etc., grasshoppers, ants, frogs, mice, lizards, snakes etc. For example, do not just write 'tiger' or 'lion' as top carnivores.

Chapter 6 - Maintenance and awareness: Ask the villagers what they do to maintain the grassland, how frequently they burn the grasses, what their opinion is about grazing too many animals on a small patch of grassland, what are the harmful effects of overgrazing, in what way does it degrade the soil, how can a degraded grassland be brought back to normalcy etc. Discuss all such aspects with them. They will learn many things from you; and you will learn many things from them. It will be a meaningful interaction.

Chapter 7 – Conclusion: ………

Chapter 8 – References:……….

8.2.4 Hill/Mountain

There may be a mountain or a hill or even a hillock not far from your college. Pay a visit to it on a field study trip, and prepare a good field report.

Chapter 1 – Introduction: ……….

Chapter 2 – Location and General Information: Give all details about locality. Then observe and note down whether the slope of the hill is steep or gradual, what type of top the hill has – flat or conical etc. If the top is flat, it is an old hill; if it is conical, it is a new hill. Observe the direction of the mountain chains, if the mountain has been formed as ranges. Observe which side has more vegetation and which side has less. Try to find the reason, considering the falling of sun's rays, exposure to wind, more rain etc.

Chapter 3 – Mapping: Make a map of the hill/mountain with the help of the observations made, like presence of protruding rocks, streams, waterfalls, forests, dry areas, scrubby vegetation, trees and other plants, springs etc.

Chapter 4 – Nature of soil on the slopes: Observe and note down the colour and texture of the soil, See whether there is a good soil cover of grasses or anything else. Mostly, soil cover is not good on slopes as the top soil is washed away in rain and so plant roots cannot establish well. Look for gully formation and erosion, signs of cattle grazing etc. Describe the degraded nature of the soil.

Chapter 5 – Flora and fauna: Note down the plants growing on the mountain slopes and the valleys below. Note the differences. On the slopes, growth is stunted due to deficiency of water and minerals, while the valley has luxuriant vegetation as it has enough water, minerals and also organic content due to large quantities of humus. Look for animals of various kinds including mammals, birds, reptiles and also insects, worms etc. on the soil.

Chapter 6 – Uses: Understand and note down in what ways the hill is useful to the local people. They may be collecting various produce from the plants and animals on the hills slopes and valleys. They may be getting drinking water from its streams. There may be enough grass for their cattle. The valleys may be good for picnics, the local people may say it can be even develop as a tourist spot. Talk to the villagers and get as much information a possible.

Chapter 7 – Awareness: Find out how far the villagers are aware of the importance of mountains and hills and valleys for maintaining the equilibrium of the biosphere. Discuss with them topics like the role of mountains in steering winds and causing rainfall in different sides and the like. Let the importance of these ecosystems sink down in their minds.

Chapter 8 – Find out from the local people whether there has been any reduction in the forest produce in the past years. Try to find out the reason, along with them. Discuss the various measure that can be adopted to stop this degradation.

Chapter 9 – Conclusion: ………….

Chapter – 10 – References: ………..

8.3 VISIT TO A LOCAL POLLUTED SITE

8.3.1 Urban Site

Students living in towns or cities can take up this project. Pollution results from undesirable changes in the environment. In urban areas, there are many types of pollution. A site of solid waste accumulation or an industrial area can be chosen.

Chapter 1 – Introduction: ……

Chapter 2 – Location and general information: Give exact details of where the town/city is situated, in which state etc. Write about nearby rivers, hills/mountains, rail, road and plane connectivity, approximate population, general livelihood of people, weather conditions, greenery of the place, and the neatness etc.

Chapter 3 – Observations at the site: You may see heaps of garbage at the sides of the roads, spilling into drains and choking them, polluting water nearby, overflowing onto the roads in the rainy season, emitting foul smell, giving shelter to disease causing rats and breeding flies and mosquitoes. Plastic plates and glasses may be scattered here and there. This and other such sights are common in some urban areas. These are places where the urban poor live in closely packed small houses or huts, giving rise to slums. Sometimes, even in areas where comparatively richer people live, garbage may pile up due to the inefficiency of the municipality or corporation. Make a note of which of these are degradable and which are non-degradable. In some other urban surroundings, industries may be emitting smoke and various gases into the air or may be emptying effluents into nearby rivers or streams or other water bodies, polluting the water. On the roads in urban areas, automobile pollution is very common. During traffic jams, large quantities of smoke and harmful gases are emitted, and the continuous honking of horns by the automobiles becomes jarring to the ears. In this way, make a note of whatever you observe in the urban area. Some urban areas on the other hand may be clean and well maintained, with clean roads and properly planned drainages and avenue trees on the sides of the roads etc. Similarly, some industries have waste water treatment plants, rainwater harvesting systems etc. and are responsible socially. Make a note of these plus points also.

Chapter 4 – Effects of pollution: Talk with the people in the surrounding areas and find out the occurrence of water-borne and air-borne diseases, various allergies, breathing problems and the like in the area. Ask them in what other ways they are affected by pollution in their area, like getting contaminated fish, lack of clean drinking water etc.

Chapter 5 – Extent of pollution: Find out whether the pollution is severe or moderate or slight or nil. Study in detail what is being polluted (air, water, soil) and what are the polluting agents. Map the area according to the severity of pollution, the water bodies, industries etc.

Chapter 6 – Remedial measures: Ask the people around what can be done to prevent and control pollution in their area. Are they depending wholly on the authorities to remedy the situation or are they doing anything substantial themselves? Do they think that they are also responsible for the situation? In what way do the rag pickers help in removing the wastes? Is there any solid waste management system in the area, like a community biogas plant? In industrial areas, can they get together and protest against pollution? In cities, can they create some awareness regarding the uselessness of honking horns in a traffic jam? Ask the people such and other questions and discuss the points with them.

Chapter 7 – Conclusion: ....

Chapter 8 – References: ....

8.3.2 Rural Site

If your college is set in rural surroundings, you can easily take up a project on pollution in villages. The type of pollution in villages is different from that in towns and cities.

Chapter 1 – Introduction: ....

Chapter 2 – Location and general information: State where the village is situated. Mention clearly the name of the state, district, village and tehsil. Write about how you reached there, and if you faced any difficulty in doing so due to bad roads. You can give description of the beauty of the place, the village scene that greeted you on your arrival etc. In this way, give some general information.

Chapter 3 – Observations at the site: Pollution in the villages is not as much as is in towns and cities. Pollution is due to excess fertilizers and pesticides used in the farms flowing into nearby water bodies, due to villagers bathing cattle in the same pond where they are taking bath, due to wastes from houses being thrown out and lack of facilities where wastes are taken away etc. Dirty water bodies may become breeding ground of mosquitoes, causing attacks of malaria, filarial, dengue fever etc. among the villagers. Poor people in the villages use cheap detergents for washing clothes, thus polluting the ponds where they themselves take bath. In this way, make a note of all the types of pollutants in the rural areas.

Chapter 4 – Effects of pollution: Find out from the villagers what types of diseases they suffer from due to pollution. In many cases, if they are illiterate, they may not even realize that pollution is the cause of some of their ailments. In such cases, you may have to explain everything in detail to them.

Chapter 5 – Remedial measures: Biogas plants, smokeless chulhas, compost pits, vermicompost pits, solar lamps, organic farming, rainwater harvesting, using masks while spraying pesticides, the importance of hygiene, etc.

Chapter 6 – Discussion with the rural folk: You have gone to a rural area on a field study visit. Let it be useful to the rural folk too. So discuss with them in detail all that you know about pollution, pollutants, prevention and control of various types of pollution.

Chapter 7 – Conclusion: ....

Chapter 8 – References ....

8.3.3 Industrial Area

Chapter 1 – Introduction: ....

Chapter 2 – Location and general information: In this section, describe where the industry you are about to study is situated – in which state, district, town/city. Write about the neighbouring villages/towns/cities, the name of the river or any other water body nearby, where the products manufactured here are transported, about the roads and railways nearby etc.

Chapter 3 – The industry or industries in the area: In some cities, there may be one main industry; but in some others, there may be many. The smaller ones may be supplying parts to the major industry. Write their names and find out what items are manufactured in them. This will give you an idea of what types of waste materials are produced in them.

Chapter 4 – Polluting, non-polluting or less polluting industries: Find out in which river or any other water body they discharge their effluents and whether these industries have sewage treatment plants so that the effluents are treated before release. Have the industries installed air pollution monitoring devices? Have they given their workers ear plugs if their machineries are making too much nose? Do they take care of their workers when they get diseases due to pollution? Are they following safety norms? Have they planted enough trees around the factory? In this way, find out as much as you can about the industry.

Chapter 5 – Interaction with local people: Get in touch with the local people and get their opinions about the pollution being spread by the industry and the remedial measures being taken by them. Ask them whether they are happy with the permits given to the company by the government. Are the industries drawing too much water from underground resources? Are they planting enough trees?

Chapter 6 – Suggestions for improvement of the situation: In this section, you can include the various suggestions given by the local people and also your own opinions.

Chapter 7 – Conclusion: ....

Chapter 8 – References: ....

8.3.4 Agricultural Area

India is a country of villages. Different crops are cultivated throughout India. In the earlier days, cultivation was environment friendly. But the teeming millions of our country had to be fed. So, when Green Revolution was proposed by Dr. Swaminathan in the 1960's, use of artificial manures and pesticides increased manifold. Thus, though Green Revolution multiplied our agricultural production, it had its own disadvantages.

Chap' 1 – Introduction: ....

Chapter 2 – Location and general information: Agricultural areas are seen in continuity in villages; they can be seen in small plots on the outskirts of urban areas also. So you can go to such an area without much difficulty wherever you are. Describe the area under cultivation, the main crops, the source of water, the availability of electricity, the general prosperity of the people, medical and educational facilities etc.

Chapter 3 – The method of cultivation adopted: Here you can describe in detail all the practices adopted by the farmers of the area in the case of each crop. From ploughing to seeding, watering, adding fertilizers, spraying pesticides, harvesting, storing, marketing etc. all must be included.

Chapter 4 – Environment unfriendly methods used at present: In this chapter, describe the unsustainable methods of cultivation practised, if any. For example, in many areas, you can see that there is wastage of a large quantity of precious water. Most of the farmers in our villagers do not know the details of drip irrigation or precision farming. They may not be giving the required attention to the dose of fertilizers. They may not be using masks while spraying pesticides or washing their hands after spraying. In this way, find out each of the unsustainable practices followed in the agricultural area.

Chapter 5 – Reasons for such practices: Reasons can vary. The farmers may be ignorant of new sustainable methods, or they may not be able to afford the new measures. Government may be neglecting their needs. There may be geographical barriers. Infrastructure may be poor. Find out the reasons and describe them in detail.

Chapter 6 – Awareness: Discuss with the farmers about the modern sustainable methods of farming like drip irrigation, lift irrigation, using better agricultural implements, reducing on artificial fertilizers, better methods of animal husbandry, setting up co-operatives for better marketing, use of IT for gaining more knowledge on all matters etc.

Chapter 7 – Conclusion: …………

Chapter 8 – References: …………..

8.4 STUDY OF COMMON PLANTS, INSECTS AND BIRDS

Various types of plants, insects and birds are seen in all places. It is only that in our busy schedule of lives, we do not observe much. Therefore this field study visit will give you some hours of relaxation, time of being one with nature and experiencing immense peace of mind. For this study, it will be ideal to carry with you a pair of binoculars and a hand lens in addition to a note-book and pen. You may also carry some old newspapers to press small plants or different types of leaves or flowers.

Chapter 1 – Introduction: ……………….

Chapter 2 – Location and general information: Give some general information regarding the location you have chosen to study.

Chapter 3 – Common plants: Observe the vegetation of the place properly. Trees may be big or small. They may be with leaves or all leaves may have been shed (if the season is autumn). Observe that if it is the beginning of autumn, trees like Palash, silk cotton etc. will be in full bloom. Later in the season, leaves would have replaced the flowers. In evergreen trees, leaves remain throughout the year and shed also at all times. See the shape of the canopy. Do they have special types of roots above ground? Like prop roots or buttress roots? If you are Botany students, you will know. Others can take help from them. Shrubs of different types are also common. Observe whether they are large sized or small, thorny or otherwise, scrambling on nearby trees or independent. Herbs are small plants, erect or creeping. Grasses are all herbs. Some climbers may be seen climbing on supporting trees. They may have various organs for supporting them. Observe the size, colour, scent etc. of the flowers. See what types of leaves and fruits the plants have. See that some plants do not produce flowers at all. Examples are mosses, ferns, mushrooms etc. In this way, gather as much information about the local plants as possible. You may know the names of some; ask the local people for the names of others. Find out which plants are more in number or spread in the locality. With the help of Botany students or teachers or a local guide, find out if there are any rare plants in the area. Describe at least 5 plants in your study report.

Chapter 4 – Common insects: Though the variety of insects is great in nature, it is difficult to notice them all even in a small locality as many of them are underground and many can fly about. So the common insects that you see will most probably be ants, mosquitoes and house flies, bees and butterflies, bugs and beetles etc. Spiders, you may know, are not typically insects as they have eight legs and scientifically, all insects are 6 legged. Observe the size, colour and behaviour of these insects and make a record of the same.

Chapter 5 – Common birds: Crow is perhaps the commonest bird in any locality in our country. Mynahs of different types are common too. Sparrows used to abound earlier, but now they are threatened in many localities. Pigeons, bulbuls, parrots etc. are also common. Parrots can be seen pecking at ripe or unripe fruits on trees. Woodpeckers may be seen. Eagles fly high in many places. Vultures have become endangered and most probably you will not see them anywhere. Near water bodies, storks and kingfishers are seen. Observe the various kinds of birds in the study area. See their flying habits and food preferences. Try to locate nests of some. Make notes of everything.

Chapter 6 – Common mammals and reptiles: Make a list of all common mammals that you see in the study area, and reptiles like lizards and describe them in brief.

Chapter 7 - The importance of biodiversity conservation: Note the variety of plants, insects, birds, mammals and reptiles you see all around. Discuss the importance of biodiversity in nature.

Chapter 8 – Conclusion: …………..

Chapter 9 – References: ………….

8.5 STUDY OF SIMPLE ECOSYSTEMS

8.5.1 Pond Ecosystem

Ponds are common in many areas, though their numbers have gone down greatly due to poor maintenance and draining for construction activities.

Chapter 1 – Introduction ……….

Chapter 2 – Location: Give the exact location of the pond under study.

Chapter 3 – Abiotic components of the pond ecosystem: Here, describe the various physical and chemical components like water temperature, light, wave action if any, pH, etc.

Chapter 4 – Biotic components of the pond ecosystem: Find out which are the producers, and consumers of various levels living in the pond. Study the various food chains and food webs in the pond.

Chapter 5 – The edge of the pond: Study the soil and the various types of living organisms present on the soil surface and just under the soil surface. Describe detritus food chain operating here.

Chapter 6 – Importance of pond ecosystem: Mention here how a pond is habitat to a variety of animals and plants, how it helps in water conservation and how it provides fish, edible plants, water etc. to the local people, and hence should be maintained well. Discuss these points with the local people and record their observations.

Chapter 7 – Conclusion: ....

Chapter 8 – References: ....

8.5.2 River

Rivers, big or small, flow through all the states of India. Visit a riverside to study the river ecosystem.

Chapter 1 – Introduction: ....

Chapter 2 – Location: Give the exact location of the river being studied.

Chapter 3 – Abiotic components of the river ecosystem: Here, describe the various physical and chemical components like water temperature, light, wave action, pH of the water etc.

Chapter 4 – Biotic components of the river ecosystem: Find out which are the producers, and consumers of various levels living in the pond. Fish can be seen at all trophic levels, except the producer level, in a river. Study the various food chains and food webs in the river.

Chapter 5 – The river bank: Study the sand and the types of plants growing on the bank. These are deep rooted sand binders near the water edge. Away from the bank, they gradually give way to plants with normal root systems. Observe crabs and other insects and amphibians on the bank.

Chapter 6 – Importance of river ecosystem: Mention here how a river provides shelter to a variety of animals and plants, how it helps in water conservation and how it provides fish, edible plants, water for drinking and irrigation etc. to the local people, and hence should be maintained well. Discuss these points with the local people and record their observations.

Chapter 7 – Conclusion: ……………….

Chapter 8 – References: ……………….

8.5.3 Hill Slopes

Hills, if not mountains, are common in many places. Small hills called hillocks may be commoner. Study a hill along with its slopes for preparing a field report.

Chapter 1 – Introduction: ……………..

Chapter 2 – Location and general information: Describe the exact location of the hill, its altitude, latitude, longitude, water courses originating from it etc.

Chapter 3 – The slopes: Describe the direction of the slopes, its exposure, nature of the soil on the slope, soil cover if any, degree of sloping – whether steep or gradual etc.

Chapter 4 – Vegetation: Give an account of the vegetation on the hill, the slopes and the valley, and compare and contrast among the three. Give reasons.

Chapter 5 – Fauna on the hill and slopes: Try to locate the various types of reptiles, insects etc. on the slopes. You may see some mammals roaming around and some birds flying above the hills and the valleys. Cattle may be grazing on the slopes. Make a note of them.

Chapter 6 – Vulnerability of slopes: Try to find out why slopes are vulnerable and unstable. Stress the necessity of protecting them.

Chapter 7 – Importance of hill ecosystems: Understand how important hill ecosystems maintain the balance of nature.

Chapter 8 – Awareness: Discuss with the local people about the benefits they get from the presence of the hill in their neighbourhood.

Chapter 9 – Conclusion: …………..

Chapter 10 – References: …………..

The format given here for field visit report can be used for dissertations, Ph.D. theses etc. Therefore, the students can benefit from it at a later stage in studies also.

Appendix 1

International Conventions and Protocols to Protect the Environment

- 1946 – International Convention for the Regulation of Whaling (ICRW), Washington
- 1954 – International Convention for the Prevention of the Sea by Oil (London)
- 1957 – European Agreement concerning the International Carriage of Dangerous Goods by Road (ADR), Geneva
- 1959 – Antarctic Treaty, Washington DC
- 1963 – Vienna Convention on Civil Liability for Nuclear Damage, Vienna
- 1966 – International Convention for the Conservation of Atlantic Tunas (ICCAT), Rio de Janeiro
- 1969 – International Convention on Civil Liability for Oil Pollution Damage 1969 (1969 CLC), Brussels, also 1976 and 1984
- 1969 – International Convention Relating to Intervention on the High Seas in Cases of Oil Pollution Casualties Intervention Convention, Brussels
- 1971 – International Convention on the Establishment of International Fund for Compensation for Oil Pollution Damage (as amended – Brussels)
- 1971 – Ramsar Convention on Wetlands of International Importance, Ramsar
- 1972 – World Heritage Convention Convention Concerning the Protection of the World Cultural and Natural Heritage, Paris
- 1972 – London Convention on the Prevention of Marine Pollution
- 1973 – Convention on the International Trade in Endangered Species of Wild Flora and Fauna, (CITES), Washington DC
- 1973 – Convention on the International Trade in Endangered Species of Wild Flora and Fauna, (CITES), Washington DC

- 1974 – World Food Conference, Rome
- 1978 – International Convention for the Prevention of Pollution from Ships (Montege Bay)
- 1979 – Convention on Long-Range Trans-boundary Air Pollution (LRTAP), Geneva
- 1979 – Convention on the Conservation of Migratory Species of Wild Animals (CMS), Bonn
- 1981 – Environmental Protection: Aircraft Engine Emissions - Convention on Civil Aviation, Montreal
- 1982 – United Nations Convention on the Law of the Sea (LOS) Convention, Montego Bay
- 1983 – FAO International Undertaking on Plant Genetic Resources, Rome
- 1985 – Vienna Convention for the Protection of the Ozone Layer
- 1985 – FAO International Code of Conduct on the distribution and use of Pesticides, Rome
- 1986 – Convention on Early Notification of a Nuclear Accident (Notification Convention), Vienna
- 1986 – Convention on Assistance in the Case of a Nuclear Accident or Radiological Emergency (Assistance Convention), Vienna
- 1987 – Vienna Convention for the Protection of the Ozone Layer, Vienna, 1985, including the Montreal Protocol on Substances that Deplete the Ozone Layer, Montreal
- 1989 – Convention on Civil Liability for Damage Caused during Carriage of Dangerous Goods by Road, Rail, and Inland Navigation Vessels (CRTD), Geneva
- 1989 – Montreal Protocol on Substances that Deplete the Ozone Layer, Montreal
- 1989 – Basel Convention on the Control of Trans-boundary Movements of Hazardous Wastes and their Disposal, Basel
- 1990 – International Convention on Oil Pollution Preparedness, Response and Co-operation (OPRC), London
- 1991 – Espoo Convention on Environmental Impact Assessment in a Trans-boundary Context, Espoo
- 1992 – The Earth Summit at Rio de Janeiro, Brazil
- 1992 – Convention on Biological Diversity (CBD), Nairobi
- 1992 – Watercourses and International Lakes (ECE Watr Convention), Helsinki
- 1992 – Convention on the Trans-boundary Effects of Industrial Accidents, Helsinki
- 1992 – Framework Convention on Climate Change (UNFCCC), New York

- 1992 – Convention on the Protection and Use of Trans-boundary Watercourses and International Lakes (ECE Water Convention), Helsinki
- 1994 – Convention to Combat Desertification (CCD), Paris
- 1994 – Convention on Nuclear Safety, Vienna
- 1994 – International Tropical Timber Agreement, (ITTA), Geneva
- 1996 – Comprehensive Test Ban Treaty (CTBT)
- 1996 – International Convention on Liability and Compensation for Damage in Connection with the Carriage of Hazardous and Noxious Substances by Sea (HNS), London
- 1997 – Kyoto Protocol on Climate Change, Kyoto
- 1998 – Aarhus Convention on Access to Information, Public Participation in Decision-making and Access to Justice in Environmental Matters
- 1998 – Rotterdam Convention on the Prior Informed Consent Procedure for Certain Hazardous Chemicals and Pesticides in International Trade, Rotterdam
- 2000 – European Agreement Concerning the International Carriage of Dangerous Goods by Inland Waterways (AND), Geneva
- 2001 – Stockholm Convention on Persistent Organic Pollutants, Stockholm
- 2002 – *International Conference* on *Environmental* Threats to the Health of Children: Hazards and Vulnerability, Bangkok, Thailand
- 2003 – International Conference on the Protection of the Environment from the Effects of Ionizing Radiation, Stockholm, Sweden
- 2003 – The World Climate Change Conference, Moscow
- 2004 – World Conference on Disaster Reduction, Kobe, Japan
- 2006 – World Climate Conference, Nairobi
- 2009 – The Copenhagen Climate Summit or Climate Conference, Copenhagen

Appendix 2

Glossary

A

Acid rain – Rain containing acids like sulphuric acid and nitric acid formed by the dissolution of sulphur dioxide and nitrogen oxides emitted from industries in the water vapour of the atmosphere

Afforestation – Planting trees in areas where there were no trees earlier

Agro forestry – A modified form of social forestry where trees are planted in suitable locations on agricultural land

AIDS (Acquired Immune Deficiency Syndrome) - Progressive destruction of the immune system of the body by HIV (Human Immunodeficiency Virus).

Allogenic succession – A succession where replacement of a community is caused largely by external conditions and not by the existing vegetation

Atmosphere – A protective covering around the earth; that component of the environment which is a mixture of different gases in different proportions; a reservoir of several elements essential for life

Anthropogenic extinction – Extinction of species due to man's activities

Aquatic ecosystem – Ecosystem in water like pond, river, ocean etc.

Autogenic succession – A succession in which a vegetation interacts with and modifies the environment, resulting in its own replacement by new-comers

Autotrophs – Photosynthetic plants and chemosynthetic bacteria that can manufacture their own food are called autotrophs

B

Biodegradable pollutants – Pollutants that can be easily decomposed by natural processes

Biodiversity – The totality of genes, the species, the ecosystems in a given region of the world; the existence of a large variety of living organisms in any locality

Biogeochemical cycle – Circulation of materials between the biotic and the abiotic components of an ecosystem

Biomass – Total amount of living organisms present in an ecosystem

Biome – Plants, animals and microorganisms of any region together, influencing one another and interacting with one another

Biosphere – The largest, almost self sufficient biological system consisting of the living organisms of the earth interacting with the non-living environment

Biosphere reserves – Special protected areas where people form a component

Biotic community – same as biome

Buffer zone – Zone of a biosphere reserve, where educational activities and research are allowed, outer to the core zone

C

Carnivores – Animals eating herbivores

Climate – The average of weather over a long period

Conservation – A constant effort to increase resources, while at the same time utilizing them for our needs

Conservation value of a plant species – The percentage of soil that would have been eroded in the absence of that species

Contours – Ridges made at the same level in farms

Contour farming – Farming where planting is done on contours, i.e., ridges made at the same level, mostly on slopes of hills

Coral reefs – Corals in continuous populations

Core zone – Undisturbed zone in the centre of a biosphere reserve

Crustose lichens – Thin and papery lichens growing on rocks

Cryopreservation – Storing germplasm at very low temperature of –196°C (temperature of nitrogen)

Cyclones – Deadly atmospheric storms blowing in a spiral manner, formed due to closed low pressure systems

D

Decomposers – Top consumers

Desertification – The process by which productive potential of a land is reduced by 10% or more

Detritus – Fallen leaves and twigs on land and dead algae and parts of other water plants that fall to the bottom of water bodies

Detritus eaters – Animals like protozoa, nematodes, snails etc. that eat detritus

Detritus food chain – A food chain that starts with dead organic matter into microbes and then into larger organisms

Domestication of plants – The process of cultivating wild plant species for human needs

Drought – A period when rainfall is nil and evaporation from the soil is higher than normal

E

Earthquake – A motion of the ground surface caused by wave motion starting from a focal point.

Ecological niche – The status or position and the role of an organism in the structure and operation of an ecosystem

Ecological pyramids – Graphic representations of the different trophic levels in any ecosystem

Ecological restoration – Scientific measures taken by experts in conservation leading to ` restoration of degraded or damaged areas surrounding protected areas

Ecological succession – The process of orderly evolution of communities from birth to maturity, leading to the establishment of a relatively stable community in total harmony with the environmental conditions

Ecological system – same as ecosystem

Ecosphere – same as biosphere

Ecosystem – A natural unit of living communities and the non-living environment interacting and exchanging materials between themselves

Endemic species – Species restricted to small areas, mostly the last remnants of a group which is becoming extinct

Environment – External conditions influencing the growth and development of living organisms in any way

Environmental disasters – Extreme events, natural or man-made, which exceed tolerable magnitudes.

Environmental hazards – same as environmental disasters

Environmental pollution – Any physical, chemical or biological change in the environment that leads to its deterioration

Environmental studies – Study of any one or more aspects of the environment and living organisms in relation to each other

Epicentre – The place on the earth's surface, right above the hypocentre

Epilimnion – Upper part of a water body with warm water

Ethics – The science of morals

Exhaustible resources – Resources in limited supply in nature and which can be exhausted if not used carefully

ex situ conservation – Conservation of genetic resources outside the natural habitats of organisms by maintaining sample populations

F

Flood – A condition of abnormally great flow of water in a river; submergence or inundation of vast areas of land under water for many days in continuation

Focus – The place of origin of an earthquake

Foliose lichens – Lichens with flat plant bodies like leaves

Food chain – The sequence or order of trophic levels through which food energy passes from the primary producers to the top carnivores

Food web – Food chains that are interconnected at some points so that an interlocking pattern is formed

Forest conservation – Protecting the present forests and preventing them from degradation

Freshwater ecosystem – Ecosystem in freshwater, like pond, river etc.

Fungicides – Chemicals that kill fungi

G

Gene bank – Institutions where seeds or other living materials are stored as genetic materials

Gene pool – The sum total of all the genes in all the wild species

Gene reserve – Same as genetic engineering – The procedure by which one or more genes from a desirable variety is introduced into or deleted from another variety which needs to be improved

Genetically modified (GM) crops – same as transgenic plants

Geographic Information System (GIS) – A system for creating, storing, analyzing and managing geographical information

Geothermal energy – Naturally occurring heat from the interior of the earth flowing to the cooler surface

Germplasm – The sum total of the reproductive cells of any organism, that later gives rise to the whole organism

Global warming – Heating up of the earth's atmosphere due to the action of greenhouse gases

Grazing food chain – A food chain that starts from living plants and goes through herbivores to carnivores

Greenhouse effect – The thick layer of CO_2 in the atmosphere functioning like gas panels of a greenhouse, allowing sunlight to pass through, but preventing its heat from radiating out

Greenhouse gases – Gases which contribute to the greenhouse effect, like CO_2 (Carbon dioxide), N_2O (Nitrous oxide), CH_4 (methane), CFCs (Chloro Fluoro Carbons), SO_2 (Sulphur dioxide), HFC's (Hydro Fluoro Carbons), and about 25 other gases

Green revolution – The most popular and common practice in trying to increase agricultural yield per unit area

Gross production – Total energy fixed in food by green plants

Ground water resources – Water found below the surface of the earth

Gully erosion – Erosion through gullies or channels formed by joining of many rills along slopes making the surface uneven

H

Halosere – Succession beginning in salty water

Herbivores – Animals and parasitic plants which feed directly on the green plants

Heterotrophs – Organisms that cannot prepare their own food, and get their food from green plants directly or indirectly, like animals, fungi and many bacteria

Hotspots – Regions in the world that have rich biodiversity and high endemism and are under threat of extinction and habitat destruction

Hydrosere – A succession beginning in water or watery habitats

Hydrosphere – All the water on the earth, i.e., the water collectively present in the rivers, lakes, ponds, seas, oceans, glaciers etc. and the water stored underground in aquifers

Hypocentre – same as focus

Hypolimnion – Deeper part of a water body with cooler water

I

Incineration – Burning (of waste materials) at a specific temperature in specific conditions

Inexhaustible resources – Resources which are in unlimited supply in nature

Information Technology (IT) – The use of technology in managing and processing information

Insecticides – Chemicals that kill insects

In situ conservation – Conservation of biological diversity in its natural surroundings

Integrated farming – Farming in which agriculture, animal husbandry, poultry, fishing etc. are all integrated

J

Jatropha – A plant of Family *Euphorbiaceae* that is being cultivated widely these days, the extract from which can be distilled into a substitute for petrol.

K

Keystone species – The dominant or the most prominent species in an area

L

Land reclamation – Bringing degraded land back to normalcy

Landslide – Tumbling down of huge landmasses is called a landslide (same as slip erosion)

Leaching – A process by which soluble minerals dissolve in rainwater and move deep down into the soil

Lichen – Symbiotic association of a specific alga and a specific fungus growing together as one plant

Limnetc zone – The zone in a water body where light is less, but enough for photosynthesis, where many rooted floating and submerged plants grow

Lithosere – A xerosere originating on a rock

Lithosphere – Component of the environment comprising of the soil, the rocks and other parts of the earth's surface

Littoral zone – The shallow plankton rich zone in a water body where there is enough light and where free floating plants grow

M

Macroconsumers – Large sized consumers like herbivores, carnivores and top carnivores

Marine ecosystem – Ecosystem in marine or saltwater conditions like sea, ocean, estuary, backwater etc.

Mass extinction – Extinction of a large number of species during the same time period due to some natural calamity

Mesosere – Succession beginning on areas with adequate water

Metalimnion – Middle part of a water body where epilimnion and hypolimnion meet

Microconsumers – Decomposers like some fungi, bacteria, protozoans etc.

Mineral – Any naturally occurring inorganic substance, having a definite structure, and found along with earthly impurities

Mineral resource – A concentration of a naturally occurring solid, liquid or gaseous material inside or on the earth's crust and whose extraction is profitable

Mining – Extraction of minerals from the earth

Monoculture – Cultivation of single crop in a field repeatedly

Mulching – Ploughing stubble as mulch under the soil

N

Natural extinction – Slow, gradual extinction of species due to environmental changes

Natural laws – Laws based on ethics and morals

Natural resources – Substances, different forms of energy etc. that living organisms can take from nature for various purposes

Net production – What remains of gross production after usage in respiration and other activities

Non-degradable pollutants – Materials that do not degrade or degrade only very slowly in nature

Noise – Unwanted or undesirable sound

Non-renewable natural resources – Natural resources which, once used, are exhausted for ever, and cannot be restored or replaced

O

Off site conservation – same as *ex situ* conservatiopn

Ores – Those minerals from which metals are extracted commercially at a comparatively low cost and with minimum effort

Orthodox seeds – Seeds that are not damaged by drying and freezing, like those of legumes and cereals

P

Parasites – Organisms which depend on other living organisms totally or partially for food

Pesticides – Chemicals that kill pests i.e., undesirable organisms like harmful insects

Photosynthesis – The making of carbohydrates by green plants from carbon dioxide and water with the help of the green pigment chlorophyll and solar energy

Plant community – An aggregate of plants belonging to different species living together in small or large patches in a particular area, having mutual relations among themselves

Pollutant – An agent which causes environmental pollution

Pollution – An undesirable change in the air, water or land of our environment that harmfully affects human, animal and plant life in different ways

Producers – Autotrophic plants in an ecosystem forming the 1st trophic level (T_1)

Profundal zone – The zone of deep water where there is no effective light penetration, where saprophytic fungi, bacteria etc. grow

Primary producers – same as producers

Prisere – same as primary succession

Protected areas – Areas, land or water, protected legally by the government or by community efforts

Psammosere – A succession beginning on sand

Pyramids of biomass – An ecological pyramid where relationships between the trophic levels of an ecosystem on the basis of living weight, fresh or dry are shown

Pyramids of energy – An ecological pyramid that shows the total energy contained in its different trophic levels

Pyramids of numbers – An ecological pyramid where the number of individuals in the different trophic levels at any given time is shown

Pyrolysis – Combustion (burning) in the absence of oxygen

Q

Quarry – An open pit for cutting off building stone, slate etc. from underground rock source

R

Radioactive isotopes – Isotopes of elements that emit radiations

Radionuclides – Same as radioactive isotopes

Recalcitrant seeds – Seeds that get killed by drying and freezing, like coconut, litchi, jackfruit and many other fruits and timber plants

Red Data Book – A book published by IUCN (International Union for Conservation of Nature and natural resources) that lists plants and animals which are in different levels of extinction

Reforestation – Bringing areas now devoid of trees back to the original form of forests

Remote sensing – The science of identification and classification of the surface features of the earth using electromagnetic radiation.

Renewable natural resources – Natural resources that can be restored and replaced

Resource ecology – Ecology of natural resources

Restoration ecology – The branch of study dealing with the conservation of species outside the protected areas

Rill erosion – Erosion where the running water moves rapidly and cuts small streams like channels on the land.

Riparian erosion – Soil erosion happening on river banks by cutting the river bank by the fast flowing water

S

Sacred grove – A patch of dense vegetation including a small shrine and a pond or stream

Saltation – Wind erosion where salty soil is carried away by wind in small quantities

Secondary succession – Succession beginning on a secondary area, i.e., an area where plant life had existed earlier, nutrients are already present in the soil and conditions are comparatively favourable, like cut out or burnt out forests, abandoned croplands, new ponds formed after heavy rains or floods etc.

Seed bank – Institutions where plant materials (genes) are stored as seeds

Seismometer – An instrument used for recording the waves of energy moving in widening circles away from the hypocentre, i.e., origin of the earthquake

Sheet erosion – Removal of soil in small but uniform amounts all over the land surface like a sheet

Silviculture – Cultivation of trees and forest regeneration

Slip erosion – Slipping down of rocks or masses of soil in the rainy season

Smog – Fog made darker and heavier by smoke and chemical fumes

Soil – The top cover of the earth in which plants can grow

Soil degradation – The fall in the quality of soil

Soil erosion – Loss of top soil or disturbance of the soil structure; removal of soil particles from the upper layers and deposition of the same in some other place

Soil reclamation – Same as land reclamation

Solar energy – Energy from the sun

Species – Individuals resembling each other in all important external characters, both vegetative and reproductive

Standing crop – The total amount of living material in an ecosystem at any given time

Standing quality – same as standing state

Standing state – The amount of inorganic substances like phosphorus, sulphur, carbon, nitrogen, hydrogen etc. present in the environment of an ecosystem at any given time

Stream bank erosion – same as riparian erosion

Stubble – Basal parts of plants left in the fields after harvest

Subsere – same as secondary succession

Succession – A natural process by which the same locality becomes successively colonized by different groups or communities of plants.

Surface creep – Wind erosion where soil particles are pushed away by strong wind

Surface water resources – Water resources found above the ground

Suspension – Wind erosion where fine dust is lifted in the air and remains there

Sustainable agriculture – Agriculture during which air, water and land are not polluted by any means and the natural gene pool is also not disturbed

Sustainable development – Development that fulfills the needs of the present generation without harming in any way the needs of the future generations

Symbionts – Pairs of organisms which live in association with each other and benefit from each other

T

Terrestrial ecosystem – Ecosystem on land, like forests, deserts etc.

Thermal pollution – Degradation of water quality by any process that changes ambient water temperature

Thermocline – Same as metalimnion

Tissue culture – Growing isolated cells or tissues or organs or embryos in an artificial medium under aseptic conditions outside the parent organism

Top carnivores – Animals that are not killed and eaten by other animals

Totipotency – The capacity of any cell to give rise to a new plant

Transgenic plants – Those plants into which one or more foreign genes have been introduced permanently

Transition zone – The outermost zone of a biosphere reserve where management of reserve interacts with the local people

Transpiration – Release of water vapour from green plants into the atmosphere

Trophic level – Each energy level in an ecosystem

Troposphere – The lowermost layer of the atmosphere, comprising a mixture of several gases, and about 20 km thick, in which man and other organisms live

True Earth Citizens – People who follow ethical guidelines

Tubectomy – Sterilization operation in females

U

Unsustainable development – Any development during which natural resources get depleted faster than their renewal

V

Values – Principles based on which we decide what is right and what is wrong.

Vasectomy – Sterilization operation in males

Vermicomposting – Making compost manure using earthworms

Vermitechnology – Technology for large scale use of earthworms for converting types of organic wastes into vermicompost and the final application of the in croplands

Vermi-wash – Water passed through vermicompost

W

Weedicides – Chemicals that kill weeds

X

Xerosere – A succession beginning in dry situations

Z

Zero waste companies – Companies which waste nothing by recycling or reusing their waste materials and by-products in various suitable ways

Appendix 3

Answers to Multiple Choice Questions

Unit – 1

1 – d. 2 – c. 3 – a. 4 – d. 5 – b. 6 – c.

Unit – 2

1 – b. 2 – c. 3 – c. 4 – c. 5 – a. 6 – c. 7 – d. 8 – a. 9 – d. 10 – d. 11 – b. 12 – c. 13 – a.

14 – d. 15 – c. 16 – c. 17 – d. 18 – c. 19 – b. 20 – d.

Unit – 3

1 – d. 2 – c. 3 – a. 4 – d. 5 – c. 6 – d. 7 – a. 8 – c. 9 – a. 10 – d.

Unit – 4

1 – b. 2 – a. 3 – d. 4 – d. 5 – c. 6 – a. 7 – d. 8 – d.

Unit – 5

1 – b. 2 – d. 3 – b. 4 – d. 5 – d. 6 – a. 7 – c. 8 – a. 9 – a. 10 – b. 11- d. 12 – d. 13 – d.

14 – c. 15 – b. 16 – d.

Unit – 6

1 – c. 2 – d. 3 – d. 4 – c. 5 – b. 6 – d. 7 – a. 8 – c. 9 – c. 10 – b. 11 – d. 12 – a. 13 – b.

14 – d. 15 – d.

Unit – 7

1 – b. 2 – c. 3 – d. 4 – c. 5 – a. 6 – b. 7 – b. 8 – c. 9 – d. 10 – d. 11 – c. 12 – b.

Unit – 8

1 – b. 2 – c. 3 – c. 4 – b. 5 – d. 6 – b. 7 – a. 8 – c. 9 – a. 10 – d.

Appendix 4

Websites and Search Engines

Suppose you want to get some information on a particular topic and would like to use the internet for that. After getting connected to the internet, type in the name of any search engine in the long box named address, and press enter. Google.com, yahoo.com, msn.com, ask.com, lycos.com, alltheweb.com, altavista.com etc. are all search engines.

As soon as the page of the search engine appears, type the most important words from the topic in the Search box. Search engines are websites that help you find information located on other websites based on the words you supply. Thus, as soon as you type the words, the search engine gives you a list of web pages to visit.

Now you can click on any website you want. Within a very short time, mostly within seconds, you will reach a world of unlimited knowledge. After you finish with the first site, you can click on 'back' to return to the previous page and click on another website. The links are unlimited.

There are different types of websites. A website ending with .edu is often the most reliable, as it is the website of any educational institution. Similarly, websites ending with .org are those of organizations. One ending with .gov is the website of the US government. Indian government websites have .nic.in at the end. Thus the last letters in the name mostly indicate the organization/institution/government etc.

If you know the name of the website, you need not go through the search engine. You can type the site name directly in the address box. For example, if you want to go to the site of the World Health Organisation (WHO), type Who.org in the address box and press enter. The site of United Nations Environment Programme is unep.org

Similarly, if you want to visit the Ministry of Environment and Forests (MoEF), Government of India, type in envfor.nic.in and press enter.

Some endings are country specific. For example, the website of the Government of Australia ends in.gov.au A website ending with .uk is that of the United Kingdom while .pk is the website of Pakistan.

Encyclopedias and dictionaries are also available on the net. For example, wikipedia is a web-based encyclopedia just like dictionary.com is a web-based dictionary. Wikipedia is an excellent encyclopedia. It appears almost in the beginning on opening many sites. The information you get in wikipedia is not obtained by searching through various other websites – it is information that experts on various topics have already entered in wikipedia itself.

Some of you must be aware of all this and more. This Appendix is meant for those students who would like to know the basics of browsing (surfing) the internet. In these days of technical advancement all over the world, knowledge of computers has become a necessity. Those of you who do not know how to search the net are losing out a lot. That is the reason for dealing with this topic briefly here. Hope it will be of help to you. Happy surfing!!

Appendix 5

Some Tips to Students

Following are some tips for students. By following these recommended suggestions, the students can improve their memory, concentration power and confidence. This will help them perform better in the classroom and examinations. It will be in fact helpful throughout the life. Tips are given on how to answer in examinations also. Facts have been presented in an informal fashion.

- *Importance of attending classes*: Hope you have started attending the third year classes right from day 1. Attendance is important not only for the sake of procuring the required percentage of attendance, but listening to your teacher with full attention in the classroom will help you a lot in the examination hall.
- *Interest in the subject*: When you are seeing a movie, I am sure you will watch the scenes and listen to the dialogues and songs with rapt attention. It is because you want to know what is happening. Similarly, if you want to really know what is being taught in class, nothing can stop you from paying attention. Develop interest in the subject consciously, in case you did not have it earlier! Enjoy the subject. There is no subject that is mundane.
- *Need of single-minded attention*: Remember the story of Arjun. Once, he and his brothers were asked by their archery teacher to hit the eye of a wooden bird placed sidewise on the branch of a tree. The boys stood in a line ready to shoot their arrows. As each boy came forward and took aim, the teacher asked him what he was seeing. Some answered they were seeing the tree; some said the branch on which the bird was kept and yet some others said they saw only the bird. Only Arjun gave the right answer. He said, "I see only the bird's eye". The teacher was happy. And Arjun hit the aim with great accuracy.... When the teacher is explaining Environmental Science, do not think of anything else, about home or anything that happened in the bus or of Chemistry, Economics or Commerce. Forget everything else and concentrate on Environmental Science. In Environmental Science, think of only that chapter, only that part of the chapter, that is being taught. Suppose

the teacher is explaining Ecosystem. Think of Biotic factors when that part is being dealt; within that, think of producers when the teacher is talking about that, if it is primary producers that is being discussed, think of primary producers. In this way, keep your mind glued to each sentence, each word being spoken in class. This is possible. It comes with practice.

- *Yoga to remove tension, improve concentration power and memory*: Follow some yoga practices every day. Yoga has become popular in many foreign countries. It is high time it became popular in its land of origin, our very own India.

Each *asana* (posture) has multiple benefits. Those of you who have learnt yoga know that inverted and balancing poses are especially good to improve concentration. As inverted poses are more difficult, and you can learn them later, let me tell you about the balancing poses. There are many different balancing poses. Let me explain the simplest of them, Pranamasana. It is also called Eka Pada Pranamasana, meaning the one legged prayer pose.

Pranamasana: Stand erect, with feet together and arms on your sides. Lift one foot and place its sole on the inside of your other thigh. Fold your hands in front of your chest. Look straight ahead, fixing your eyes on one point, like a black dot made on a wall. Do this asana for as long as possible. For balancing your body, you have to remain concentrated. Therefore, it improves your power of concentration. It also removes stress, anxiety and worries. If you become tense at any time due to any reason, you can do Pranamasana for some time. Additional benefits are strengthening the leg, foot and ankles.

Trataka: Practice of Trataka can help to improve concentration to any extent. There are many variations to this technique. I shall explain the simplest form here: Place a lighted candle on a table or stool in such a way that the flame will be at the level of your eyes when you sit down for the practice, about one and a half to two feet away. Sit in a convenient pose and make yourself comfortable. Any broad based pose will be ideal for this. Keep your back straight. Relax your whole body and close your eyes. Soon, you will become aware of your body only and you will be still like a statue. Now open your eyes and look at the brightest spot of the flame, just above the tip of the burning wick, with total concentration. Do not blink or move your eyeballs. Do not change the position of your body even slightly.

When the eyes become tired, close your eyes. Now you can see the image of the flame in front of your eyes. Practice Trataka on this image. When the image begins to fade, open your eyes and gaze at the real flame again.

Trataka can be done for about 15 minutes. In the beginning, your eyes may pain. So increase the time to 15 minutes only gradually, within a few days or weeks. Soon, you will be practising Trataka with ease.

Additional effects of Trataka are that it stabilises nerves, relaxes troubled minds and corrects weakness as well as some defects of eyes like short-sightedness.

Pranayama: Pranayama is excellent to increase the power of concentration. When we breathe normally, the lowermost part of our lungs is never emptied. Therefore some stale air always remains there. By forcefully expelling this stale air, like during the Kapalbhati Pranayama, we can increase the efficiency of our lungs. Pure air filling the lungs cleans all cells in the body during blood circulation. Brain cells also become more active. This leads to improved memory and concentration.

It is best to do Pranayama early in the morning. However, it can be done at other times also, but not immediately after meals. A gap of about 3 to 4 hours can be left. There are various types of Pranayama. Of these, Anulom Vilom, Nadishodhan etc. are very good to improve memory. For more details on pranayama, various balancing *asanas* (postures), other *asanas* for general health and well-being, and to know about exercises for eyes, refer any good yoga book.

- *Eye care*: Reading or writing for a long time continuously, especially before examinations, tires the eyes a lot. Therefore, give your eyes some rest in between. Close the eyes and just relax, physically and mentally for a minute or so after about every hour of reading. You can also rub your palms to make them warm and place them on your closed eyes during this period. Putting one or two drops of pure rose water (*gulab jal*) in each eye once or twice a day also refreshes the eyes.
- *Regularity and planning*: Study daily. Do not keep everything for the last minute. Make a plan daily and try to follow it. Every night, before going to sleep, check how much of it you have been able to finish. This will help you to know your strengths and weaknesses. You will be able to make better plans for the next day. In the weekends, revise what was taught in the whole week. Similarly, in long vacations, revise everything taught in the term.
- *Display method*: Write equations, formulae, important points etc. on big sheets and display them in prominent places. You can read them while walking past them.
- *Revising while relaxing*: Do not think that you are studying only when you have a book in your hands. You can try to recollect what you read recently while taking bath or brushing your teeth or walking to the market. This is also studying.
- *Drawing and writing*: If there are pictures, formulae etc. in the chapter, draw or write them down while studying. This can be done after studies also, as a self test.
- *The importance of getting concepts clear*: Do not just mug up facts without understanding the meaning. Studying only to get good marks is not advisable. Study to learn new things. That makes your concepts clear. This will invariably translate into good marks. When your concepts are clear, you will answer the examination with confidence.

- *Thinking positive*: A positive mindset is crucial in every walk of life. As far as examinations are concerned, if you know the lessons, you will be in a positive frame of mind. That helps a lot in the examination hall.
- *Be relaxed*: Do not be all tensed up while you enter the examination hall. You will feel that you have forgotten all that you have learnt. This happens to everybody. There is no need to worry. Once you start writing, you will remember everything.
- *Carefulness in the examination hall*: While writing your roll number and other particulars on the answer book, be very careful. If you forget to write your roll number or enter a wrong number, you may fail.
- *Reading and understanding the questions*: Read the question paper thoroughly. Understand fully what has been asked. See the marks allotted to each question. Decide which questions you plan to answer. Find out how much time is available for each answer. Try to follow this plan as much as possible. Time management is very important. Many students write the 1st and sometimes 2nd answers for a very long time and then get very little time to answer the rest. This reduces the overall marks drastically.
- *Order of writing*: Write the one you know best in the beginning. This will boost up your morale. However, follow the time schedule while answering.
- *Revision*: If there is time for revision, do that. Otherwise, just check the entries on the first page before handing over the paper.

Go home happily. Come prepared for the next examination. Remember, 'God helps those who help themselves". Wish you all the best! God bless you!

Index